International Perspectives on Geographical Education

Series Editors

Clare Brooks, UCL Institute of Education, London, UK
J. A. van der Schee, Faculty of Earth and Life Sciences (FALW),
Vrije Universiteit Amsterdam, Amsterdam, The Netherlands

This series is under the editorial supervision of the International Geography Union's Commission on Geographical Education. Led by the priorities and criteria set out in the Commission's Declaration on Geography Education Research, the series plays an important role in making geography education research accessible to the global community. Publications within the series are drawn from meetings, conferences and symposiums supported by the Commission. Individual book editors are selected for special editions that correspond to the Commission's ongoing programme of work and from suitable submissions to the series editors. In this way, research published represents immediate developments within the international geography education community. The series seeks to support the development of early career researchers in publishing high quality, high impact research accounts.

More information about this series at http://www.springer.com/series/15101

Graham Butt

Geography Education Research in the UK: Retrospect and Prospect

The UK Case, Within the Global Context

 Springer

Graham Butt
School of Education
Oxford Brookes University
Oxford, UK

ISSN 2367-2773 ISSN 2367-2781 (electronic)
International Perspectives on Geographical Education
ISBN 978-3-030-25953-2 ISBN 978-3-030-25954-9 (eBook)
https://doi.org/10.1007/978-3-030-25954-9

This Springer imprint is published by the registered company Springer Nature Switzerland AG
The registered company address is: Gewerbestrasse 11, 6330 Cham, Switzerland

Acknowledgements

Writing a book is a process that is pleasurable, inspiring and rewarding—but also, at times, cathartic, frustrating and painful. This is the case not only for the author himself (or herself), but also for those around him.

There are always people to thank—especially those who have supported the production of the publication in different ways. I am particularly grateful to Clare Brooks and Joop van der Schee—the commissioning editors for the small collection of books on geography education research of which this publication is a part. The series is supported by the International Geographical Union Commission on Geography Education (IGU-CGE) and by the series publisher, Springer. I was delighted to work as an editor with Clare, and her colleague Mary Fargher, on the first book in this series—*The Power of Geographical Thinking*—in 2017. I hope my current offering will be as well received as that publication has been.

Special thanks must go to two of the leading lights in geography education research, nationally and internationally, who generously gave their time to be interviewed by me—namely Clare Brooks and David Lambert, Institute of Education, University College London. David has been something of a 'fellow traveller' with me during my career in geography education: he has also always been a source of level-headed, intellectually secure and inspirational thinking in the field. I spent the first three happy years of my career teaching in a state school in Hertfordshire under his direction as Head of Geography—a position I took over from him when he moved to work at the Institute for the first time in 1986. Clare and David have also been close colleagues of mine in the Geography Education Research Collective (GEReCo), which Clare and I set up in 2007. Also interviewed for this book were two leading education academics from beyond the field of geography education: Christine Counsell, Ex-Lecturer in History Education at Cambridge University and now Member of Ofsted's curriculum advisory group; and Alis Oancea, Director of Educational Research at the University of Oxford, Department of Education. Both were characteristically generous with their time and thoughts—I owe them many thanks.

I must also acknowledge the receipt of one of Oxford Brookes University's first Research Excellence Awards, which funded a sabbatical of one semester's research leave in 2017. This afforded me valuable time away from other duties, during which much of the groundwork for this project was completed.

Finally, my thanks to those who have been affected most by my efforts to write—to my wife, Cathy, who has always been wise enough to know when to shut the door and 'leave me to it', especially if the writing process has become frustrating. And, finally, to my much loved Spanador (Google it!)—Bramble—who joined me for long walks in Cannon Hill Park in Birmingham, and latterly in Aughton Wood in the Lune Valley, when I was trying to 'think things through' during the writing process.

As always, errors of omission and any mistakes in the text—for which I apologise—are all my responsibility.

Brookhouse, Lancashire, UK   Graham Butt
January 2019

Contents

Chapter 1
Scene Setting

1.1 Context

The introductory chapter of *Geography Education Research: Retrospect and Prospect* starts by raising the important question of whether geography education research, as a 'specialist field of intellectual endeavour' (Lambert 2010), actually exists. Although such a consideration may appear rather incongruous—research obviously occurs in this area, whether 'specialist' or not—it does lead us towards a discussion of the nature of research in geography education, how it might be described, whether it constitutes a field, and why (or indeed if) it matters. This approach soon raises further questions which form part of a much larger debate: concerning theory and thought in geography education, the progress made in pursuit of our research questions, the impact on research of changes in education policy and practice, and the positioning of geography education research within and beyond higher education institutions. Many of these questions are equally valid both in the context of the UK and further afield.

One of the reasons for writing this book has been to open up—for geography educationists, as well as for those from other disciplinary and subject backgrounds who are interested in the role of research in education—a challenging account of past and future research endeavours in geography education. To do so successfully a clear focus has to be achieved on the bigger picture of research in education. There is some agreement—driven by a considerable amount of evidence, some anecdotal, but much that is research-driven—that education research in the UK is nearing a state of crisis (Whitty 2005; Furlong 2013). This is not to say that in every institution where research is undertaken, whether these are universities or schools, there are crippling problems—this is far from the case. But the general 'health' of education research, and of the outcomes it produces, is patchy. Some individuals and institutions that regularly engage in research thrive, while others struggle to survive: importantly most researchers have concerns about their continued productivity, about future funding for their research, and about the high stakes assessment of their publications. Reviews of

© Springer Nature Switzerland AG 2020
G. Butt, *Geography Education Research in the UK: Retrospect and Prospect*, International Perspectives on Geographical Education,
https://doi.org/10.1007/978-3-030-25954-9_1

the state of research in education in general, and of geography education research in particular—both at the national and international scales—suggest that these problems are widespread (Butt and Lambert 2014; Lidstone and Williams 2006a, b). This situation encourages those of us who have dedicated much of our professional lives to the advancement of education research to consider just how we got to where we are now. More significantly, it also encourages us to think about how we might get to where we want to be in the future. This is the essence of this book—with its dual focus on both the 'retrospect and prospect' of geography education research.

On the surface things may not appear to be too bad. It is certainly not my intention to over inflate, sensationalise, or dramatise the current 'state of play' in education research: rather I strive to provide evidence that is open to critical assessment and reflection, evidence which offers the reader the opportunity to consider all sides of the arguments. To start on a positive note, the volume of research undertaken in geography education in the UK, and internationally, has substantially increased over the past 15 or so years. There are many drivers that have caused this increase—not least the expectation that almost all academics in universities should now be engaged in some form of meaningful research and publication, and that teachers who wish to gain education qualifications (within and beyond those offered in pre-service preparation for the profession) should also undertake some substantive research. Research in geography education is now venturing into spaces and ideas that take us beyond narrow, small-scale, 'what works' enquiries based solely in classrooms.[1] There is also a greater realisation that change in the practice of education is best achieved, and more long-lasting, if it is made from a basis of reliable evidence—rather than from an acceptance of commonly held beliefs, or opinions. The importance of evidence-based, or evidence-informed, practice in education—with such evidence being provided by research—is now more widely accepted than ever before. The need to generate research evidence to help drive future developments in education policy and practice is therefore more firmly established, although there are inevitable disagreements about what constitutes valid evidence.

1.2 Introduction

We must note that the number of contributors to high quality research in geography education (be they academics, postgraduate researchers, students, practitioner researchers, or others) is relatively modest when compared with other research fields in education (Butt 2018). This is true not only in the UK, where the traditions of educational research in geography are considered to be strong, but also across most of the globe. There are particular dangers associated with being a university-based researcher in a field where the scale of research activity is small and outputs are

---

[1]This may sound pejorative, but see Sect. 1.7 for a more detailed consideration of why an over-reliance on 'what works' research may not be an entirely healthy foundation for educational research in geography, or indeed in any other subject.

modest, particularly in relation to achieving a critical mass of high-quality research publications. The implications for funding, academic tenure, professional status and (of increasing importance) impact are also substantial. But geography education researchers, and researchers in many other subject areas, are not alone in facing such issues. These are difficulties that most researchers in education encounter—their status in the overall hierarchy of research endeavours in academia is often unexceptional, particularly if their research is subject-based. The issues relate to the scale of their research efforts, the possibilities of achieving funding for research, the methodological rigour of the research work carried out, and the expected culmination of the research. These are real and pervasive issues (Whitty 2005). There are also pertinent and troubling questions to answer regarding whether most geography education research is merely 'education research with a geographical hue' (Lambert 2010). To summarise: in many higher education institutions, nationally and internationally, education research and researchers find themselves positioned near the base of a pyramid—where those involved in subject-based education research are usually on the lowest level (Butt 2018). This chapter seeks to 'set the scene' and introduce some of the main questions pursued within the book by highlighting why it is essential for geography educators to articulate clearly the need to establish a strong research foundation within their field—one which expressly justifies a theorized contribution to education and geography curriculum thinking. This can be encapsulated into a question which Roberts (2000), with characteristic clarity, articulated as follows: 'What is the purpose of geography education and what is worth researching?'

This introductory chapter offers a brief overview of the current state (or, if you prefer, 'condition' or 'health') of geography education research, in the UK and elsewhere. It illustrates the comparative fragility and weakness of such research, as part of a research sector that is not considered to be one of the 'big players', either within universities or beyond. But, at the same time, it offers positive suggestions for 'ways forward' which will hopefully see geography education research sustain its current position and subsequently grow. The chapter is therefore structured in such a way as to 'flag up' questions, which are explored further in the text that follows—essentially these are questions about what is significant, original and really matters in the field of geography education research. Importantly, space is also given to explain the methods and methodology employed in researching the book, to describe the education systems within the four jurisdictions that make up the *UK case*, and to celebrate the importance of research into primary geography education.

1.3 Education—A Discipline?

John Furlong begins his influential book '*Education—an anatomy of the discipline: Rescuing the University project?*' (Furlong 2013), with a chapter titled 'Education—a discipline?' This is a helpful question to ask for, as Pelikan (1992) explains, the positioning of the 'professional schools' within universities (such as education, medicine, law) witness them continually striving to maintain a delicate balance between being part of the university sector on one hand, and part of wider society on the other. These disciplines are broad, intellectually coherent fields of study, connected to everyday matters and part of the wider world; given the nature of their work they are also often the focus of political debate and action. Their position, status and function within the public domain has an impact on the ways in which the disciplines have developed—and on their changing research emphases, their epistemological stance(s), and the directions in which their theoretical and methodological foundations shift. The long-established connections between the discipline of education (as opposed to its conception as a 'field'[2]) and the professional preparation of teachers is, for Furlong (2013) at least, a source of its political weakness. This is compounded by the fact that although universities in the UK have been involved in teacher education since the end of the nineteenth century, their schools, departments and faculties of education have remained largely peripheral to the main university project (often being physically located outside universities, in teacher training colleges and former polytechnics[3]). Long established universities in the UK, many having eventually accepted the arguments that education should be represented within their faculty structures, now find themselves at the whim of increasingly centralized political control of teacher education and research. The hardest won victory—of ensuring that education was recognised as a discipline in universities in the UK, and indeed elsewhere—was essentially pyrrhic, given the extension of government control over universities from the end of the twentieth century.

If the weakness of education, as it is represented within universities, is structural or systemic then this problem is possible to resolve. But of greater concern is the argument that education, as a discipline, is also 'epistemologically weak' (Keiner 2010; Becher 1989). This is because education researchers often seek to answer epistemological questions in largely pragmatic and professionally oriented ways, utilising 'soft' knowledge derived from (re)interpretations of information gathered in the humanities and social sciences. This has created a seemingly endless debate, and long-standing disagreement, over the form, content and control of knowledge—leading to contentions about what educational knowledge is and what it should be, what the role of education research is, and also about how knowledge should be represented in professional education and practice (Furlong 2013; see also Ryle 1949). Geography education, and research into geography education, exists within a sub-set of this larger conundrum, both nationally and internationally. As a brief aside, it is encouraging to note that Furlong (2013)—who we must remind ourselves is *not* a geographer—recognises that although many of the problems that societies face across the globe go far beyond the traditional purview of schools, a significant

---

[2]Furlong (2013) explains that 'field' may arguably be the preferred term for education, as it covers so many different contexts (from early years education to lifelong learning); topics (from the teaching of reading to management of HE); other disciplinary perspectives (from neuroscience to philosophy); and approaches to research and scholarship (from literary studies to Randomised Controlled Trials). As such, for some, education fails the first test of a discipline: with respect to coherence, distinctiveness and epistemological rigour.

[3]In the UK change in legislation enabled polytechnics to become universities after 1992. Subsequently all polytechnics chose to change their status, giving rise to the term 'post 1992 universities'.

number of substantive and enduring problems do nevertheless fall within the realm of geographers' and geography educationists' interests:

> The environment, our aging population, migration, economic recession, global competitiveness, all of these issues have important educational dimensions that raise powerfully important questions that educationalists, in collaboration with others, need to respond to (Furlong 2013, p. 5)

Moreover, the nature of these issues highlight the importance of educating teachers broadly within the liberal education traditions, to instruct and inform them about what Arnold (1869) referred to as the 'best that has been thought and said'. In part, a consequence of such reasoning was the publication of the Robbins Report[4] (1963), which recommended that all teachers in the UK be educated in the higher education sector.[5] It was also recognised at the time that there was a distinct moral element to teacher education—specifically regarding the instruction of teachers as to the right ways to act in a professional community, in addition to ensuring their intellectual development. This emphasis is still seen in educational research, as opposed to education research, where research is often about 'making a difference' rather than simply discovering more about pedagogical practice (Whitty 2005).

1.4 What Is the Status of Research in Geography Education?

The purpose of undertaking this project has been to explore four broad objectives. These are best expressed in the form of questions to pursue, which are briefly stated as follows:

- What has been successfully achieved during geography education research's short life?
- Why has geography education research failed to have more of an impact, within the remit of subject-based education research, in the wider panoply of education research in general, and most importantly on practice in schools?
- What have been the main foci for research in geography education and why have the results largely been modest, or indeed non-existent. Why have some areas yielded to research efforts (such as enquiry), while others have proved resistant (such as progression)?

---

[4] As a result of the Robbins Report educational research diversified into 'sub disciplines' (sociology, philosophy, economics, psychology, etc.), and into subject-based, often practitioner-led, research. What Robbins proposed, and universities were ready to accept, was a 'highly academic, highly theoretical model of teacher education' (Furlong 2013, p. 32). This still remains part of the policy debate, as governments struggle with the concept of teacher education, particularly initial teacher education (ITE), being located in research driven universities.

[5] Day Training Colleges having already been established by universities for training teachers in the UK at the end of the nineteenth century.

- Why has geography education research struggled to establish links with disciplinary research in geography, resulting in an inability to answer the 'big questions' about what we should teach in our rapidly changing world?

Each of these questions asks something significant—not only about the current status of research in geography education, but also about the importance of education(al) research in general. It is apparent that criticism of the quality and relevance of education research in the UK has a long history, stretching back at least to the 1990s. These criticisms include concerns about the overall quality, ideological bias, lack of rigour, and poor user engagement of much of the research produced in education. Critics, particularly in government circles but not exclusively so, have also indicated their disquiet about the lack of culmination of such research, its low value for money and its theoretical incoherence (see, Whitty 2005; Hargreaves 1996; Hillage et al. 1998; Tooley and Darby 1998; Maclure 2003; Thomas 2007). It is important to recognise this backdrop, for we cannot truly understand the struggles that geography education research and researchers have faced, and still face, domestically and internationally, without a sense of the landscape in which their endeavours sit.

1.5   Why Does Research in Geography Education Matter?

Reservations about the overall usefulness of research which narrowly pursues questions of 'what works?' in the classroom are easily understood (see Sect. 1.7 below). We must recognise, as a matter of course, the two chief components of our interest in geography education research: specifically, an intellectual pursuit of knowledge creation in 'geography' and of the process of 'education'. Much current research work in geography education is, understandably, situated in classrooms—research that is taking place on a regular basis, formally or informally, among geography teachers and their students. This research is not necessarily always narrowly focussed on the continuing advancement of students' learning of geography, although this is indeed common. In the classroom context practitioner researchers, usually understood to be teachers researching their own practice, may have numerous motivations for undertaking their research. This may be to better support children's learning, or to help solve a particular pedagogical problem that the teacher and his or her learners face—research may, perhaps, aim to enhance the development of one's professional practice, or improve the perceived status of the teacher in career terms. For many practitioner researchers, the driving force for undertaking research is simply to improve the geographical learning experiences of their students, by pursuing a question or issue that the researcher believes to be important or which drives their intellectual curiosity; such research is often embedded within the structures of award bearing qualifications (such as a Post Graduate Certificate of Education (PGCE), Post Graduate Diploma in Education (PGDip Ed), Masters or doctoral programme). Here we are provided with one answer to the question of why research in geography education matters, and indeed to whom it matters. To the researcher, the value of this research matters not only in terms of their students' intellectual gain, but also because it serves as a support to their own professional practice, status and possibly career progression. It has the potential to make them better classroom practitioners and enhances the opportunities their students have to learn. Indeed, Butt (2015a), when referring to the contributions made to an edited book by a number of researchers in geography education—some of whom were Masters' students who had qualified to teach geography barely two years earlier, while others were academics who had international research profiles stretching back over 40 years—notes almost universal agreement that teacher-researchers are:

> encouraging greater critical engagement with teaching and learning, supporting reflexivity, aiding problem solving and enhancing professional learning (p. 4)

Nonetheless, we must acknowledge the reasons why research in geography education matters *differently* to different people. Almost certainly most classroom-based practitioners' views about why research matters (if indeed they do believe this) are slightly different to those of university-based academics. This is not to simplistically valorise one type of research or researcher above another; but to acknowledge that the foci of research—the scale, the methods applied, the theoretical frameworks employed and the nature of the questions pursued—may not always be co terminus. They almost certainly represent different types of enterprise, with distinctly different expectations of end results. Therefore, the question of why research in geography education matters can be a slippery one: it will be regarded by those *external* to the field through very different lenses and against very different criteria to those applied *internally* (Kent 1999a). What we perhaps need to be more aware of, as geography education researchers, is the way in which external agents value our research. In short, why should others—who are often in more influential and powerful positions than researchers in geography education, whose work they judge—believe that our research matters? In such a small field—as geography education research inevitably is, wherever it is undertaken—we cannot afford to be too inward looking.[6] Understanding and enhancing our credibility, utility and impact beyond the cosy groves of academe is surely what counts. If we are found wanting in our efforts to convince others, or indeed ourselves, that geography education research really matters, then the game is over.

Let us briefly return to the importance of research for teachers and the types of research that are necessary to support teaching and learning. Teachers are 'knowl-

---

[6]The dangers of introspection within the geography education research community are approached in an interesting way by Albert and Owens (2018), who pose the intriguing question: 'Who is listening to us from geography education? Is anyone out there?' Their research sought to assess the interchange between geographers, geography educationists and other disciplines based on a bibliometric study of citations of journal articles—its conclusions are remarkably sanguine about the influence of research work in geography education on both our own, and on other, disciplines and fields: 'The flow of information is two-way; … geography education is engaged at all levels, from its own geography education circle to the larger geography community and substantially to scholars in the field of education'. However, most significantly, the authors also reveal that 'our attempts to find research from geography education filtering into applied circles failed to generate much' (sic).

edge workers'—they exchange knowledge with learners, they help their students create knowledge, as well as assisting them in developing understanding and skills. The Chartered College of Teaching (CCT) captures the importance of pedagogical research effectively when it speaks of the importance of 'teacher knowledge'—the broad range of skills, strategies and knowledge(s) that enable teachers to function effectively in diverse classroom settings and with a wide range of student needs. The CCT asserts that teaching:

> has to be informed by a body of rigorous, high quality research and evidence rather than based on taken-for-granted assumptions, routines and habits (https://chartered.college/membership/knowledge-and-research)

Teachers possess a detailed and complex knowledge of pedagogic practice(s) which enables them to function as educators on a daily basis. But they must also strive to update and extend the knowledge they have of the subject they teach (Brooks 2015). Pedagogic content knowledge (PCK), as first described by Shulman (1987), is central to their effectiveness as educators. Their possession, and regular reinvigoration, of PCK implies a need for teachers to engage in subject-based research and scholarship. This is a prime reason why geography education research, and indeed research into all subject specialisms, should contribute to the initial teacher education and continuing professional development of teachers.

In their increasingly pressured professional lives, teachers need to be able to quickly differentiate between information and data that is truly important and meaningful—this includes aspects of educational research—from that which is merely spurious decoration. Teachers must therefore hone their abilities to undertake what Postman and Weingartner (1971) famously referred to as 'crap detecting'. They must critically sift between new policies and directives from government, pronouncements from higher education (both on education and about advances in their parent discipline), shifts in pedagogic methods, research outputs and new curricula and syllabuses. To do so we expect teachers to be secure in their handling of many different forms of knowledge and data; essentially to be constructively critical of all that is presented before them. This requires agency and an understanding of what research, in its myriad forms, is capable of offering (and more importantly what it is *not* capable of providing). Increasingly this is a 'part and parcel' expectation of the modern, professional teacher and connects to the point about teachers being 'knowledge workers' who need to regularly refresh their subject and disciplinary knowledge. It would be unthinkable to believe that a teacher could, or indeed should, rely wholesale on the subject knowledge they acquired as an undergraduate for 20, 30, 40 or more years of working in the classroom.

Academic research seeks to move disciplinary and subject knowledge forward and must be effectively conveyed to, and received by, teachers. It is

> more than the simple backdrop to the acts of teaching and learning; it is central to the whole process of geography education in schools (Butt 2015a, p. 12)

But there is a danger that regular, day-to-day, practice in classrooms becomes entrenched, habitual and therefore largely unquestioned by the teachers who engage in it. If this *is* the case, then we have a set of circumstances where research can enliven and challenge practice—whether this research is carried out by an individual, critically reflective teacher who simply wants to know and understand more about their own classroom practice, or involves reference to a more substantial, established body of existing research work. Here teachers may either find themselves undertaking formal research for an award bearing course or qualification, or may be considering work that has previously been carried out as part of a large, funded project. According to the nature of the problem researched, this may involve enquiry into appropriate research methods and methodologies to employ, relevant concepts and theories to understand, literature reviews to complete, and data to collect—but in all honesty it may not result in truly significant additions to the sum total of our research intelligence. There are also considerations to make about the audience(s) for these research findings—which may be modest in reach, or involve the communication of results to a much wider community of practice (Lave and Wenger 1991) through blogs, websites, professional or academic publications, newsletters, seminars and conferences.

I therefore take seriously the debates about how education research can inform practice in classrooms—indeed this is a major theme that runs throughout the book—but do not consider classroom-based research to be the only, or indeed always the most important, type of research that geography educators can undertake.

1.6 What Is the Current 'State of Play'?

The locus of decision making, influence and power in education has shifted significantly in the UK, and elsewhere, over the past fifty years. It has moved away from the previous post war consensus on education held among the major political parties and towards something much more directive and centralised. In England we have shifted from an era when the 'secret garden' of the curriculum—as observed by the Minister of Education, David Eccles, in 1960—was the sole preserve of academics and education professionals. The pendulum has swung, such that the views of education 'experts' (wherever they may reside), and the research on which they may base their views, are not to be trusted—or often even considered—in the drive to devise new education policies and practices. Politicians and the general public are no longer happy to leave things to the professionals, or to those who proclaim themselves to be 'experts'; witness the comments of one of the longest serving Secretaries of State for Education in England, Michael Gove, who in 2014 referred to academics in education as 'the blob'. Clearly there are many stakeholders in education whose opinions should be considered and valued. The 'laity' must be allowed to state its views—no bad thing, if kept within reasonable parameters—but we ultimately need to agree on which forms of external accountability are key. The current state of play with respect to much education research is that it does not appear to be highly valued beyond the academy—even the research commissioned by governments themselves risks being pilloried if it fails to deliver 'expected' messages. Claims by government ministers to support 'evidence-based policy and practice' therefore only take us so far—frequently up to the point at which research outcomes support the original intentions of the policy maker. This is clearly a profound misuse of research evidence, full of bias and redolent of philosophical and methodological insecurity. What appears more sensible for educational research is to strive for a balance between central direction—inevitably often politically driven—secure educational purpose, and a clear intention to improve the quality of students' educational experience.

In recent years UK governments, in their attempts to modernise and remodel the teaching workforce, have noted that teachers need to be more comfortable with handling data and interrogating research findings. This led to a period in the early twenty-first century when there was a reification of evidence-based practice in education, associated as it was with an expected re-professionalization of the teaching workforce and the growth of a new generation of teachers who would be more comfortable with research activity (Beck 2009; Butt and Gunter 2007). But the goal of achieving such a research-engaged profession no longer seems to be a priority in UK government circles—particularly having now moved initial teacher education away from university partnerships towards wholly school-based preparation. Indeed, the weakening of requirements for new teachers to have been prepared and certificated by higher education institutions—to a process of teacher preparation based on classroom observation and 'trial and error'—suggests that research-informed practice now has little place. The onus for new teachers to undertake at least *some* research as part of their initial preparation for the classroom has also been removed. I would argue that basic competency training is an inadequate preparation for a genuinely modern, research-driven profession (Butt 2015b). The intention of creating a research competent teaching workforce was previously pushed, in part, by UK governments enviously viewing the Programme for International Student Assessment (PISA) scores of other nations and attempting to ape the ways in which they appeared to achieve academic success for their young people. Such policy borrowing is not only professionally dangerous, but profoundly wrong-headed. Rarely, if ever, are the social, political, cultural, historical and philosophical foundations of different nations and societies closely aligned, such that attempting to replicate their educational structures and systems rarely produces similar results. Hence, intentions to refashion the UK teaching profession into a 'Masters only' workforce—similar to that of certain Scandinavian countries—have been abandoned in England. As a consequence, the associated benefits to teachers of engaging with research activities have also been denied.

The wider context is important—Beauchamp et al. (2013) have noted that, both nationally and internationally, there is now a growing momentum towards recognizing, in Lambert's (2018) words, 'a research base that might inform teaching and teacher education' (p. 359). In this context geography education researchers must be ready to make the case for the important contribution of practical and conceptual research within their field.

1.7 Are There Dangers in 'What Works' Research?

It is apparent that there are inherent disadvantages in narrowly pursuing pragmatic 'what works' research—although at first glance this statement may appear to be counter intuitive. One might ask: why shouldn't we prioritise the types of research that seek answers to the specific pedagogical problems that teachers face? However, this is to misunderstand the nature of research in education and, more importantly, what research is realistically capable of achieving. Let us consider why seeking 'off the shelf' solutions to classroom problems—whether or not these solutions have been generated using research evidence—is not necessarily the best way forward. Firstly, the methods, techniques and approaches employed in successful teaching and learning are often highly specific to a particular context, classroom, group of students and teacher (Lambert 2015; Butt 2010). The transferability of research 'evidence' in education (or, if we are bold enough, its claims to success) is notoriously problematic—what works for *this* teacher and *these* learners, on *this* particular day, may not work for *those*. Secondly, although striving to achieve evidence-based practice is still very attractive, recent reactions *against* the sharing of 'expert knowledge'—from the inspectorate the Office for Standards in Education, Children's Services and Skills (OfSTED), the Department for Education (DfE), the former National College for Teaching and Leadership (NCTL), advisory services and some academics—has had a negative impact on education research and researchers. Thirdly, there is a danger in narrowly pursuing 'what works' research for, as Roberts (2000, 2015) points out, just because something 'works' does not necessarily mean it's worth doing. This issue is underpinned by assumptions about what things teachers, politicians and other stakeholders believe are important. Gaining a definitive view on all this is difficult—as there are competing ideas about what is significant in education, what education is for, and what it ought to be doing. These questions also have profound implications for our choice of research methods. The rise of Randomised Controlled Trails (RCTs), and reactions against qualitative research, have underlined an official drive towards finding empirical *evidence*, particularly where policy formulation is an aim. On the one hand, it is difficult to argue against basing practice on hard evidence; however, the overheated claims of objectivity and replicability with respect to some quantitative research are often unrealistic and should be countered. Education research is certainly *not* the same as research into dentistry, medicine, law or accountancy.

In all honesty teachers tend to be sceptical about most educational research. As many have previously observed, teachers do not tend to value research as they believe that it merely tells them what they already know, or that it has been conducted 'by academics, for academics'. The consequence is that classroom practitioners either choose not to access research evidence, or would not (through little fault of their own) be able to usefully engage with what research is telling them (Naish 1993). This plays into the hands of those (often politicians and policy makers) who push for 'what works' research—which they consider to be better value for public money, with its (false) promise of producing easily applicable solutions to intransigent problems. Teachers, for understandable reasons, are often not considered to be the main audience for education research—themselves being the object of study, or tangential to much of the research work undertaken by education researchers (see Cochran Smith and Lytle 1999; Roberts 2000). In conclusion, Slater (2003) observes that:

teachers do not read research, generally speaking, and do not use its findings in their teaching practices. Teachers can be disdainful about and even hostile to research and research jargon… many teachers see themselves as practitioners dealing with the practical, not the theoretical (p. 285)

Additionally, as Lambert (2018) warns, it is unhelpful:

to expect the role of the teacher to expand to become a researcher too … To heap active research onto teachers is an unreasonable and unnecessary expectation, and a distraction from teachers' core professional responsibility for curriculum making (p. 366)

1.8 'Education Research with a Geographical Hue'?

In many respects classroom-based research, when practitioner-led, may simply be a slightly more formalised extension of the problem solving teachers do in their daily work. Essentially, practitioner research may:

build upon their regular process of reflection – which all teachers engage with on a day-to-day basis, often when tired, at the end of frenetic working days – to achieve more considered, systematic and structured forms of sustained enquiry (Butt 2015a, p. 6)

There is a significant amount of such research and scholarship in geography education, much of it resulting from formal postgraduate study for award bearing courses (Brooks 2018; Catling and Butt 2016). However, there is an issue as to whether most, indeed any, of this research ultimately has a profound impact on education policy and practice. This begs a larger question: if the bulk of research is practitioner based, how valuable is educational research? I started this chapter with Lambert's (2015) observation about whether geography education research constitutes a 'specialist field', involving 'educational research conducted within the context of geography as a school subject' (p. 21). Such research, he acknowledges, takes in 'geography' and 'education'—but also has to consider a much larger backdrop of previous work on the disciplinary nature of geography and education, often involving consideration of what constitutes subject knowledge and what ought to be taught.

Essentially, what is described above is a problem of scale and focus. Roberts (2015) helpfully reminds us that although researchers in geography education can draw on their own (albeit small) field of research, this is not the only starting point for their research endeavours:

it is worth drawing on three other fields: generic research, academic geography research and research related to other school subjects (p. 49)

This is an extremely pertinent point, as too much research in geography education is, for wont of a better phrase, 'self-regarding'—often using the same sources of research evidence, possibly ignoring similar (better?) work done elsewhere, or seeking to answer questions that have already been successfully pursued in other contexts. Perhaps this criticism is too harsh? Let us not lose sight of what *has* been achieved within the field of geography education research, especially where it has created specific links to research within the discipline of geography. Over the years we have seen some publications that have, with varying degrees of success, attempted to bring together the thoughts of academic geographers and geography educationists. Often such publications have offered separate sections for the work of academic geographers and geography educationists, maybe with additional overview comments for the reader that draw out links and connections (see, for example, Kent 2000; Daugherty and Rawling 1996). Few publications have examples of geographers and geography educationists writing together (although see Kirby and Lambert 1984, 1985a, b; Bradford and Kent 1977); or have academic geographers and educationists from beyond the field of geography education commenting on the work of geography educators (although see Butt 2011). As such, the connections between research in geography and in geography education tend to run *parallel*—with only a few researchers explicitly trying to forge enduring links and understandings between the two. It is my contention that striving to create stronger connections between the research of geographers, and that of geography educationists, would have mutually beneficial outcomes.

1.9 How Healthy Is Research in Geography Education?

Slater's (1994) introduction to a research monograph featuring three pieces of geography education research, each of which drew on a different approach to enquiry (positivist, interpretivist, postmodern), questions whether the achievement of 'pure' research in geography education is either possible, or helpful. She therefore questions whether our definitions of education research exist. By encouraging boundary crossing in research activity, drawing on methods and techniques from other paradigms, Slater argues that advances in geography education research could be made. This is both an attractive proposition and somewhat dangerous—witness, for example, the recent criticisms of the rise of mixed method approaches in research in education (Giddings 2006). Whilst Slater raises what are essentially methodological and philosophical questions about research into geography education, we must also be aware of issues that are more prosaic and practical. Unfortunately, given the criticisms targeted at much educational research, we find that research in geography education is not strongly positioned:

> It has not traditionally attracted much government, funding council or project money—large scale, generously funded research projects in geography education are unusual, with most research being produced as small-scale, practitioner-based, unfunded work often reflecting the interests of a few individuals (Butt 2015a, p. 7)

The intellectual and practical contribution of much of our research work is, arguably, limited. It may support the professional development of individual teachers, enable small groups of teachers (and learners) to move forward in their pedagogic endeavours, or help someone complete a project or dissertation—but not contribute substantially to the advancement of thought and deed in geography education. Because our research tends to be small scale and practitioner-based our expectations must be tempered—how can we expect significant impacts and substantial shifts in professional practice from such research? However, both Brooks (2010), with reference to practitioner research in geography education, and Finney (2013) in music education, make similar important points—if individual teachers can benefit from practitioner research surely this may be 'enough': not all research will cause seismic shifts in government education policy (nor is it designed to). This does not invalidate the research undertaken by teachers or academics, but we must recognise its local and context bound nature.

Lambert (2015) asserts that making a case for a research base for the teaching profession is 'unremarkable':

> surely any professional body is anchored by a specialist knowledge base, and this is created and refined through research and scholarship (p. 15)

He acknowledges that some politicians, and even some educationists, argue that the need for training and professional development for teachers is largely redundant—teachers, in their view, simply need an enthusiasm to impart subject knowledge. This position, if believed, strongly undermines the case for the teaching profession to be research informed. Moving away from the everyday pragmatics of 'what works' research, to more nuanced discussions about the core principles that underpin effective teaching, must (for Lambert) introduce consideration of the curriculum and how it is taught, linked to an appreciation of the contribution of the subject discipline.

In conclusion, there is clearly a need for robust research evidence in geography education—evidence which advances our knowledge and understanding and which can stand up to outside scrutiny. But research in geography education (and indeed in all other subjects) regularly finds itself questioned, externally, concerning its significance and rigour. The dangers of small scale, non-generalizable research—or research that is, at best, limited to making 'fuzzy generalisations' (Bassey 2001)—are apparent to practitioners, academics and politicians alike. Research based in single classrooms and on unique encounters has some value—but rarely is it robust enough to ensure confidence if we expect strong recommendations, or policy, to be built on the back of it.

1.10 What Are the Ways Forward?

I have a firm conviction, in agreement with Hargreaves (1996), that teaching is a research-based profession. With Lambert (2010), I believe that geography education research is concerned with exploring the relationship between two 'big ideas'—

those of 'geography' and 'education'. However, it is an inescapable fact that the political influences on research in higher education, affecting both geography and geography education, are substantial and very unlikely to go away. Referencing the political turmoil at the time of the creation of the geography national curriculum in England—with a comment which aptly suits the act of policy making throughout the quarter century that followed—Walford (1996) stated:

> One year's apparent certainties have turned into the waste paper of the next; that which was praised by one Minister was frequently overturned by a successor (p. 131)

Noting that in the eight years from its inception to its 'final revision' the passage and development of the geography national curriculum was influenced by no fewer than five Secretaries of State for Education, Walford's point is a good one. It recognises that political change is often rapid and contradictory, cutting across well intentioned plans for research in the most dramatic ways. Given that the vagaries of political 'interference' will always be present—no matter what the area of research, or the national context—perhaps we should focus more directly on questions closer to home, about things we still have some degree of control over. These might include considerations of past, current and future research priorities for geography education; about what we have already discovered (or think we have discovered); about what we can be confident in saying about teaching and learning in geography; and about what we need to research next. Roberts (2015) outlines some 'significant gaps' in research in geography education—often where access issues exist for academic researchers, making them more applicable to practitioner researchers—including:

> 'the influence of school culture and location on geographical education within a school; how geography teachers think about and practice 'curriculum making'; how students use and make sense of geographical source materials found in textbooks and on computers; how students make sense of statistical information; students' understanding and misconceptions of key geographical concepts; how classroom discussion and dialogic teaching can contribute to geographical education and how students learn to 'think geographically' (pp. 49–50).

A considerable job has to be done to make education research, both in geography and elsewhere, more relevant to all the stakeholders who should engage with it. Two points occur: firstly, that this research has to be good enough; secondly, it has to be communicated effectively. One of our biggest challenges is to make high quality research findings the cornerstone of classroom practice.

1.11   Research Methods and Methodology

This book is largely a personal, reflective piece about the changes witnessed in geography education research in the UK over the past 30, or so, years—with additional thoughts on, and comparisons with, similar research at the national and international scales. It does, however, have a structure and methodology that hopefully elevates it from being merely a subjective and introspective piece of writing. A significant element of the work involved in producing this book has entailed reviewing extant literature—mainly in the fields of geography, geography education and education. This body of literature included articles published in professional and academic journal, single authored and edited books, research papers and pamphlets, conference proceedings, transcripts of public lectures (including inaugural professorial lectures), newsletters and, occasionally, blogs and webcasts. The main themes of interest were pursued methodically across these texts. Given that the focus of this book is predominantly on geography education research in the UK, but with the inclusion of selected examples of comparable research at the international scale, the primary sources of information were often linked to the UK context. I, like others, note the difficulties faced in attempting to achieve a truly systematic, global account of geography education research (or indeed any form of subject-based research), as well as detailing the inherent dangers in attempting to do so (Butt and Lambert 2014; Gerber 2001, 2003; Rawling 2004).

The outcomes of my research have inevitably been influenced by discussions with colleagues in the field. Such discussions are often informal and self-interested, however they were—on four occasions—elevated into more coordinated, formalised and research-oriented, semi structured interviews. These interviews were conducted with leading figures in contemporary education research, two from the field of geography education and two from beyond—Professor David Lambert, Dr. Clare Brooks, Professor Alis Oancea and Christine Counsell—each respondent was UK-based, but with a national and international reputation for excellence in their chosen areas of expertise. The text explains the reasons why these four were selected to illuminate subsequent chapters with their thoughts, as well as through reference to their previously published work. The research and publications of Lambert and Brooks will already be known to many in the field of geography education research; Oancea's and Counsell's contributions perhaps less so. The latter were specifically chosen for their respective expertise in the national assessment of research outputs (such as the Research Excellence Framework in the UK) and for ideas about the future shape and location of subject-based education research.

The nature of this predominantly personal, reflective piece of research means that it inevitably draws on a variety of publications that I produced over the last 30 years, or so, of my academic career. This book has been strongly influenced by my previous role as a secondary school geography teacher, and subsequently as an academic with a keen interest in research in geography education working in higher education institutions in England. As a consequence it contains something approaching a re-evaluation, or indeed a re-purposing, of much of my published work, including at times an element of exegesis—'a retrospective critical analysis of the works which locates them in broader fields of theoretical enquiry and scholarly literature, and establishes the contribution the publications have made to knowledge' (Biddulph 2016, p. 1). The personal and professional contexts in which any research is produced are important—they need to be recognised and reported. This serves to help the reader understand the researcher's situation and motivation, but also identifies sources of potential bias—from which all research, whatever its methodological framing, will suffer. As Bourke (2014) articulates, bias is an inevitable element of research endeavour—a consequence of the particular beliefs and 'identity' of the researcher, shaped by their class, race, gender and educational background. One has to be honest and (hopefully) principled about the bias that affects the researcher's work, despite our aspirations to achieve objectivity in the face of the influence of subjectivity (Biddulph 2016).

The sub-title of the book—*The UK case, within the global context*—indicates that although the text is UK-focused, it also embraces significant reference to geography education research produced elsewhere. Historically research into geography education produced in the UK, particularly in England, has had an important international sway. However, some would argue that its impact has not necessarily always been benign (Bagoly-Simó 2014; Lidstone and Gerber 1999; Lidstone and Stoltman 2002). Indeed, the previous editors of *International Research in Geographical and Environmental Education*, reflecting on their stewardship of the journal during the first eight years of its publication, stated:

> We are left with the disturbing feeling that the result of all our endeavours favours both those whose first language is English and, perhaps more importantly, those whose cultural background – both personal and academic – rests in the English tradition (Lidstone and Gerber 1999, p. 219)

The productivity of UK-based researchers in geography education alone might justify the parameters drawn for this book, but it would be very wrong to imply that equally, or more, important research in the field does not occur elsewhere (Chalmers 2011). Indeed, it is probable that the future directions and outputs of geography education research and researchers may well be shaped principally beyond the shores of the UK. The UK influence on geography education research, in terms of productivity and organisational structures, is therefore historic, deep-seated and profound—but not without notable parallels and prospects in other countries. British researchers in geography education have traditionally been prominent in advancing education theory and in influencing educational policy and practice—often helping to model ways forward for organisations and institutions involved in geography education and research, internationally (Solem and Boehm 2018). Their outputs should not, however, be considered in isolation; the canon of research produced by geography education researchers that make up *the UK case*[7] is therefore placed here within the broader international context. However, it is to a consideration of schooling in the UK context that I now turn, as the backcloth to much geography education research undertaken across its four jurisdictions.

1.12 The UK Case

To international audiences the nature of education within the UK can sometimes be confusing, often being conflated solely with the English national setting. While

---

[7]The term 'case' used in the sub-title is important: the book is not technically a case *study*, but nonetheless draws a firm focus on geography education research within the parameters of education in the UK.

English influences remain strong across the UK—and form the starting point for much of the analysis and commentary within this book—we must acknowledge that the four countries that comprise the UK have different, devolved systems of governance which affect their schooling, education policies and research. Devolution—the statutory delegation of central government powers in the UK to its four jurisdictions since 1997—has meant that education policy and legislation has developed differently in each nation for the past 20 years. As Lingard (2013) observes, the variance in education policy and practice across the four jurisdictions reflect specific historical and political dissimilarities, with each system now being distinctive in nature and scale (see also, Brock 2015). The consequences of political devolution therefore have cultural, historical and political impacts affecting both how subject-based education is conceptualised and how subjects are taught. Any blanket reference to the influence of the 'British government' with respect to educational matters is therefore dangerous, given the devolved nature of education systems in the UK (Butt et al. 2017). Curricula in the four jurisdictions are clearly related, but nonetheless contain their own, very distinctive, distinguishing elements.

England

Education in England is overseen by the Department for Education (DfE). Central government increasingly exercises direct control over state education, as Local Authorities (LA) continue to decline in importance with respect to the implementation of national education policies and practices. The majority of children attend state schools, including comprehensive and selective 'grammar schools' at secondary level, with around 7% attending fee paying independent or 'public' schools—which are not required to teach the national curriculum, nor employ specialist trained teachers. Comprehensive education, enjoyed by the majority of pupils, is further divided into free schools and academies, including religious schools.[8] All schools are inspected by the Office for Standards in Education, Children's Services and Skills (OfSTED). A national curriculum exists, first introduced in 1988, which divides learning phases into Key Stage 1 (KS1: ages 5–7, or infants), Key Stage 2 (KS2: ages 7–11, or juniors), Key Stage 3 (KS3: ages 11–14) and Key Stage 4 ((KS4: ages 14–16). Key Stage 5 is for 'post 16 education' (16–18)—typically within a school 'sixth form', a sixth form college, or Further Education (FE) College—with 'tertiary education' describing provision for students who wish to continue their education at 18+. Education is compulsory until 18, following the passing of the Education and Skills Act 2008, although *schooling* ends at 16. Post-16 education is either academic or vocational, including apprenticeships, traineeships, and volunteering. Students typically take examinations for the General Certificate of Secondary Education (GCSE) at 16 and Advanced (or 'A') Levels at 18. Geography remains a popular

---

[8]More precisely, there are six types of state funded (or 'maintained') schools in England: free schools, academies, community schools, foundation schools, Voluntary Aided schools and Voluntary Controlled schools. Secondary schools are mostly comprehensive schools requiring no entrance examination. The majority of comprehensives are also 'specialist' schools—receiving additional state finance, and with the option of applying partial selection of intake, to one or more subjects in which the school chooses to specialise.

GCSE option for many students, encouraged by its inclusion in the so-called English Baccalaureate (EBacc)—not a qualification in its own right, but 'a combination of GCSE subjects, including a language, that offer an important range of knowledge and skills to young people' (DfE 2017). Higher education includes bachelor's degrees, master's degrees and doctorates, with tuition fees for the former currently at £9250 per year for English, Welsh and European Union students.[9] Barnes and Scoffham (2017) indicate that in England 'schools in both urban and rural areas operate in rapidly changing social, economic and political conditions' (p. 298); the English population is also ethnically more diverse than in the other jurisdictions (Office for National Statistics 2014; Scottish Government 2012). Since the introduction of the national curriculum, English schools function in a landscape where a stream of initiatives and government regulations have meant that there are 'few certainties, apart from an ever-increasing demand for accountability and ever more stringent targets' (Barnes and Scoffham 2017, p. 298).

Curricula in English schools are commonly divided into separate subjects, as opposed to themes or topics, with three subjects regarded as 'core' (English, mathematics and science) and the rest 'foundation' (including geography) (Catling 1999; Kent 1999b; Rawling 1999). This arrangement has obvious implications for the prioritisation and status of subjects in primary and secondary schools. Cross curricular links are therefore limited, as knowledge tends to be contained within discrete subject blocks.[10] Guidance on the constitution of the school curriculum is rather unspecific, with the National Curriculum in state schools having been gradually 'watered down' since its inception in the late 1980s. The 2014 version of the National Curriculum simply states 'every state-funded school must offer a curriculum which is balanced and broadly based' (DfE 2014). However, in recent years there has been an increased government emphasis on the provision of factual knowledge and on the role of subject disciplines. As Barnes and Scoffham (2017) report, with particular reference to primary education in England, contrasts with the other UK jurisdictions are notable:

> In history and geography, for example, the statements on subject contents and skills all start with the words 'pupils should be taught to' or 'pupils should be taught about'. In geography, pupils are required almost without exception to 'name', 'locate', 'identify' and 'use'. The term 'understand' is only used once in KS1 and twice at KS2. The Scottish 'Curriculum for Excellence', by contrast, commonly uses terms like discuss, consider, explore, appreciate and explain (p. 300)

Foundation subjects, such as geography, therefore have smaller programmes of study, are subjected to inspection frameworks from the Office for Standards in Education (OfSTED)—where performance in individual subjects is subsumed into reports

---

[9]At the time of writing, before BREXIT, national academic examinations and vocational qualifications in schools and higher education in England met European Qualifications Framework requirements which aligned them to assessment regimes across the European Union.

[10]This is not the case in the curriculum for children under the age of 5 where the Early Years Foundation Stage (EYFS) curriculum contains seven areas of learning, including one titled 'understanding the world'. Evidence of the amount of time children spend studying the humanities subjects at primary school (ages 5–11) is limited, with most schools spending about 4% of teaching time (about an hour each week) on each of the three humanities subjects.

on 'work across the curriculum'—and are appraised by non-subject specialist inspectors. This situation is paralleled in the inspection and review of initial teacher education (ITE) courses—such as the Carter Review (2015)—within which few discrete mentions of geography, or history, occur. The allocation of time for ITE in geography in English university-based training programmes has been reduced steadily since the late 1990s, with foundation subjects at primary level being provided with just a few hours for the preparation of new teachers—often resulting in teachers entering service with poor or inadequate subject knowledge in geography (Catling 2016; Catling and Morley 2013).

According to the World Health Organisation's report on 'Health Behaviour in School-Aged children' (WHO 2016), English schoolchildren are not as happy as most of their European counterparts—children report higher tendencies for disliking school, often feel pressurised by school work, and do not perceive their peers as being kind and helpful. The report presents a picture of declining mental and social well-being among English children, and of their relatively poor psychological, physical and social development—particularly among children from economically deprived backgrounds. This state of affairs is further substantiated by children's charities working in England. There is an inevitable concern about any education system that labels children as successes or failures at an early age, creating dissatisfaction among students about their formal education and schooling.

Northern Ireland

The education system in Northern Ireland is arguably characterised by the most significant and distinctive levels of academic and religious difference, indeed segregation, throughout the UK jurisdictions. Here the national context—with its particularly complex, troubled history and record of sectarian and cultural division—has had a particularly strong influence on schooling, revealing a tendency to respond to economic and social change in distinctive ways (Clarke 2006). Selective education still exists in Northern Ireland, with pupils tested at 11 to determine whether they will progress to grammar or secondary schools, and with faith schools—Protestant and Catholic—even now exerting a hold on education policy and practice. There are currently comparatively few 'integrated' schools in Northern Ireland—that is, schools that are non-denominational and unaffected by religious affiliation—creating a distinct and enduring separateness in educational provision. A handful of Irish language medium schools, and English-speaking schools that have Irish language units, also exist. State schools in Northern Ireland are referred to as 'Controlled Schools', under the management of the Education Authority (EA) and elected governing bodies. The Northern Ireland Curriculum (NIC), which is based on the national curriculum used in England and Wales, has entered a phase of greater flexibility and local decision making since the Good Friday power-sharing agreement was signed in 1998. This followed long periods of direct control from Westminster—indeed, in the two decades following the introduction of the national curriculum in the late 1980s:

> Northern Ireland followed fairly closely the pattern of curriculum prescription by government in England and Wales (Greenwood et al. 2017, p. 309)

The majority of examinations taken by students in Northern Ireland are set by the Council for the Curriculum, Examinations and Assessment (CCEA), although these broadly follow the format of English and Welsh GCSEs and A Levels. The performance of students in public examinations are consistently the highest of all the four home nations in the UK.

Schools in Northern Ireland contrast with those in England regarding geography teaching—particularly with respect to the adoption of issues-based approaches to geographical investigations and a willingness to engage in cross curricular approaches, which has achieved statutory recognition. This reflects a growing willingness among teachers in the province to recognise, and in part address, the historically divided society in which young people are growing up. Here, as in other UK jurisdictions, Religious Education (RE) is a separate subject that stands outside the national curriculum, but controversially in Northern Ireland the responsibility for the 'Core Syllabus' for RE remains with the four major Christian denominations (Greenwood et al. 2017). The importance of Northern Irish schools reflecting the desires for a shared future, tolerance at every level of society, and preparation for life in a diverse, inter-cultural community led to radical proposals for changing the Common Curriculum in 1999 which, according to Gallagher (2012), had by 2004:

> transformed the original content-heavy, strongly regulated format of the original NIC into as more flexible, framework-style curriculum which gave schools much greater freedom to interpret minimum requirements in ways that would meet the needs of their pupils (Greenwood et al. 2017)

The 2007 revision of the NIC primary curriculum (CCEA 2007) has created the most radical curriculum framework in the UK—it being structured into 'Areas of Learning', rather than separate subjects as in other jurisdictions. At primary level this has meant that teachers have had to adopt 'a pragmatic, balanced and reflective approach to their planning, something which requires a high skill level in 'curriculum making' which, in turn, can be enhanced by effective in-service training' (Greenwood 2013, p. 443). The majority of teachers in Northern Ireland appear positive towards geography, history and science being integrated within one Area of Learning (Greenwood 2007, 2013).

Scotland

The Scottish education system is distinctly different to those which exist across the rest of the UK, with the Scottish Government having full legislative control of education following devolution at the end of the 1990s. The Scottish Curriculum is dissimilar to the variants of the national curriculum applied in schools in England, Wales and Northern Ireland having greater breadth and consisting of eight curricular areas of study—previously encapsulated in a '5-14 Programme'. Partly as a consequence of the structure of the school curriculum, most university degrees in Scotland are four years long, compared to three in England, Wales and Northern Ireland. Data from the Office for National Statistics (ONS 2014) reveals that Scotland's tertiary education system contributes significantly towards making Scotland one of the most highly educated countries in the world, with attainment levels for 16–64 year olds above those of people living in Finland, Ireland and Luxembourg. Just as elsewhere in the UK, local authorities are involved in the management and operation of state schools, supported by Education Scotland. Less than 5% of students in Scotland are privately educated.

There is an emphasis in the current Curriculum for Excellence (CfE), introduced into Scottish secondary schools in 2012, on cross curricular learning, as well as on the promotion of literacy, numeracy, health and well-being. The curriculum is broad, flexible and gives emphasis to the Scottish contexts for learning, within a national and global setting (Robertson et al. 2017). The CfE consists of Experiences and Outcomes in the curricular areas, with Social Studies encapsulating elements of geography education alongside history, Business Studies and Modern Studies. State secondary schools in Scotland are comprehensive, non-selective and mainly non-denominational in nature. Predominantly, these are referred to as High Schools and Academies, alongside a much smaller number of Secondary Schools, Grammar Schools and Colleges. Vocational education is provided by Further Education colleges and through apprenticeships up to graduate level; some schools provide Scottish Gaelic medium education. All universities are funded by the Scottish Government, with students from Scotland and EU countries (at the time of writing) not required to pay tuition fees for their first undergraduate degree, although fees are exacted for students from across the rest of the UK.

The public examination system and qualifications in Scotland differ from those in other UK jurisdictions. At 16 Scottish students sit the Scottish Qualifications Certificate, usually comprising between 6 and 8 subjects, including compulsory English and Mathematics. At the school leaving age of 16 students can opt to study for Higher and Advanced Higher examinations, with the latter recognised as being of broadly equivalent status to A Levels studied elsewhere in the UK.

Wales

Just as in the Scottish case, the Welsh government has sought to develop a strong sense of national identity through its school curricula, and has encouraged the growth of community-based schools—in contrast to England, where reform has promoted parental choice of schools, with a corresponding variety of school types (Marsden 1999). Geography and history are part of the National Curriculum for Wales which, like other foundation subjects, are often delivered through a topic-based approach. The Welsh national identity, and the promotion of 'Welshness', are important curricular considerations—involving teaching of the Welsh language and the inclusion of the national context in subject content (Jones and Whitehouse 2017). Around 15% of students are educated either wholly, or partly, through the medium of Welsh, with one in ten students attending schools where a significant proportion of the curriculum is bilingual. This has been encouraged through Curriculum Cwmreig, which aims to foster an understanding of what is distinctive about living and learning in Wales (ACCAC 2003). Following devolution in 1997 the Welsh curriculum has promoted Education for Sustainable Development (ESD) and Global Citizenship (ESDGC), as well as a skills' framework and early years' pedagogies. The introduction of a Foundation Phase, extending early years practice for 'under-fives' to KS1, was phased in from 2008—to increase well-being, health and enjoyment, rather than promoting discrete subject disciplines. A more radical curriculum—set out in 'A Curriculum for Wales' (WG 2015)—is planned for implementation by 2021, to further understanding of the country's heritage, culture and geography. In 2015 the Donaldson Review's recommendations were fully implemented in the new 3-16 Curriculum for Wales, mindful that successive changes to the curriculum over the previous 20 years had led to stakeholders' concerns about curriculum and assessment overload, greater curricular prescription, increased system complexity, and decreasing relevance of education outcomes (Jones and Whitehouse 2017; Donaldson 2015). This new curriculum is organised into six Areas of Learning and Experience (AoLE), including Humanities, with three cross curricular responsibilities (literacy, numeracy and digital competence).

The importance of promoting a sense of 'Welshness' is highlighted by Jones and Whitehouse (2017) who comment that:

> statistics from the Office for National Statistics (2002) ... reported 87% of people born in Wales identify themselves as being Welsh as opposed to being British. This was significantly different from for those born in England, where only 15% identified with being English as opposed to being British. These statistics suggest that the concept of national identity is complex, multi-dimensional and subject to change, for national identity is not fixed and is dependent on a range of factors which are country specific (p. 338).

New curriculum and assessment arrangements, introduced as a consequence of the acceptance of the findings of the Donaldson Report, have been paralleled by changes to Initial Teacher Education and Training (ITET) in Wales, and the introduction of measures under the so-called New Deal for the Education Workforce. The latter seeks to identify and deliver upon the training and Continuing Professional Development (CPD) needs of education professionals.

1.13   Geography Teaching in Primary Schools in the UK

Having discussed the structure of education in schools in the four jurisdictions of the UK, with brief mention of geography teaching and assessment, it is important to recognise the specific contribution to young learners' development of experiencing geography education in primary schools. Traditionally, much research into geography education has been linked to teaching in secondary schools—however, primary education stands as the foundation on which all subsequent subject-based education is built and is, in itself, an important focus for education research (Catling 2013a, b). Valuing research into primary geography education is therefore key to achieving a more complete understanding of the field. We must also remember that many of the themes addressed by geography education researchers have equal interest both to those who work in the primary *and* secondary sectors, often with direct relevance to each.[11] But many aspects of primary education provision, particularly the educational contribution made by individual subjects, are complex to research. The Cambridge Primary Review (Alexander 2010) drew attention to the 'muddled discourse of the primary curriculum in relation to terms such as subjects, knowledge and skills' (Eaude et al. 2017). Alexander (2010) emphasised the importance of educating the 'whole child'—as opposed to focusing too narrowly on the development of subject-based conceptual, procedural and propositional knowledge—while urging that literacy and numeracy skills should be developed through the teaching of motivational and engaging subject content. In its way, the Review also argued that subject-based learning can help children to refine:

> different types of knowledge and skills related to enquiry, interpretation, reasoning, and formulating an argument. While such opportunities are not limited to History, Geography and RE, these disciplines offer fertile opportunities to explore human culture and personal and collective identity. Without such opportunities children will not be fully equipped with what is necessary to cope confidently and thoughtfully with change now and in the future (Eaude et al. 2017, p. 393).

However, as Catling (2017b) reminds us, in UK primary schools the teaching of geography as a discrete subject is rare. Geography is more commonly taught within what might be termed 'humanities education' (Catling 1999). The UK's four Education Departments rarely use the term 'humanities'—inspection reports also use this term only sparingly—nonetheless, primary schools usually deploy 'humanities subject leaders, giving the impression of a discrete curriculum area' (p. 354). Such leaders have responsibilities to oversee the teaching of the subjects of geography, history and (possibly) Religious Education[12]—each of which, in England, have their own national curriculum orders and are individually referenced in inspection reports 'reflecting a policy approach to distinguish and differentiate between these subjects' (Catling 2017b, p. 354). The situation in the other jurisdictions varies slightly:

> In Scotland geography and history are named within the 'social studies' curriculum alongside societal, economic and political studies, but Religious and Moral Education is separate. In Northern Ireland 'The World Around Us' area of learning covers geography, history, science and technology, with Religious Education identified separately. The term 'humanities' is being introduced in Wales as an area of learning and experience encompassing geography, history and Religious Education, with politics, economics, society, culture and beliefs…

---

[11] For a helpful collection of personal and academic reflections specifically on *primary* geography education and research, see Simon Catling's edited 'Reflections on Primary Geography', published in 2017 (Catling 2017a). Written by participants at the 20th Charney Manor Geography Conference it provides an overview of influences and impacts on primary geography teaching and initial teacher education spanning back to the start of the annual Charney conferences in 1995. This series of brief reflections, many research-based but some scholarly or personal, also benefits from contributions made by a number of international authors and secondary school experts in geography education research.

[12] Conceptions of humanities education can extend beyond what are often considered to be its constituent subjects of geography, history and Religious Education. Some would seek to include other areas: such as citizenship, as well as a modern foreign (or native) language, art, music, drama and literature. Nonetheless, boundaries are necessary and these are currently often drawn using existing subjects, or disciplines.

Each nation identifies geography, history and Religious Education either singly or linked, at times connected with other subjects and areas of learning (pp. 354–5).

It is obvious that the individual humanities subjects, such as geography, have a particular role to play in delivering the curricular breadth and balance required of primary education, but that too narrow a focus on prepositional subject knowledge can be limiting (Eaude et al. 2017). This issue is reflected in other countries. What has been referred to as the 'commandeering of the curriculum' (Owens 2013) by literacy and numeracy has also occurred in the US, where studying geography and history is deemed less important than mathematics and reading in elementary and middle schools (Leming et al. 2006). Schmeinck (2013) concurs, reporting that 'in recent decades in many primary school systems in Europe, geographical aspects have only played a marginal role' (p. 399). Based on his investigation of high-quality teaching of humanities subjects in primary schools, using evidence provided by UK school inspectorate reports, Catling (2017b) nevertheless concluded that the best primary curricula were founded on subject and cross-subject ideas and connections, where learning about separate subjects—such as geography—was clearly visible. Other, more generic, components of high-quality teaching and learning—such as high levels of teacher enthusiasm; strong commitment to, and expectations of, children's performance; effective teacher-pupil relationships; use of a diverse range of resources and approaches to teaching; sound curriculum leadership; and the presence of enquiry learning—were also essential.

1.14 Primary Geography Education in England

The quality of teaching and learning of primary geography in England, according to Catling et al. (2007), showed some marked improvements from the 1980s into the 2000s—predominantly as a consequence of its inclusion in the National Curriculum. Despite some notable negative comments about the provision of geography education in primary schools—not least within the 2003 OfSTED report that identified geography as the 'worst taught subject' (OfSTED 2003; Bell 2004; Marsden 2005) in the sector—there are clearly pockets of exceptional practice in geography education in English primary schools (Catling 1999). Nonetheless, many schools still teach only modest amounts of geography, quality of provision relies closely on the interest and involvement of specific staff, meaningful fieldwork experience remains scarce, and teacher subject knowledge in geography is often weak. Subject leaders in schools therefore need targeted support and professional development opportunities to ensure that geography education remains strong, or improves. Additionally, OfSTED reports, when commenting specifically on geography, tend to target pupil underachievement in the subject, declining subject provision, and poor curriculum and staff leadership (OfSTED 2008, 2009, 2011, 2013).

When well-taught, primary geography exhibits a range of teaching and learning approaches, explores topical issues, and engages with children's interests—ensuring excitement, enjoyment and enthusiasm among learners (Catling and Willy 2009). This point is extended by Martin (2008, 2013a), and Catling and Martin (2011), who refer to the 'ethnogeographies' of young learners, which draw on the child's experiences and personal geographies. Others note the:

> rich, contested meanings to be made of and through geography, implying a range of knowledge and skills as well as emotional experiences (Owens 2013, p. 382).

The emphasis in many countries on outcomes in literacy and numeracy at primary level, leading to a loss of curricular breadth and balance, is clearly damaging for young learners. Not surprisingly, English primary schools tend to focus their time and energy on areas of provision that are inspected and tested, especially towards the older end of primary schooling, putting pressure on time, resources and professional development opportunities for the foundation subjects—such as geography and history (Eaude et al. 2017).

1.15 Research into Primary Geography Education[13]

Catling (2011), Martin (2008) and Morley (2012) each identify that many teachers in primary schools stopped studying geography at the age of 14, with the result that their conceptions of the subject are arguably similar to those of early secondary pupils (in contrast to the majority of specialist secondary school geography teachers in the UK who must either have attained a degree in geography, or in a geography-related discipline, before training to teach). Greater investment in professional development and initial teacher education in geography—to increase teachers' pedagogical and subject knowledge, expertise and confidence in delivering the subject, and to encourage geography's inclusion in interdisciplinary and cross curricular forms of education—is regularly demanded by geography educationists (Catling 2016; Catling and Morley 2013). Deficiencies in teachers' subject knowledge will not be addressed by placing the onus on individual teachers to improve their own practice, but by recognising a collective lack of expertise and the need to invest in professional development and initial teacher education (Slater 1999). The importance of research providing an evidence base, alongside the support of subject associations and central government, may still not be enough to exact change, for:

> In troubled times, different ways of thinking are needed, without discarding the wisdom of traditions and experience (Eaude et al. 2017, p. 395).

Geography education in primary schools is therefore often associated with poor teacher subject expertise, but also suffers from low expectation of pupil performance in the subject, lack of curricular prioritisation, and poor allocations of curricular

---

[13] A useful source of research evidence on primary geography in the UK is the Register of Research in Primary Geography, coordinated and edited, at the time of writing, by Rachel Bowles. See also, Bowles (1999, 2005) and the associated *Forum* papers in the respective editions of the journal *International Research in Geographical and Environmental Education*.

time and resource (Catling 2011). The issue of poor teacher subject knowledge is increasingly pressing—particularly following a 'knowledge turn' in the subject, and the political drive and renewed emphasis on the value of 'core knowledge' (DfE 2010). Debate in geography education research has consequently focused on the nature of pupils' development of core knowledge, possibly at the expense of their understanding and skills (Martin and Owens 2011), on how knowledge is organised in the geography curriculum (Firth 2011), on conceptual thinking (Martin 2011), and on core concepts (such as place and space) (Matthews and Herbert 2004).

The set of circumstances facing primary geography education in the UK is common elsewhere: Jo and Bednarz (2014) outline similarities in the US context, while Blankman (2016), in the Netherlands, notes that pre-service teachers are non-specialists, usually possessing only a limited knowledge of geography. The amount of time assigned to train primary teachers in geography education in England is, at its maximum, some 16 hours, but pre-service teachers may receive no geography training at all during postgraduate teacher education (Catling et al. 2007). The amount of time allocated to training in the foundation subjects has steadily decreased in recent years, directly contrasting with increased provision for numeracy and literacy training (Catling and Morley 2013). This leads to the production of teachers who may be well prepared in terms of their pedagogical knowledge and skills, but less so in terms of their subject knowledge—a concern when the majority of primary teachers are expected to teach the *whole* range of curriculum subjects in key stages 1 and 2 of the national curriculum.[14] Martin (2013b) believes that there is also a deeper issue about how primary teachers perceive and construct their geographical knowledge; while Catling and Morley (2013) note that the geographical subject knowledge of primary teachers has not been explored as rigorously as their knowledge of, say, mathematics or the sciences. Based on their interviews with 'geography educators', identified by their schools as 'enthusiasts' for the subject, they outline five common subject-related themes for further research. These themes indicate that primary teachers have an understanding that geographical knowledge is 'informational' and 'conceptual', as well as recognising the educational significance of 'geographical thinking'—teachers appreciate geography's breadth, its function of conveying knowledge about the world, and the notion of it being a 'living subject' that should be made accessible to all children. Many also suggest the need for geography to be more visible in the curriculum.

In conclusion, the 'knowledge turn' experienced in education policy thinking—which has affected subject teaching in the UK and elsewhere since the first decade of the twenty first century—arguably heightens the need for research into primary teachers' conceptions of geography and geographical knowledge (see also, Brooks 2011; Firth 2011; Lambert 2011; Lambert and Jones 2013 for a similar discussion in the context of secondary geography education). Analysis of international research concludes that comparatively little is known about teachers' geographical subject

---

[14]Catling (2013b) notes that primary ITE courses from 2002 to 2007 were given the option by the government of dropping either geography, or history, from their programmes; although few did so, those that did tended to drop geography.

knowledge, either within the primary or secondary school sectors (Bednarz et al. 2013). The issues of poor teacher preparation in geography, particularly for primary schooling, are clear—these can be listed as: a lack of confidence in subject knowledge and understanding, difficulty in interpreting the nature of the subject in curriculum orders, poor subsequent structuring of learning units, lack of clear and sequential structure in geographical learning, poor progression, superficiality of taught content, over reliance on commercially produced materials, poor planning, and the resultant fragmentation of learners' experiences of geography (Martin 2013a; Catling and Morley 2013; OfSTED 2008, 2009, 2011). This has led to research, at both primary and secondary levels, not only into teachers' subject knowledge (Puttick et al. 2018), but also about how teachers use this knowledge to develop their pupils' understanding of geography (see Martin 2008; Brooks 2011).

1.16 Primary Geography and Creativity

Providing high-quality, effective and motivational learning experiences in geography in primary schools is challenging (Bent et al. 2014; Blankman 2016). For some, this is chiefly a matter for primary initial teacher education to address (see, in the Dutch context, Blankman et al. 2015, 2016), for others it is a matter of subsequent professional development. Scoffham (2013) explores how primary geography teaching and learning can be enhanced by engaging with 'creative practice', but acknowledges that the concept of creativity is both complex and contested. Truly creative pedagogy goes beyond the construction of imaginative geography lessons, putting children at the heart of the learning process and avoiding the problems associated with the (supposedly) simple transmission of knowledge. Owens (2013) refers to creativity as a concept to be applied across the curriculum, with students engaging with 'problem-solving, lateral thinking and hypothesising' (p. 395). This leads to the promotion of learning through enjoyment, with associated benefits to health and emotional well-being, and the potential:

> to redefine the boundaries of the formal curriculum and encourage flexibility and innovation at a time when the curriculum … runs the danger of becoming rigid or inert (p. 369)

Government and inspection reports (DCSF 2007; OfSTED 2006) have stated that enhancing creativity is strongly linked to raising standards, and that it contributes positively to the cognitive and emotional development of young people. Many teachers in primary schools have also taken a lead from other key reports (DfES 2003, 2004) to introduce creativity into the curriculum, enabling a more flexible, cross curricular approach to teaching and learning. There is, however, a tension between inspectors encouraging creativity on the one hand, while simultaneously expecting the delivery of an accountable curriculum, demonstrably based on raising standards, on the other.

For some, good geography education and 'thinking creatively' are inseparable (Scoffham 2013; Barnes and Scoffham 2013; Catling 2009; Craft 2001; Pickering 2013; Lambert and Owens 2013). Indeed, Lambert and Owens (2013) refer to 'empathic geographies' where the development of children's imaginations and creativity are essential to their consideration of others' views and values. However, many would acknowledge that the link between creativity, geography and geography education is not straightforward and that more research is needed into its connection with improving geographical understanding in young people. Catling (2011) asserts that the thoughts, feelings and perceptions children have about the world are integral to their geographical imaginations, which are in turn heightened by engaging with creativity (see also Lyle 1997). As Scoffham (2013) explains, connecting with creativity entails much more than teachers and learners simply interpreting a given curriculum in an imaginative way:

> a pedagogy which sees creativity as the dynamo at the heart of joyful and ethical learning has much more to offer than this. In particular it:
>
> - respects the autonomy and agency of the child, thereby enhancing their well-being;
> - acknowledges that children learn in different ways at different speeds and with different motivations;
> - recognises the complexity and messiness of ideation, and
> - attributes value to emotional and existential knowledge alongside more visible cognitive achievements (p. 379)

This suggests a need for research into creativity to move beyond accepted wisdoms, case studies and anecdote, towards the employment of theorised, empirical studies—possibly involving neuroscientific research.

1.17 Primary Geography, Space and Place

Schmeinck (2013) explains how the 'European dimension' has been a prominent theme for primary and secondary geography education in schools across Europe for some time, despite particular shortcomings in primary provision. At primary level, the concentration on geographical data such as place names, facts and figures; a lack of consideration of children's ideas, perceptions and pre conceptions about Europe; and limited teaching about Europe and the wider world, has affected children's overall understanding of space and place. Schmeinck (2007) divides research broadly about 'space and place', in primary and secondary geography education, as follows:

i. cartographic knowledge of pupils (e.g. Wastl 2000; Livni and Bar 2001; Umek 2003)
ii. awareness and understanding of topographic terms (e.g. Kelly 2004)
iii. spatial concepts and knowledge of pupils concerning Europe and the world (e.g. Lambrinos 2001; Buker 2001; Schniotalle 2003)

iv. perceptions of proximity and foreign countries (e.g. Barrett 2007), and
v.   impacts of in-school and out-of-school knowledge and experiences on learners'
     perceptions (e.g. Harwood and Rawlings 2001; Schmeinck 2007; Schmeinck
     et al. 2010; Holl-Giese 2004; Huttermann and Schade 2004; Schmeinck and
     Thurston 2007)

(after Schmeinck 2013)

However, research conducted across Europe concerning children's place and space
knowledge is currently limited—such that 'solid empirical data and analyses that
allow for well-founded conclusions about children's perceptions concerning Europe
and the world remain rather scarce' (Schmeinck 2013, p. 399). Germany has taken
steps to encourage the study of the European dimension since the 1970s, actions
that are 'illustrative of decisions that have been taken by many European nations
over the same period' (p. 400). Nonetheless, the teaching of place and space in
primary geography education is still based largely on traditional practice, rather
than being informed by research evidence—which has encouraged teaching about
the local and/or home region, rather than locations further afield. Schmeinck et al.
(2010) report that the situation in many European countries is similar to that in
Germany: for example, in English primary schools there was no requirement to
include specific elements of the European dimension as a distinct focus until 2014,
while in Greek schools Europe is mentioned as a topic—but often appears simply
through the provision of maps of European Union members and potential members
(Flouris and Ivrideli 2002). In an era of globalisation and, for most member states
of the European Union, increasing international collaboration, the focus on local
and regional understanding at the expense of more international and global learning
seems restrictive and limiting. This is recognised by many geography education
researchers at primary level (see Catling 2003; Catling and Willy 2009; Martin 2006).
Schmeinck (2013) further justifies the inclusion of a European dimension in the
primary curriculum for five reasons: the increasingly multicultural dimension of life
in most European states, with experiences based on living together with peoples of
different countries and cultures; improved mobility and increasing travel activities;
the influence of the mass media; access to a wide range of consumer products; and the
existence of official programmes to support the integration and 'coming together of
Europe' (p. 402). This last point, especially for those living in the UK in the shadow
of BREXIT, is an interesting and salutary one.

Martin (2013a) notes how the English Geography National Curriculum has traditionally encouraged teachers to focus on 'similarities and differences' when teaching
about distant places—a particular problem with regard to the Global South, as stereotypical attitudes may be promoted by teaching that essentially 'others' people from
further afield. Based on research undertaken in Gambia and Southern India for an
Economic and Social Research Council (ESRC) project, Martin (2013b) explored
what teachers and student teachers learned from undertaking study visits to contrasting locations. She concludes that although poor geographical subject knowledge is
often stated as an issue for primary teachers, the problem goes far deeper—to the
epistemological basis of their geographical knowledge. The assigning of peoples

and cultures to different categories in primary geography teaching leads to a consideration of what is 'similar' or 'different', unwittingly creating a situation where difference is judged as the negative deviation from a Western cultural norm.

1.18 Primary Geography and Fieldwork

The importance of fieldwork, and all educational work outside the classroom, in helping primary school children develop their observational, investigative, recording and analytical skills—alongside their inquisitiveness—is viewed as the essence of enquiry learning (Catling 2013a). Outdoor learning, including the provision of Forest Schools (Pickering 2017), has resource and professional development implications which connect to mainstream, as well as discovery and sensory, fieldwork activity (Bowles 2017; Witt 2017). There are links to creativity in geography education (see Sect. 1.16), and show that when given appropriate professional and mentoring support—through projects such as 'Young Geographers: A Living Geography Project for Primary Schools' (see Catling 2011)—primary teachers soon embrace the need for outdoor education and geographical fieldwork.

1.19 Conclusions

Geography education research is highly specialized, but relatively modest in terms of its number of proponents and outputs—it can therefore never be expected 'to grow and mature into anything other than an important niche interest' (Lambert 2015, p. 25). Research into almost all subject-based teaching and learning is youthful and small scale, even in the 'core' subjects of maths, sciences and English which attract greater political interest and larger research funding. To maintain relevance it is therefore important that those who have responsibilities to research into geography education regularly remind themselves of the nature of their role—for researchers occupy privileged positions, most of which are (at least to some extent) publicly funded. The general public therefore has a right to ask what researchers are doing, and to what end. We must also be aware of the dangers of simply undertaking 'research for research's sake'—a problematic waste of time, effort, and resource which runs the risk of opening up our work to legitimate criticisms regarding its relevance, validity and fitness for purpose.

The 'backdrop' of where geography education, and geography education research, resides within the academy is important. Marginson and Considine (2000) have previously indicated how university-based research sits inside institutions that are essential self-organising and, to a greater or lesser extent, self-regulating. In the modern world these institutions can appear archaic and oddly configured, with their 'monastic and cathedral symbolism of rectors, deans and readers' (p. 1). Universities inhabit an important position, which can easily be taken for granted, with their demands for 'academic freedom' and the rights to either defend or criticise the prevailing government, society and culture. But there is a fragility here: particularly in education institutions that find themselves on the peripheries of the universities they inhabit. With only a weak standing in their host institutions, a high reliance on teaching income, and purveying a wide variety of approaches to research they are rarely regarded as being in the vanguard.

While not wishing to outline a distinct set of research priorities for geography education—which would be presumptive, counter-productive and somewhat dictatorial—we must have an awareness of the dangers of simply letting 'a thousand flowers bloom'. Some of our research efforts will inevitably lead us down 'blind alleys'—this is a function of the messy, tentative, 'backwards and forwards' nature of all research—but these steps still need to be taken. What is indefensible is spending time and resource unthinkingly and uncritically retracing previous steps to see whether they really lead us nowhere (again!) All research involves risks and opportunity costs. Lambert (2015) refers to researchers occupying a 'space of ideas'—but recognises that although these ideas often appear to hold promise, they may eventually 'run into the sand'. His point is slightly different to that above, but it alerts us to the issues of undertaking research that does not take risks and which merely amasses more data, simply because it is easy to do so. This begs the question 'what is the consequence of this enquiry?' or, in more prosaic terms, 'so what?'

Aligning subject-based research with the 'bigger' issues and problems faced in education is not without risks. In a position paper written for the Geography Education Research Collective (GEReCo) in 2017, David Lambert is clear about the dangers of research endeavours in education that 'drift' towards solving issues that have a school-wide, often generic, focus. Here the emphasis on subject specialism tends to get lost, subsumed beneath an insistence that research should be conducted into the priorities identified by school senior leadership teams (SLTs)—unsurprisingly, these priorities have a clear management focus, regularly linked to raising pupil attainment, meeting the strategic aims of the school development plan, or supporting the drive to improve school effectiveness. All this is important, of course, but takes attention away from the primacy of subject-based research and its questions about what should be taught and learned. Recently subject-based research has gained a fillip from a slightly unexpected source—Ofsted's Chief Inspector, Amanda Spielman—who in England in 2017 acknowledged that curriculum and subject matters were not afforded an important enough position in the inspection and evaluation of schools:

> Given the importance of the curriculum, it's surprising just how little attention is paid by our accountability system to exactly what it is pupils are learning in schools …. the taught curriculum is in fact just one among 18 matters for consideration in reaching the leadership and management judgement, making it somewhat of a needle in a haystack. I believe that lack of focus has had very real consequences (cited in Lambert 2017a, b)

Ideally, geography education research should be significant not only in advancing the teaching and learning of our subject, but also in having something to offer other subject-based pedagogical research, both national and internationally.

References

ACCAC. (2003). *Awdurdod Cymwysterau, Cwricwlwm ac Asesu Cymru – Qualifications, Curriculum and Assessment Authority for Wales) Developing the Curriculum Cymreig*. Thames Ditton: ACCAC.

Albert, D., & Owens, E. (2018). Who is listening to us from geography education? Is anyone out there? *Review of International Geographical Education Online (**RIGEO**)*. Accessible at www.rigeo.org.

Alexander, R. (Ed.). (2010). *Children, their world, their education—Final report and recommendations of the Cambridge Primary Review*. Abingdon: Routledge.

Arnold, M. (1869). Culture and anarchy. In R. Super (Ed.), *The complete prose works of Matthew Arnold* (Vol. V). Ann Arbor: University of Michigan Press.

Bagoly-Simó, P. (2014, August 22). *Coordinates of research in geography education. A longitudinal analysis of research publications between 1900–2014 in international comparison*. Paper delivered at IGU Regional Conference, Krakow.

Barnes, J., & Scoffham, S. (2013). Geography, creativity and the future. In S. Scoffham (Ed.), *Teaching geography creatively* (pp. 180–193). London: Routledge.

Barnes, J., & Scoffham, S. (2017). The humanities in English primary schools: Struggling to survive. *Education 3-13, 45*(3), 298–308.

Barrett, M. (2007). *Children's knowledge, beliefs and feelings about nations and national groups*. Hove: Psychology Press.

Bassey, M. (2001). A solution to the problem of generalization in education: Fuzzy prediction. *Oxford Review of Education, 27*(1), 5–22.

Beauchamp, G., Clarke, L., Hulme, M., & Murray, J. (2013). *Policy and practice within the United Kingdom: Research and teacher education: The BERA RSA enquiry*. London: BERA RSA.

Becher, T. (1989). *Academic tribes and territories: Intellectual enquiry and the cultures of disciplines*. Buckingham: Open University Press.

Beck, J. (2009). Appropriating professionalism: Restructuring the official knowledge base of England's modernized teaching profession. *British Journal of Sociology of Education, 30*(1), 314.

Bednarz, S., Heffron, S., & Huynh, N. (Eds.). (2013). *A roadmap for 21st century geography education research. A report from the Geography Education Research Committee*. Washington DC: AAG.

Bell, D. (2004). David Bell urges schools and the geography community to help reverse a decline in the subject. OfSTED Press release, 24th November.

Bent, G., Bakx, A., & Den Brok, P. (2014). Pupils' perceptions of geography in Dutch primary schools: Goals, outcomes, classroom environments and teacher knowledge and performance. *Journal of Geography, 113*(1), 20–34.

Biddulph, M. (2016). *What does it mean to be a teacher of geography? Investigating the teacher's relationship with the curriculum* (Unpublished Ph.D.). University of Nottingham.

Blankman, M. (2016). *Teaching about Teaching Geography: developing primary student teachers' pedagogical content knowledge for the subject of geography*. Haarlem: University of Applied Sciences.

Blankman, M., van der Schee, J., Volman, M., & Boogaard, M. (2015). Primary teacher educators' perception of desired and achieved pedagogical content knowledge in geography education in primary teacher training. *International Research in Geographical and Environmental Education., 24*(1), 80–94.

Blankman, M., van der Schee, J., Volman, M., & Boogaard, M. (2016). Design and development of a geography curriculum for first year primary student teachers. *Curriculum and Teaching, 31*(2), 5–25.

Bourke, B. (2014). Positionality: Reflecting on the research process. *The Qualitative Report, 19*(33), 1–9.

Bowles, R. (1999). Research in UK primary geography. *International Research in Geographical and Environmental Education, 8*(1), 59 (Special Forum).

Bowles, R. (2005). Forum: Children's voices: Younger children versus pedagogy. *International Research in Geographical and Environmental Education, 14*(4), 295–296.

Bowles, R. (2017). Changing perspectives on fieldwork. In S. Catling (Ed.), *Reflections of Primary Geography. Conference participants' perspectives on aspects of primary geography* (pp. 151–156). Sheffield: Register of Research in Primary Geography/GA.

Bradford, M., & Kent, A. (1977). *Human geography: Theories and their application.* Oxford: Oxford University Press.

Brock, C. (Ed.). (2015). *Education in the United Kingdom.* London: Bloomsbury.

Brooks, C. (2010). How does one become a researcher in geography education? *International Research in Geographical and Environmental Education, 19*(2), 115–118.

Brooks, C. (2011). Geographical knowledge and professional development. In G. Butt (Ed.), *Geography, education and the future* (pp. 165–180). London: Continuum.

Brooks, C. (2015). Research and professional practice. In G. Butt (Ed.), *Masterclass in geography education* (pp. 31–44). London: Bloomsbury.

Brooks, C. (2018). Insights on the field of geography education from a review of master's level practitioner research. *International Research in Geographical and Environmental Education, 27*(1), 5–23.

Buker, K. (2001). Europa-(k)ein Thema fur die Grundschule? *Grundschule, 33*(4), 34–40.

Butt, G. (2010). Perspectives on research in geography education. *International Research in Geographical and Environmental Education, 19*(2), 79–82.

Butt, G. (Ed.). (2011). *Geography, education and the future.* London: Continuum.

Butt, G. (2015a). Introduction. In G. Butt (Ed.), *Masterclass in geography education* (pp. 3–14). London: Bloomsbury.

Butt, G. (2015b). What is the role of theory? In G. Butt (Ed.), *MasterClass in geography education* (pp. 81–93). London: Bloomsbury.

Butt, G. (2018, October 22). *What is the future for subject-based education research?* Public seminar delivered at the Oxford University Department of Education. www.podcasts.ox.ac.uk.

Butt, G., Catling, S., Eaude, T., & Vass, P. (2017). Why focus on the primary humanities now? *Education 3-13, 45*(3), 295–297.

Butt, G., & Gunter, H. (Eds.). (2007). *Modernising schools; people, learning and organizations.* London: Continuum.

Butt, G., & Lambert, D. (2014). International perspectives on the future of geography education and the role of national standards. *International Research in Geographical and Environmental Education., 23*(1), 1–12.

Carter, A. (2015). *The Carter review of Initial Teacher Training (ITT).* London: HMSO.

Catling, S. (1999). Geography in primary education in England. *International Research in Geographical and Environmental Education, 8*(3), 283–286.

Catling, S. (2003). Curriculum contested: Primary geography and social justice. *Geography, 92*(3), 164–210.

Catling, S. (2009). Creativity in primary geography. In A. Wilson (Ed.), *Creativity in primary education* (pp. 189–198). Exeter: Learning Matters.

Catling, S. (2011). Children's geographies in the primary school. In G. Butt (Ed.), *Geography, education and the future* (pp. 15–29). London: Continuum.

Catling, S. (2013a). The need to develop research into primary children's and schools' geography. *International Research in Geographical and Environmental Education, 22*(3), 177–182.

Catling, S. (2013b). Editorial—Optimism for a revised primary geography curriculum. *Education 3-13, 41*(4), 361–367.

Catling, S. (2016, February 26). *Geography for generalists in England's University-based primary initial teacher education.* Unpublished Paper. Charney Manor Primary Geography Education Conference.

Catling, S. (2017a). High quality in primary humanities: Insights from the UK's school inspectorates. *Education 3-13, 45*(3), 354–364.

Catling, S. (Ed.). (2017b). Reflections of primary geography. In *Conference Participants' Perspectives on Aspects of Primary Geography*. Sheffield: Register of Research in Primary Geography/GA.

Catling, S., Bowles, R., Halocha, J., Martin, S., & Rawlinson, S. (2007). The state of geography in English primary schools. *Geography, 92*(2), 118–136.

Catling, S., & Butt, G. (2016). Innovation, originality and contribution to knowledge—Building a record of doctoral research in geography and environmental education. *International Research in Geographical and Environmental Education, 25*(4), 277–293.

Catling, S., & Martin, F. (2011). Constructing powerful knowledge: The primary geography curriculum as an articulation between academic and children's (ethno-) geographies. *The Curriculum Journal, 22*(3), 317–335.

Catling, S., & Morley, E. (2013). Enquiring into primary teachers' geographical knowledge. *Education 3-13, 41*(4), 425–442.

Catling, S., & Willy, T. (2009). *Achieving QTS: Teaching primary geography*. Exeter: Learning Matters.

CCEA (Council for the Curriculum, Examinations and Assessment). (2007). *The Northern Ireland curriculum: Primary*. Belfast: CCEA. www.ccea.org.uk.

Chalmers, L. (2011). Counting for everything: Research productivity in geographical and environmental education. *International Research in Geographical and Environmental Education, 20*(2), 87–89.

Clarke, L. (2006). A single transferable geography? Teaching geography in a contested landscape. *International Research in Geographical and Environmental Education, 15*(1), 77–91.

Cochran Smith, M., & Lytle, S. (1999). The teacher research movement: A decade later. *Educational Researcher., 28*(7), 15–25.

Craft, A. (2001). Little c creativity. In A. Craft, B. Jeffrey, & M. Leibling (Eds.), *Creativity in education* (pp. 45–61). London: Continuum.

Daugherty, R., & Rawling, E. (1996). New perspectives for geography: An agenda for action. In E. Rawling & R. Daugherty (Eds.), *Geography into the twenty-first century* (pp. 1–15). Chichester: Wiley.

Department for Children, Schools and Families (DCSF). (2007). *The children's plan. Building brighter futures*. London: HMSO.

Department for Education (DfE). (2010). *The importance of teaching: The schools white paper*. London: The Stationery Office.

Department for Education (DfE). (2014). *The national curriculum in England*. London: HMSO.

Department for Education (DfE). (2017). *Help your child make the best GCSE choices*. Available at https://assets.publishing.service.gov.uk/government/uploads/system/uploads/attachment_data/file/761031/DfE_EBacc_Leaflet.pdf.

Department for Education and Skills (DfES). (2003). *Excellence and enjoyment: A strategy for primary schools*. London: DfES.

Department for Education and Skills (DfES). (2004). *Every child matters*. London: DfES.

Donaldson, G. (2015). *Successful futures: Independent review of curriculum and assessment arrangements in Wales*. WG23258. Crown Copyright.

Eaude, T., Butt, G., Catling, S., & Vass, P. (2017). The future of humanities in primary schools—Reflections in troubled times. *Education 3-13, 45*(3), 386–395.

Finney, J. (2013). Music teachers as researchers. In J. Finney & F. Laurence (Eds.), *Masterclass in music education* (pp. 3–12). London: Bloomsbury.

Firth, R. (2011). Debates about knowledge and the curriculum: Some implications for geography education. In G. Butt (Ed.), *Geography, education and the future* (pp. 141–164). London: Continuum.

Flouris, G., & Ivrideli, M. (2002). The image of Europe in the curriculum of Greek primary education: Comparative perspective of the last three decades of 20th century (1976–77, 1982, 2000). In S. Bouzakis (Ed.), *Contemporary issues of the history of education* (pp. 473–510). Athens: Gutenberg.

Furlong, J. (2013). *Education: An anatomy of the discipline*. London: Routledge.

Gallagher, C. (2012). *Curriculum past, present and future: A study of the development of the Northern Ireland curriculum and its assessment in a UK and international context* (Unpublished Ph.D. thesis). University of Ulster.

Gerber, R. (2001). The state of geographical education around the world. *International Research in Geographical and Environmental Education, 10*(4), 349–362.

Gerber, R. (2003). The global scene for geographical education. In R. Gerber (Ed.), *International handbook on geographical education* (pp. 3–18). Dordrecht: Kluwer.

Giddings, L. (2006). Mixed methods research—Positivism dressed up in drag? *Journal of Research in Nursing, 11*(3), 195–203.

Greenwood, R. (2007). Geography teaching in Northern Ireland primary schools: A survey of content and cross-curricularity. *International Research in Geographical and Environmental Education., 16*(4), 380–398.

Greenwood, R. (2013). Subject-based ad cross-curricular approaches within the revised primary curriculum in Northern Ireland: Teachers' concerns and preferred approaches. *Education 3-13, 41*(4), 443–458.

Greenwood, R., Richardson, N., & Gracie, A. (2017). Primary humanities—A perspective from Northern Ireland. *Education 3-13, 45*(3), 309–319.

Hargreaves, D. (1996). *Teaching as a research-based profession: possibilities and prospects*. London: TTA.

Harwood, D., & Rawlings, K. (2001). Assessing young children's freehand sketch maps of the world. *International Research in Geographical and Environmental Education., 10*(1), 20–45.

Hillage, J., Pearson, R., Anderson, A., & Tamkin, P. (1998). *Excellence in research in schools*. London: TTA.

Holl-Giese, W. (2004). Geographisches Weltbildes aus Grundschulsicht. In A. Huttermann (Ed.), *Untersuchungen zum Aufbau eines geographischen Weltbildes bei Schulerinnen und Schulern. Ergebnisse des 'Weltbild' Projektes an der Padagogischen Hochschule Ludwigsburg* (pp. 18–34). Ludwigsburg: Padagogische Hochschule.

Huttermann, A., & Schade, U. (2004). Das Weltbild-Projekt des Fachs Geographie an der Padagogischen Hochschule Ludwigsberg. In A. Huttermann (Ed.), *Untersuchungen zum Aufbau eines geographischen Weltbildes bei Schulerinnen und Schulern. Ergebnisse des 'Weltbild' Projektes an der Padagogischen Hochschule Ludwigsburg* (pp. 3–9). Ludwigsburg: Padagogische Hochschule.

Jo, I., & Bednarz, S. (2014). Developing pre-service teachers' pedagogical content knowledge for teaching spatial thinking in geography. *Journal of Geography in Higher Education, 38,* 301–313.

Jones, M., & Whitehouse, S. (2017). Primary humanities: A perspective from Wales. *Education 3-13, 45*(3), 332–342.

Keiner, K. (2010). Disciplines of education: The value of disciplinary observation. In J. Furlong & M. Lawn (Eds.), *The disciplines of education: Their role in the future of education research*. London: Routledge.

Kelly, A. (2004). 'It's Geography Jim, but not as we know it': Exploring Children's Geographies at Key Stage 2. In R. Bowles (Ed.), *Register of Research in Primary Geography: Occasional paper 4. Place and Space*. London: Register of Research in Primary Geography.

Kent, A. (1999a). Guest editorial: Image and reality—How do others see us? *International Research in Geographical and Environmental Education, 8*(2), 103–107.

Kent, A. (1999b). Geography in secondary education in England. *International Research in Geographical and Environmental Education, 8*(3), 287–290.

Kent, A. (Ed.). (2000). *Reflective practice in geography teaching*. London: Paul Chapman Publishing.

Kirby, A., & Lambert, D. (1984). *The region: Space and society*. Harlow: Longmans.

Kirby, A., & Lambert, D. (1985a). *Land use and development: Space and society*. Harlow: Longmans.

Kirby, A., & Lambert, D. (1985b). *The city: Space and society*. Harlow: Longmans.

Lambert, D. (2010). Geography education research and why it matters. *International Research in Geographical and Environmental Education, 19*(2), 83–86.

Lambert, D. (2011). Reviewing the case for geography, and the 'knowledge turn' in the English National Curriculum. *The Curriculum Journal, 22*(2), 243–264.

Lambert, D. (2015). Research in geography education. In G. Butt (Ed.), *MasterClass in geography education* (pp. 15–30). London: Bloomsbury.

Lambert, D. (2017a). Position Paper for GEReCo. Unpublished.

Lambert, D. (2017b, May 23). Interview with David Lambert at UCL Institute of Education, London.

Lambert, D. (2018). Editorial: Teaching as a research-engaged profession. Uncovering a blind spot and revealing new possibilities. *London Review of Education, 16*(3), 357–370.

Lambert, D., & Jones, M. (2013). Introduction: geography education, questions and choices. In D. Lambert & M. Jones (Eds.), *Debates in geography education* (pp. 1–14). Abingdon: Routledge.

Lambert, D., & Owens, P. (2013). Geography. In R. Jones & D. Wise (Eds.), *Creativity in the primary curriculum* (pp. 98–115). London: Routledge.

Lambrinos, N. (2001). World maps: A pupils' approach. In *Proceedings Third International Conference on Science Education Research in Knowledge-Based Society* (Vol. 11, pp. 505–507). Thessaloniki.

Lave, J., & Wenger, E. (1991). *Situated learning*. Cambridge: Cambridge University Press.

Leming, J., Ellington, L., & Schug, M. (2006). The state of social studies: A national random survey of elementary and middle school social studies teachers. *Social Education, 70*(5), 322–327.

Lidstone, J., & Gerber, R. (1999). What does the word 'International' in international research in geographical and environmental education really mean? *International Research in Geographical and Environmental Education, 8*(3), 219–221.

Lidstone, J., & Stoltman, J. (2002). Spreading the word—The geographical distribution of IRGEE authors. *International Research in Geographical and Environmental Education, 11*(2), 99–101.

Lidstone, J., & Williams, M. (2006a). *Geographical education in a changing world*. Dordrecht: Springer.

Lidstone, J., & Williams, M. (2006b). Researching change and changing research in geographical education. In J. Lidstone & M. Williams (Eds.), *Geographical education in a changing world* (pp. 1–17). Dordrecht: Springer.

Lingard, B. (2013). Reshaping the message systems of schooling in the UK: A critical reflection. In D. Wyse, V. Baumfield, D. Egan, C. Gallagher, L. Hayward, M. Hulme, R. Leitch, K. Livingston, I. Menter, & B. Lingard (Eds.), *Creating the curriculum* (pp. 1–12). London: Routledge.

Livni, S., & Bar, V. (2001). A controlled experiment in teaching physical map skills to grade 4 pupils in elementary schools. *International Research in Geographical and Environmental Education, 10*(2), 149–167.

Lyle, S. (1997). Children age 9–11 making meaning in geography. *International Research in Geographical and Environmental Education, 6*(2), 111–123.

Maclure, M. (2003). *Discourse in educational and social research*. London: McGraw Hill.

Marginson, S., & Considine, M. (2000). *The enterprise university: Power, governance and reinvention in Australia*. Cambridge: Cambridge University Press.

Marsden, W. (1999). Forum: Geographical education in England and Wales: The state of play at the end of the millennium. *International Research in Geographical and Environmental Education, 8*(3), 268–272.

Marsden, W. (2005). Guest editorial reflections on geography: The worst taught subject? *International Research in Geographical and Environmental Education., 14*(1), 1–4.

Martin, F. (2006). *Teaching geography in primary education: Learning to live in the world*. Cambridge: Chris Kington Publishing.

Martin, F. (2008). Knowledge bases for effective teaching: beginning teachers' development as teachers of primary geography. *International Research in Geographical and Environmental Education, 17*(1), 13–39.

Martin, F. (2011). Global ethics, sustainability and partnerships. In G. Butt (Ed.), *Geography, education and the future* (pp. 206–224). London: Continuum.

Martin, F. (2013a). What is geography's place in the primary school curriculum. In D. Lambert & M. Jones (Eds.), *Debates in geography education* (pp. 17–27). London: Routledge.

Martin, F. (2013b). Same old story: The problem of object-based thinking as a basis for teaching distant places. *Education 3-13, 41*(4), 410–424.

Martin, F., & Owens, P. (2011). Well what do you know? The forthcoming primary review. *Primary Geography, 75,* 28–29.

Matthews, J., & Herbert, D. (2004). Unity in geography: Prospects for the discipline. In J. Matthews & D. Herbert (Eds.), *Unifying geography: Common heritage, shared future* (pp. 369–393). London: Routledge.

Morley, E. (2012). English trainee teachers' perceptions of geography. *International Research in Geographical and Environmental Education, 21*(2), 123–137.

Naish, M. (1993). 'Never mind the quality—Feel the width'—How shall we judge the quality of research in geographical and environmental education? *International Research in Geographical and Environmental Education., 2*(1), 64–65.

Office for National Statistics (ONS). (2014). *Compendium of UK statistics.* Accessed January 2, 2019. www.ons.gov.uk/ons/guide-method/compendiums/compendium-of-the-uk-statistics/index.html.

OfSTED (Office for Standards in Education). (2003). *Geography in primary schools: OfSTED subject report series.* London: HMSO.

OfSTED (Office for Standards in Education). (2006). *Creative partnerships: Initiative and impacts.* London: HMSO.

OfSTED (Office for Standards in Education). (2008). *Geography in schools: Changing practice.* London: HMSO.

OfSTED (Office for Standards in Education). (2009). *Improving primary teachers' subject knowledge across the curriculum.* London: HMSO.

OfSTED (Office for Standards in Education). (2011). *Geography: Learning to make a world of difference.* London: HMSO.

OfSTED (Office for Standards in Education). (2013). *Geography survey visits.* London: HMSO.

Owens, P. (2013). More than just core knowledge? A framework for effective and high-quality primary geography. *Education 3-13, 41*(4), 382–397.

Pelikan, J. (1992). *The Idea of the university: A re-examination.* New Haven: Yale University Press.

Pickering, S. (2013). Keeping geography messy. In S. Scoffham (Ed.), *Teaching geography creatively* (pp. 168–179). London: Routledge.

Pickering, S. (2017). Without walls. In S. Catling (Ed.), *Reflections of primary geography. Conference participants' perspectives on aspects of primary geography* (pp. 147–150). Sheffield: Register of Research in Primary Geography/GA.

Postman, N., & Weingartner, C. (1971). *Teaching as a subversive activity.* London: Penguin.

Puttick, S., Paramore, J., & Gee, N. (2018). A critical account of what 'geography' means to primary trainee teachers in England. *International Research in Geographical and Environmental Education, 27*(2), 165–178.

Rawling, R. (1999). Geography in England 1988–98: Costs and benefits of national curriculum change. *International Research in Geographical and Environmental Education, 8*(3), 73–278.

Rawling, E. (2004). Introduction: School geography around the world. In A. Kent, E. Rawling, & A. Robinson (Eds.), *Geographical education: Expanding horizons in a shrinking world.* SAGT with CGE: Glasgow.

Robbins, L. (1963). *Higher education. Committee on higher education (The Robbins Report).* London: HMSO.

Roberts, M. (2000). The role of research in supporting teaching and learning. In A. Kent (Ed.), *Reflective practice in geography teaching* (pp. 287–295). London: Paul Chapman.

Roberts, M. (2015). Discussion to part 1. In G. Butt (Ed.), *Masterclass in geography education* (pp. 45–50). London: Bloomsbury.

Robertson, L., Hepburn, L., McLaughlan, A., & Walker, J. (2017). The humanities in the primary school—Where are we and in which direction should we be heading? A perspective from Scotland. *Education 3-13, 45*(3), 320–331.

Ryle, G. (1949). *The concept of mind*. Chicago: University of Chicago Press.

Schmeinck, D. (2007). *Wie Kinder die Welt sehen – Eine empirische Landervergleichsstudie uber die raumliche Vorstellung von Grundschulkindern*. Bad Heilbrunn: Kilnkhardt.

Schmeinck, D. (2013). 'They are like us'—Teaching about Europe through the eyes of children. *Education 3-13, 41*(4), 398–409.

Schmeinck, D., Knecht, P., Kosack, W., Lambrinos, N., Musumeci, M., & Gatt, S. (2010). *Europe through the eyes of Children. The implementation of a European dimension by peer learning in primary school*. Berlin: Mensch and Buch.

Schmeinck, D., & Thurston, A. (2007). The influence of travel experiences and exposure to cartographic media on the ability of 10-year-old children to draw cognitive maps of the world. *Scottish Geographical Journal, 123*(1), 1–15.

Schniotalle, M. (2003). *Raumliche Schulervorstellungen von Europa. Ein Unterrichtsexperiment zur Bedeutung kartographischer Medien fur den Aufbau raumlicher Orientierung im Sachunterricht der Grundschule*. Berlin: Tenea.

Scoffham, S. (2013). Geography and creativity: Developing joyful and imaginative learners. *Education 3-3, 41*(4), 368–381.

Scottish Government. (2012). *Ethnic group demographics*. http://www.gov.scot. Accessed March 18, 2019.

Shulman, L. (1987). Knowledge and teaching: Foundations of the new reform. *Harvard Educational Review, 57*(1), 1–23.

Slater, F. (1994). *Do our definitions exist? Reporting research in geography education*. Monograph No. 1, pp. 5–8. London: Institute of Education, University of London.

Slater, F. (1999). Notes on geography education at higher degree level in the United Kingdom. *International Research in Geographical and Environmental Education, 8*(3), 295–300.

Slater, F. (2003). Exploring relationships between teaching and research in geography education. In R. Gerber (Ed.), *International handbook on geographical education* (pp. 285–300). Kluwer.

Solem, M., & Boehm, R. (2018). Research in geography education: Moving from declarations and road maps to actions. *International Research in Geographical and Environmental Education, 27*(3), 191–198.

Thomas, G. (2007). *Education and theory: Strangers in paradigms*. Maidenhead: Open University Press.

Tooley, J., & Darby, D. (1998). *Educational research: A critique*. London: OfSTED.

Umek, M. (2003). A comparison of the effectiveness of drawing maps and reading maps in beginning map teaching. *International Research in Geographical and Environmental Education, 12*(1), 18–31.

Walford, R. (1996). Geography 5-19: Retrospect and prospect. In E. Rawling & R. Daugherty (Eds.), *Geography into the twenty-first century*. London: Wiley.

Wastl, R. (2000). *Orientierung und Raumvorstellung. Evaluierung unterschiedlicher kartographischer Darstellungsarten* (Vol. 20). Klagenfurt: Klagenfurter Geographische Schriften.

Welsh Government (WG). (2015). *Curriculum for Wales, a curriculum for life*. Cardiff: Welsh Government.

Whitty, G. (2005). *Education(al) research and education policy making: is conflict inevitable?* Presidential address to BERA Conference, University of Glamorgan, Pontypridd.

Witt, S. (2017). Fostering geographical wisdom in fieldwork spaces—Discovery fieldwork, paying close attention through sensory experience and slow pedagogy. In S. Catling (Ed.), *Reflections of Primary Geography. Conference participants' perspectives on aspects of primary geography* (pp. 159–164). Sheffield: Register of Research in Primary Geography/Geographical Association.

World Health Organisation (WHO). (2016). *Growing up unequal. HBSC 2016 survey*. Available at www.euro.who.int.

Chapter 2
Re-contextualising Knowledge: The Connection Between Academic Geography, School Geography and Geography Education Research

2.1 Context

Researching geography education is, in many respects, a recent phenomenon. This chapter takes a historical turn by reflecting on how geography education research has grown over time, contrasting research experiences in geography and education departments in universities to track and compare developments. Geography education research in England expanded as a field post 1945, partly as a result of an 'eminently practical concern' to support the pre-service preparation of teachers—which required findings from educational research to assist the training process (Morgan and Firth 2010). By the mid-1960s the Robbins Report (Robbins 1963) had brought teacher education more firmly into the higher education fold, following debate about the possibility of locating initial teacher education (ITE) in universities dating back to the Cross Committee in 1888. University-based teacher education affected the production of research in the field, particularly following the absorption of older-style teacher training colleges into higher education institutes (HEIs) where greater expectations of sustained research performance were prevalent. The dramatic growth in the numbers of academics and students in higher education in the 1960s—alongside the founding of new universities, and the creation of a number of dedicated University Departments of Education (UDEs)—generated a positive stimulus for education research. However, the status of schools and departments of education in higher education has remained low, only slowly gaining grudging acceptance in many British universities. The dominance of teacher training, rather than research activity, is a reputational issue which continues to hold back the development of education research to the present day. UDEs and teacher training colleges were founded primarily to meet the growing need for the professional preparation of teachers—but their different origins, aims and expectations help to explain the divisions between them that are still evident. Their distinct raisons d'etre make clear the divergent emphasis they place on research and teaching: stereotypically the older, elite, Russell Group universities in the UK prioritising research excellence in their schools, departments

© Springer Nature Switzerland AG 2020

G. Butt, *Geography Education Research in the UK: Retrospect and Prospect*, International Perspectives on Geographical Education, https://doi.org/10.1007/978-3-030-25954-9_2

and institutes of Education [with less concern for the practicalities of initial teacher training (ITT)], while the remaining teacher training colleges and school-centred ITT programmes focus on teacher preparation (with more limited expectations of research productivity). By extension, this also clarifies what might loosely be described as a 'theory-practice divide' in the types of education research conducted by these institutions. Through exploring the origins and locations of geography education research, and by tracking its recent development, the intention is to examine the continuing importance of applying research to classroom practise, and beyond (Butt 2018). This is as significant for geography teachers, educationists and practitioners as it is for other subject specialists. A loose chronology of the growth of geography education research in England is offered—from the immediate post war period to the present day—against which international readers may choose to compare and contrast the research progress made in their own countries. Unfortunately, offering a range of detailed examples of how geography education research has developed in other jurisdictions is unfeasible given restrictions of space, however later chapters widen the analysis to the international scale. It is noteworthy that during the recent era of 'doubt, reflection and reconstruction' (Hartley 2000) in initial teacher education, geography education research has often explored areas associated with teacher development, classroom practice, pedagogy and curriculum making—all of which have distinctly practical and utilitarian applications. This type of research contrasts with the more theoretical work carried out in other areas of educational enquiry.

It is important to acknowledge from the outset the importance of the connection between the development of the academic discipline of geography in the academy; the types of geography taught in primary and secondary schools; and research in geography education in higher education institutions. One criticism of this chapter might be that it chooses to focus on the transformation, or re-contextualisation, of geographical knowledge produced in the academy into the geography curricula taught in schools. As such, it might be argued, it fails to prioritise the centrality of geography education *research* in this process—indeed, that such research is largely unimportant, if the nature of the transition of knowledge from academy to school curriculum is essentially unproblematic, or irrelevant. This ignores two significant, related points: firstly, that in the UK, the inclusion of disciplinary and subject knowledge in the school curriculum has achieved a renewed importance in recent years (DfE 2010)—prompting a marked, if belated, response from geography education researchers (Lambert 2011; Morgan 2014; Firth 2011; Mitchell 2017; Butt 2018). Shifts in education policy have encouraged geography education researchers to consider more fully the role and place of knowledge, the function of 'powerful knowledge' (Young 2008, 2013; Young and Muller 2010; Brooks et al. 2017), and the pedagogical implication of a knowledge-led curriculum (Young et al. 2014). Secondly, that the connections between geographical research in the academy, and its expression in the content of geography curricula taught in schools, are extremely important. At times these links are strong, at others weak. Geography education research should clearly have a role to play in helping to understand and facilitate the connection between academic geography and the forms of geography children learn in schools. As Stoddart (1986) reminds us, 'what is taught in schools, in a sense, codifies our accepted knowledge and, in a large degree, the teacher is thus the custodian of the truth' (p. 9).

The chapter closes by re-examining Haubrich's (1992b, 1996a) rather over optimistic assertions—made a quarter of a century ago, and even at the time considered by many to be somewhat 'gilded'—that since the mid-1970s research in geographical and environmental education has 'increased in quality and quantity', making 'substantial progress' in terms of its scale, coverage and methods (Haubrich 1996a). The extent to which such progress has been made, with particular consideration of the negative impact of our poor engagement with research and theory from *outside* the field of geography education, is also noted.

2.2 Introduction

It is perhaps appropriate to begin by considering the post war consensus on education in the UK from 1945, and its influence on geography education and research up to the new millennium. Within this framework, the origins and outcomes of the Madingley conferences in Cambridgeshire (from 1963 to the mid-1970s), and of the Charney Manor conferences in Oxfordshire (from 1970[1] to the present day), are significant. In addition, the Schools Council Geography Projects also require examination given their noteworthy influence on geography teaching and assessment in schools and their connection to research in the field. This approach affords us a window onto the ways in which academic geographers, teacher educators and teachers have responded to developments in society, their cognate disciplines, and to the enduring question of 'what to teach?' While it is important not to over-state the significance of events that occurred over half a century ago—some of which only directly affected a small, rather select, group of educationists—particular influences continue to the present day. We should certainly recognise the mutual benefits of academic geographers and educationists combining in different ways, and the stimulus this creates for research in geography education (Butt and Collins 2013, 2018). With this in mind, it is also important to consider the forums within which academic geographers, educationists, researchers and practitioners have traditionally come together—such as conferences, seminars and symposia, and also in special interest groups, professional associations and dedicated collectives. Geography teacher educators in England, predominantly from higher education institutions, have for decades met annually at Geography Teacher Educator (GTE) conferences[2]; researchers, practitioners and special interest groups from the Geographical Association (GA) meet for an annual conference—

---

[1]The Charney Manor conferences were organised by Rex Walford in 1970, 1980, 1990 and 1993, with a predominant focus on secondary school geography. Subsequently, from 1995 to 2017, they were organised by Simon Catling, on twenty occasions, with an emphasis on primary geography.

[2]The Geography Teacher Educator (GTE) conferences, which attract around 40 people each year, have developed from the previous University Departments of Education (UDE) conferences in England. These are now more 'open' affairs, with a wider range of delegates, including geography educators from across Europe.

which also attracts a range of publishers, examination bodies and others with broad interests in geography education; the Royal Geographical Society with Institute of British Geographers (RGS-IBG), the Council of British Geographers (CoBRIG), and the Geography Education Research Collective (GEReCo) similarly gather for various purposes—to conduct business, to offer research forums, to provide opportunities for open debate, and to give doctoral students a chance to present their work.

The population of active geography education researchers in the UK, and indeed globally, has always been small—indeed, many of the same people (in slightly different guises) are members of the same organisations and forums within which geography educators meet. However, it is important to recognise that many university-based geography education researchers in England, and elsewhere, have professional profiles that enjoy an international reach. They publish their research, and present findings, to geographers and educationists at various conferences worldwide—not least those of the International Geographical Union (IGU), the IGU's Commission on Geographical Education (IGU-CGE), as well as meetings attended by academic geographers (such as, say, the Association of American Geographers (AAG) conferences), and gatherings of educationists (such as the education conferences held by the British Education Research Association (BERA), American Educational Research Association (AERA), etc.).

2.3   Education in the Post War Period

The post war consensus in the UK—which affected society, economy, politics, culture and education—promised that the future would avoid any prospect of returning to previous conditions of social injustice, conflict and poverty. The national drive for full employment and a welfare state, as well as the expectation of achieving better standards of housing, finance, health care and education, were widely supported politically. Shifts accomplished in education policy and practice during this period were significant, resulting in the provision of new schools, more and better trained teachers, and higher educational standards. From the 1960s the population became more affluent—in the main—and increasingly consumerist, a trend that inevitably also affected young people. Expectations that greater numbers of young people would now progress from 'basic' schooling into higher education created pressures for curricular and educational changes in schools: to service the wants and needs of a new generation growing up in post war Britain. Nonetheless, substantial societal and political divisions still existed in the UK with regard to class, income, gender and region.

With our focus primarily on research in geography education, it is however important to consider not only what was happening in schools and university departments of education during this period, but also in academic geography. In the 1960s and early 1970s the discipline was experiencing paradigmatic shifts towards the development of a 'new geography', state schools in the UK[3] were beginning to face up to the demands of comprehensivization, and larger numbers of school students were progressing into higher education. This was a period when many teachers and educationists believed that the promotion of academic subject content in schools was less important than nurturing children's development through 'child centred' education—with its focus on values, attitudes, character and beliefs. The re-organisation of secondary education into comprehensive schools stimulated the provision of tailored educational programmes for *all* abilities and aptitudes of pupils—in many schools this required a radical re-alignment of educational aims and a firmer realisation of an obligation to educate 'the whole child'. At the same time, shifts in academic geography—away from a regional paradigm towards one which prioritised a systematic, 'new' and more conceptual form of geography, which required the application of quantitative methods—gradually began to impact on geography education in schools. Schools and universities, in an increasingly pluralistic society, had become a focus for reform: to accommodate changes in educational theories and ideas, to differentiate the knowledge and skill sets that children experienced, and to adjust the ways in which learning was organised (Ross 2000). It is to the manner in which these shifts affected geography education, and specifically geography education research, that we now turn.

2.4 Madingley and Charney Manor Conferences

In the early 1960s two young academic geographers at Cambridge—Richard Chorley and Peter Haggett—sought to explore how new ideas in geography could be transmitted into schools. Ray Pahl, a geographer who had been appointed as a tutor in the university's Extra Mural Studies Department, was approached to organise a residential course for 25 teachers on 'Modern geography'—to update them on recent developments in the subject at university level (Haggett 2015). He invited Chorley and Haggett, who were Demonstrators in the Cambridge geography department—and without doctorates, fellowships or full-time contracts—to lead the course. There was an urgency, at least within some school and university geography departments, to discover how the sectors could work together more closely to ensure that new research findings in geography were effectively transferred between the two. But, as Rawling (1996) recalls, the first Madingley conference revealed a paternalistic relationship between the sectors, in providing an opportunity:

> for geographers from higher education to update their schoolteacher colleagues about the new trends and ideas in the subject, particularly about the so-called 'New Geography'. It was an occasion for those in universities to offer 'pearls of wisdom' to those in schools. Although there was plenty of mutual respect, schoolteachers were still viewed as the junior partners in this relationship, not surprisingly since the focus was on research findings and new ideas which had been developed in the university sector (pp. 3–4)

---

[3]Although comprehensivisation was not a significant force for reforming secondary schools in Northern Ireland.

This uneven relationship still largely exists today—although there are now examples of more 'balanced' research and publication endeavours in geography education across the academic and school 'divide'.[4] The Madingley conferences helped to create what the eminent geography educator, Rex Walford, came to refer to as a 'new model army' (Walford 2001, p. 158)—the size of which was always rather modest—of academic geographers and teachers who could together explore the implications of developments in their subject (Haggett and Chorley 1989). But their direction of travel was essentially one way—with academic geographers 'handing down' research findings and techniques to the 'junior partners in this relationship' (Rawling 1996, p. 3). Nonetheless, Unwin (1996) comments positively on the influence of Chorley and Haggett's activities on 'a generation of geography teachers'—particularly with regard to how quantitative approaches, modelling and theory building subsequently 'filtered down into school textbooks and examination syllabuses' (p. 21). Indeed, he asserts that a small number of geography teachers remained heavily influenced by the research agendas of university geographers into the 1970s and early 1980s, although we must be sceptical about the magnitude of this impact across English schools. In hindsight the influence of the Madingley conferences may be regarded as rather more symbolic, than actual.

At a Council for British Geographers (COBRIG) seminar held in Oxford in 1994, Rawling (1996) playfully questioned whether the gathering of academic geographers and geography educators might represent 'Madingley revisited'—but concluded that a rather different relationship now existed between the two sectors. Unquestionably, the seminar's aims reflected an attempt to realign a previous hierarchy: the organisers expressly sought opportunities for academics and educationists 'to share experiences and to update each other about new approaches in geography and about significant changes in the school curriculum' (Rawling 1996, p. 3). Additionally, there was an intention to clarify the 'character and contribution of geographical education from the secondary years to higher education' and to 'plan future strategies which will ensure continuing dialogue and well-being' (Rawling 1996, p. 4). More equal partnership would not only encourage intellectual exchanges from the research frontiers of the academy into schools, but also a sharing of ideas about curriculum change and educational objectives from school geography into the academy. Rawling asserts that the COBRIG seminar was therefore 'not a revisiting of Madingley, but a reformulation of the event in a way better suited to the 1990s' (Rawling 1996, p. 4). Interestingly, the contribution of geography education *researchers* to this exchange process was not explicitly mentioned by Rawling—the emphasis clearly being on investigating research from academic geography departments, and receiving updates from schools. Recognition of the research endeavours and potential influence of geography educationists was therefore tacit. (Although, in fairness, the delegate list from the conference reveals that education researchers were well represented. The

---

[4]It is significant that the leading academic journal of the GA, 'Geography', made a dramatic shift in 2008 towards being more inclusive of educationists and educational ideas in geography. The professional backgrounds and interests of those who make up the editorial collective for the journal reflect this shift.

subsequent seminar publication—Rawling and Daugherty (1996)—included chapters from seven geography educationists, out of a total of 23 contributors). Tellingly Tim Unwin, one of the academic geographers who attended the seminar, recognised that university-based geography still remained separated from 'other kinds of geography':

> between geography *in the school curriculum*, which is seen as being an *educational* perspective, and geography in *higher education*, which is considered as being the *academic* or *research* perspective (Unwin 1996, p. 20; emphasis in the original)

At the 1994 COBRIG seminar, Peter Haggett was asked to reflect on the relationship between the academy and schools over the previous three decades, and to compare how the Madingley conferences differed from the current Oxford gathering. Peter's contribution to research at the interface between geography and geography education in the 1960s, alongside Richard Chorley, is a matter of record—culminating in the publication of *Frontiers in Geographical Teaching* (Chorley and Haggett 1965) and *Models in Geography*[5] (Chorley and Haggett 1967). But by the mid-1990s, Haggett believed, 'times had changed'. Following a period of unprecedented growth in the university sector from the early 1960s—a consequence of a higher proportion of GNP being directed into higher education, particularly in the UK and US—undergraduate student numbers had risen, research outputs had increased and new, younger researchers had been employed. By the start of the 1970s the growth, and subsequent decline, of an 'optimistic and confident' positivist paradigm in geography was still playing out in many universities and schools. Commenting on the changing relationship between school and academic geography, Haggett noted that in the 1960s it was 'the university dons who taught and the schoolteachers who listened'—but in the 1990s 'this hierarchic situation would be unthinkable!' (Haggett 1996, p. 13). Daugherty and Rawling (1996) concur, noting that the 1960s were 'simpler times':

> The 1960s saw the heyday of positivism in university geography, with the emphasis on model building, theory and quantification. Great enthusiasm for the new ideas and approaches which these made possible was communicated to schools by means of conferences, courses and eventually textbooks. The relationship with schools was close, if paternalistic, with academics gladly sharing the 'new geography' with their junior partners in schools. It was also a relatively simple functional relationship since schools provided the students for geography degree courses and for the next generation of geography teachers, and higher education geographers were closely involved in designing the A level syllabuses, setting the A level examination papers and writing the appropriate textbooks (pp. 361–362).

However, from the perspective of the mid-1990s, Unwin (1996) observed that:

> From a position of considerable influence on the school curriculum in the 1960s, those in university departments of geography now have little effect on what is taught at the secondary or primary level (p. 23)

---

[5]It is interesting to reflect on a comment made by Chorley and Haggett in this text, that it would be 'better that geography should explode in an excess of reform than bask in the watery sunset of its former glories' (p. 377). Not everyone was of the same opinion—an anonymous reviewer, 'P.R.C', commented that *Models in Geography* appeared to have been written by the authors 'for one another' and contained 'barbarous and repulsive jargon'.

Some aspects of research from university geography departments, which subsequently filtered down into schools in the 1960s, 70s and 80s, *did* have a profound effect on geography education. Unwin (1996) highlights the particular impact of publications which emerged from the Madingley conferences, or which were subsequently inspired by these gatherings, as vehicles for introducing geography teachers in schools to cutting edge research ideas from academic geography:

> the continued publication of theory- and model-based texts, such as Bradford and Kent's (1977) *Human Geography: Theories and Applications*, into the late 1970s and early 1980s suggests that the teaching of secondary geography in the latter part of the 1960s and throughout the 1970s and early 1980s remained heavily influenced by the research agendas of university geographers'. (p. 22)

The longer-term, more widespread effect on mainstream geography education in UK schools is questionable, however. Unwin is careful to state that it is difficult to ascertain, with any degree of certainty, the *magnitude* of influence of research findings in academic geography on geography teachers across Britain. For example, the growth of positivist geography in the academy in the late 1960s—and the subsequent shift towards behavioural, radical, welfare and humanistic geographies in the 1970s—saw barely any immediate impact on the types of geography taught in schools. Understandably, a time lag exists between the period when research ideas become established in universities and when some might filter down into schools, but by the late 1970s, most examination boards were only just re-aligning their syllabuses to reflect the positivistic approaches adopted by academic geographers during the quantitative revolution back in the 1960s. Academic geography continued to reposition itself: recognising that most geographical questions could not be answered through the pursuit of objective, scientific facts encouraged the growth of subjective, personal, perceptual and experiential approaches to geographical enquiry. In schools, many geography teachers also believed that developing the learner's 'sense of place' would help them to achieve a better comprehension of the world. Existential and phenomenological approaches appealed, encompassing not only an explanation of place, space and environment but also embracing the development of the individual child. In contrast to much of the 'new' geography, mental maps, 'private geographies', children's place perceptions and spatial preferences could all be readily incorporated into school geography curricula, bringing together the frontiers of geographical research and educational theory (Slater 1982). Academic geographers, such as Harvey (1969, 1973) and Peet (1977), attacked the assumptions of positivistic geography, pointing out its conceptual flaws and poor applicability across much of the discipline. As Jackson (2018) recalls:

> The 1960s and 1970s was a period of disciplinary ferment, symbolised by the publication of David Harvey's rigorously theoretical 'Explanation in Human Geography' (1969), followed just four years later by his passionately Marxist 'Social Justice and the City' (1973). These debates followed the discipline's so-called quantitative revolution that sought to reposition 'Geography as a Spatial Science' (p. 119)

Radical geography advocated the fundamental reform of society, in part by putting the academic disciplines to work in challenging existing social, economic and political norms. This movement subsequently found an expression in the work of geography educators such as John Huckle and David Pepper[6] and, to a more limited extent, John Morgan and Roger Firth.

The Madingley conferences—with their principal concern about the links between academic geography and geography education in secondary schools—drew to a conclusion in the mid-1970s, at around the same time as the Charney Manor conferences began. These eventually became annual events at which primary geography educators could meet to exchange ideas (having originated with a *secondary* school focus), although in contrast to Madingley the input of academic geographers was always limited. At the Charney conference in 1990, Rex Walford reflected—perhaps rather optimistically, as a serving member of the Geography Working Group fashioning the first geography national curriculum for England and Wales—that 'Geography's place in the school curriculum is taken as assured and its rationale self-evident' (Walford 1991, p. 3). The general tone of the early Charney conferences tended to be less buoyant, however, reflecting a belief that geography's influence and status in schools still had to grow in order to attract 'more and better-quality pupils' (p. 3). Indeed, following the inaugural Charney conference in 1970—and recognising the need for geography in secondary schools to keep pace with changes at the academic frontiers of the discipline—Walford himself had written: 'if geography is to survive in the school curriculum it will have to be more than a convenient examination pass for those who seek only to memorise a jumble of facts and sketch maps' (Walford 1973, p. 2). Reflecting on the themes prevalent in the twenty primary geography conferences he organised between 1995 and 2017, Simon Catling comments:

> contributions have … included incisive consideration of the state of primary geography, offered positive and optimistic ways to take geography forwards for primary teachers and children, taken critical stances, been theoretical and practical, demonstrated innovative, creative and engaging ways to work with children, teachers and students, reported on research into and scholarly thinking about many aspects of and influences on primary geography, been challenging and enjoyable, provided solace and been thought-provoking (Catling 2017, pp. xiii–xiv)

What is less apparent is the contribution made by academic geographers to these events, although the likelihood is that similar to the GTE conferences these contributions have been relatively modest in number.

When considering the period of major change in academic geography in the 1960s and 70s, and on the subsequent influence of events on geography teaching in schools, it is tempting to question the limited involvement of geography educators and *researchers* in helping to promulgate new ideas from the academy into school geography curricula. As we have seen, radical disciplinary changes in geography were 'handed down' to geography teachers in secondary schools—in a fashion that largely bypassed geography educators and researchers from HEIs. The dangers of this circumvention are perhaps better appreciated today, as geography education

---

[6]In the 1980s the journal '*Contemporary Issues in Geography and Education*' was considered the publication of choice for radical geography educators to present their work.

researchers strive to engage more directly with ideas from their parent academic discipline (Lambert 2011, 2015; Mitchell 2011, 2017).

2.5   Geography Curriculum Development in Schools

It is recognised that not all developments at the frontiers of research in academic geography should, or indeed could, ever be transferred into school syllabuses and curricula (see Walford 1981)—nonetheless, maintaining some connection between the two is vital. This raises the question of how curriculum development in schools should best occur, and which agents should be involved in the process. The Schools Council was established in 1964 to create curricula for schools which would mirror shifts in economy, society and culture—but which would also encapsulate some of the advances made at the academic frontiers of disciplines. The 1970s saw significant changes in the ways in which educational achievement in schools—primarily measured through the award of Advanced level ('A' level) examinations, and progression to university education—was viewed as a motivator for children's learning. The raising of the school leaving age ('RoSLA') in 1972 meant that a larger proportion of young people now 'stayed on' for a further year of state education before entering the workplace, necessitating changes in curricular provision in schools. This was a time when the post war consensus in the UK finally broke down: a consequence of the election of Conservative governments from the late 1970s that were bent on fomenting radical educational, economic and social change. The realisation that many students either would not, could not, or chose not to progress educationally past the statutory age for state schooling helped to break a hierarchical link between university-based education and subject teaching in schools. Education researchers elected to explore aspects of curriculum, assessment, attainment and pedagogy in schools—rather than the contribution of disciplinary knowledge to subject content—with concomitant advances in the application of educational theory. The educational needs of the majority of learners who would *not* progress into higher education came to the fore, further weakening the connection between the academy and schools.

Clearly, it is unhelpful to conceptualise the relationships between academic and school geography as being simple, straightforward and linear—the transfer of any research ideas into schools requires careful selection and mediation. This process is, of course, often complicated by events in both sectors. From the 1980s onwards, geography departments in universities faced significant pressures: increases in undergraduate student numbers were not matched by a proportionate increase in revenue, such that many academic departments sought new sources of funding through contract research, consultancies and entrepreneurial operations. Growing competition for Research Council grants tended to skew the types of research undertaken in geography departments—with academics bemoaning their inability to direct their own work, the rise of external controls and accountability, and the enforced diminution of original, open, critical enquiry. This 'brave new world' was understandably resisted by many academics who saw their previous academic freedoms disappearing before their eyes. Ever increasing 'market place' competition for students and funding, less altruistic decision making within departments, and a greater demand for academics to display entrepreneurial flair led to less cooperation between academic staff. The decline in contact between academic geographers, and teachers and students in schools, is not coincidental—indeed, the relationship between university geography departments, schools and departments of education in higher education institutes (HEIs), and geography teachers in schools has always been somewhat problematic in England. This situation has resonance in other countries and jurisdictions. Gaining an understanding of the nature of the interface between these sectors is useful when striving to appreciate how research in the academy might best be utilised to encourage curriculum development in schools. Michael Bradford's reflections on the ways in which these inter-relationships developed from the 1970s to the mid-1990s help us to appreciate such links (Bradford 1996). Like others around the same time (Goudie 1993; Clifford 2002; Thrift 2002; Bonnett 2003), Bradford focused on a widening *division* between geographers in universities and schools, represented by:

> 'a gap or discontinuity in methods and content … Academic geographers and secondary teachers, who seem to communicate less than they did in the past, need to recognise the gap, partly so that students can be helped to bridge it and partly to ensure the future health of the subject'. (p. 277)

Writing during a period when higher education had recently experienced radical organisational shifts as a consequence of the passing of the Education Reform Act (DES 1988)—a massive piece of education legislation which not only allowed all higher education institutions to apply to become universities, but also introduced the first national curriculum into state schools in England and Wales—Bradford recognised increasing tension at the interface between secondary and tertiary education, and witnessed its impact on curriculum development. Diversity in the higher education sector was being dramatically reduced. Former polytechnics now grasped the opportunity to re-cast themselves as universities, with all institutions being funded under the same auspices as those for traditional universities in England and Wales (namely, the Higher Education Funding Councils for England (HEFCE), and in Wales the Higher Education Funding Councils for Wales (HEFCW)). University teaching and research began to be assessed, examined and evaluated in the same ways across the entire sector. Importantly, one of the few key differences between institutions concerned their attitudes towards research. Here Bradford's (1996) observations were prescient:

> There seems to be a move, however, to produce a hierarchy of universities, with varying emphases on research. For some universities, research and postgraduate teaching may have much greater prominence; while there is a threat that some others will become teaching-only institutions. The superficial unifying tendency of recent years may simply be the prelude to further diversification' (p. 279)

We shall see how the preoccupation with change in the academy and the 'sheep and goats' division of university-based research and researchers—both in academic geography and geography education—has played out in the following chapters. However, it is worthwhile noting here that some aspects of research in geography and geography education have, in recent years, advanced in parallel. In areas where their conceptual and disciplinary connections overlap—for example, say, in the study of children's geographies (Holloway and Valentine 2004), the geography *of* education (Brock 2015), transitions from schools to universities (Finn 2018; Tate and Swords 2013)—there have been mutually beneficial developments. Nonetheless, it is hard to evidence a significant body of research in geography education which directly connects developments in academic geography with those in geography education in schools (although see Rawding 2013; Goodson 1983; Livingstone 1992; Marsden 1997; Walford 2001).

2.6   The Schools Council Geography Projects

Naish (1987, 2000), like Graves (1975), views the period from 1965 until the early 1980s as one of major change in the fortunes of both school and university-based geography education. Although Graves (1975) refers to this as a time of 'laissez faire' curriculum development in school geography, he also observed a looming 'crisis in geographical education in Britain' (p. 61)—initiated by conceptual developments within academic geography in universities, advancement of educational theory, and changes in the structure of secondary schooling. Additionally, a series of curriculum crises occurred concerning timetabled provision for geography education in schools, the nature of the subject's instrumental value, and the quality of teaching and learning experienced by pupils. Marsden (1997) echoes these observations, also referring to a number of 'unhealthy stresses' that existed between school and university geography at the time.

Research and curriculum development in geography education in schools has often been stimulated by major changes in state education—rather than by narrower, more esoteric, developments in the parent academic discipline. For example, the expansion of comprehensivization from the mid-1960s in England, and the raising of the school leaving age in the early 1970s, each directly affected how subjects were taught and the content they conveyed—stimulating research into curriculum development and 'what to teach'. This was a period of weak consensus, even confusion, about what directions geography education in schools and universities should take—partly as a consequence of the paradigm shift in geographical thought in higher education, with its introduction of theoretical modelling, quantitative techniques, conceptual frameworks and spatial analysis. These changes suggested that school geography should also move away from studying what was unique—as typified within the paradigm of regional geography—towards studying concepts and content that were believed to be more generalizable. Geographers in universities, and eventually in schools, therefore moved from simply describing places to the more systematic examination of pattern and process (see Tidswell 1977). But this 'passing down' of new geographical knowledge and methods to schools was uneven and sporadic; many schoolteachers were reluctant to change, having themselves learnt geography only within the traditional regional paradigm, which they subsequently taught for a number of years. Newby

(1980), Robinson (1981) and Butt (1997) recognised that the impacts of change on many geography teachers were dramatic:

> especially since the drive towards introducing new quantitative methods and theories within schools were also combined with increasing comprehensivization, the growth of integrated curricula, greater concept-based learning and the influence of curriculum development projects such as the American High School Geography Project (HSGP)' (Butt 1997, p. 14)

Hore (1973), who investigated the impact of quantitative geography on schools, found teacher conservatism a major resistance to change.[7] Regionalism remained a strong focus for many involved in geography education in schools—despite its widespread rejection by academics, he found that geography teachers in the 1970s were:

> increasingly suspicious of persons in ivory towered universities and colleges of education, who throw out wonderful suggestions, without testing them in the white heat of classrooms composed of, say, thirty aggressive youths from a twilight urban area (p. 132).

Considerations of the broader aims, objectives and purposes of education came to the fore—a growing 'educational focus' encouraged many geography teachers to step back from contemplating geography's subject content to explore its wider *educational* benefits. Caught up in debates about vocational education and curriculum centralisation, following the rise of the New Right in the 1980s, teachers often appeared to lose touch with the disciplinary roots of their subject. This problem was exacerbated because most of the Schools Council's curriculum development projects in the 1970s had prioritised the consideration of *pupil* needs, rather than what *subjects* might offer. The 16–19 Geography project—a Schools' Council curriculum development project based at the University of London's Institute of Education from 1976, resulting in the creation of an extremely popular 'A' level geography course— was innovative in that it embraced aspects of educational theory, notions of societal change, and elements of humanistic geography from universities. However, the contribution of academic geographers was treated very selectively, as one of the Project's leaders later commented:

> since geography is to be used as a medium for education, there is no requirement that all new academic developments necessarily be translated into the school context (Naish et al. 1987, pp. 26–27)

The growing interest in transferable skills, political literacy, values education, and vocationalism in schools were a direct consequence of such thinking. As Butt and Collins (2018) explain:

> The Schools Council actively supported curriculum reform and development, sponsoring three major geography curriculum development projects in the 1970s: Geography for the Young School Leaver (GYSL); Geography 14–18 (Bristol Project) and the Geography 16–19 Project. Geography curriculum development was also influenced by a project from abroad

---

[7]Conservatism was, of course, also apparent among some geography educators and geography education researchers in UDEs. Long and Roberson (1966), for example, stated that they had 'nailed (their) flag to the regional mast'.

– the American High Schools Geography Project (HSGP, 1971) – which instructed teachers how to incorporate new ideas, content and techniques from the 'quantitative revolution' in academic geography into their schemes of work. However, most curriculum development projects focused more on how geography could contribute to the fulfilment of the needs of young people, than on considerations of academic subject content. The Geography 16–19 Project, examined at A level from 1982, achieved great popularity, experiencing a near-exponential growth in student numbers during the 1980s. This project had an impact on the teaching of geography within universities – for incoming undergraduates who had studied 16–19 Geography had been taught through a 'route to enquiry' approach, acquiring geographical content, skills, techniques and values very different from those provided by more 'traditional' geography syllabuses. The change was not universally welcomed by university geographers, many of whom criticised the (supposed) superficiality of content covered by the 16–19 syllabus, particularly of physical geography. (Naish et al. 1987, pp. 26–7)

Most geography curriculum development projects have traditionally been based within departments of education in higher education institutions (HEIs)—significantly neither schools, nor academic geography departments, regularly hosted or organised such projects (Boardman 1988). This had a major impact on the school geography curricula that were developed, where the involvement of teachers and academic geographers were often rather tangential to project outcomes—as indeed were the contributions made by geography education *researchers*. Curriculum development in school geography tended to reflect *some* elements of change from the academic frontiers of the discipline, but were not strongly influenced by them, being heavily mediated by teacher educators—many of whom were arguably more concerned about the advancement of pedagogic and curriculum theory, than innovations in disciplinary geography. Rawling (1991), herself a major contributor to one of the Schools Council projects, reflected on the impact that curriculum development projects had on school geography, asking:

• why did large scale curriculum development in geography education 'take off' in the 1970s?
• were the curriculum developments of the 1980s and 1990s different, or a continuum?
• how did the projects contribute to geography education? and
• where does the national curriculum fit into this picture?

She concludes that although the general re-organisation of state education in England from the 1970s was the main driver of change—chiefly as a result of the comprehensivization of secondary schools following the introduction of Circular 10/65—there was also a strong influence from new curriculum thinking and educational research. Rawling recognises that the Schools Council geography projects in England and curriculum development projects in the United States—specifically the High School Geography Project (HSGP)—were both led by educationists and curriculum experts based in higher education institutions, rather than academic geographers. University-based curriculum developers were therefore significant agents of change for geography teaching in schools. Disappointingly, the disbanding of the Schools Council in 1984, to be replaced by the School Curriculum Development Committee (SCDC) and Secondary Examinations Council (SEC), saw a decline in subject-based curriculum development projects across all school subjects, including geography.

2.7 The Geography National Curriculum

The development of the national curriculum at the end of the 1980s saw increasing centralisation and politicisation of the school curriculum, as a consequence of the passing of the most significant piece of education legislation in England since the Butler Act of 1944—the 1988 Education Reform Act (DES 1988). The first iteration of the Geography National Curriculum (GNC) (DES 1991) for England and Wales was widely considered to be 'traditional' and 'restorationist' (Rawling 2001), with Lambert (2011) commenting:

> The Schools Council projects introduced the idea that subject knowledge was not an end point in education, but a vehicle contributing towards educational ends (geography as a 'medium of education'). The 1991 National Curriculum can be interpreted as an attempt to restore subject knowledge (p. 248)

For some, such as Walford (1993), geography had won the 'status battle' by achieving a secure curriculum place—arguably at the expense of previously hard won educational, ideological and conceptual gains. Morgan (2008), in his review of school geography curriculum development since the 1970s, recognises the creation of the national curriculum as causing a further 'uncoupling' of school and university geography during this period. Butt (1997) also notes the limited involvement of academic geographers in the creation of the first geography national curriculum: only two were chosen to sit on the Geography Working Group; few made significant submissions to the curriculum-making process; whilst only modest numbers lobbied the Group, or politicians, either directly or through their professional associations. This indicates the minor influence university geographers had in shaping the content of geography taught in schools at this time. Indeed, by the late 1990s, academic geographers rarely crossed the school–university divide: few made contributions to the work of awarding bodies, to the creation of geography syllabuses, or to the professional development of teachers. Just as schools were enduring huge changes in education policy and practice, higher education was being subjected to increased bureaucratisation, marketisation and rising demands for accountability. The limited involvement of most academic geographers in debates about the content of public examinations in school geography, their general unwillingness to write for school teachers and students, and their lack of engagement in the professional development of teachers may be attributable to the pressures they faced to publish high-quality research and prepare for high-stakes audits (see Castree et al. 2007). According to Bradford (1996), the lack of a major conceptual revolution in university geography since the 1960s might also explain the modest impact of more recent academic ideas on schools:

during the 1980s and 1990s there has not been one major trend affecting as many areas of the subject as did either the scientific revolution of the 1950s and early 1960s, or the radical geography movement of the early 1970s. The absence of such major changes may partly account for the reduced impact of higher education on the geography taught in secondary education (p. 282)

However, it is notable that most geography educators and researchers also made little attempt to 'bridge the divide' between academic geography and geography education in schools—possibly for similar reasons.

In conclusion, the 'gap or discontinuity in methods and content' (Bradford 1996, p. 277) between school and university geography—notwithstanding recent collaborations in England through the 'A' Level Content Advisory Board (ALCAB)—has been a long-standing feature of the relationship between the two. This explains a division in pedagogy and skills that has existed for over quarter of a century—a fundamental, epistemological divide between the aims, rationale and scope of academic and school geography. With respect to research in both geography and geography education we should perhaps not be surprised by this division—as Butt (2019) points out, disciplinary geography advances in ways that school geography never could given the more restrictive demands on the subject in state education. However, the advancement of geography education in schools owes allegiance to both fields—with the geography taught in schools being *informed* by the knowledge gains achieved by academic geographers, but requiring more objectivity, stability and certainty about the content it conveys. As ever, today's teachers also need to be keenly aware of the contribution of educational research to their practice.

2.8   The Problematic Link Between School and University Geography

It is apparent that conceptualising the relationship between geography as a school subject and as an academic discipline is important, but not straightforward. There are obvious connections between the two, but also predictable and expected differences. It is important that if both sectors are to remain healthy we recognise, understand and acknowledge what unites and divides them—and seek resolution to those issues that are mutually damaging. Here the role of academic geographers, geography education researchers and others in promoting and supporting geography education in schools requires close consideration (Butt 2018).

The scholarly foundations which have previously underpinned both academic and school geography have traditionally been weak. At the start of the twentieth century university-based geography was 'dogged by the impression that it was intellectually suspect—a subject for school children rather than university academics' (Butt 1997). Indeed, university geography departments in England developed initially as a consequence of the growth of geography teaching in schools, the status of which was subsequently raised by the subject finding a place in the academy (Jackson 2018). In the 1960s—with the emergence of a 'new' geography buoyed by the acceptance of positivistic, scientific methodology—geography's academic status improved. The positive impact of the conceptual revolution on the popularity and standing of geography in schools is also often claimed (Walford 1981). However, this shift in status is perhaps rather illusory. Gregory (1978) and Goodson (1983), in their commentaries on the development of geography in schools and the academy from the mid to late 20th century, observe that ensuring the survival of the subject—rather than striving for its intellectual progression—has been the key driver for many geographers and geography educators. This, according to Smith (1973), moved dangerously close to opportunism when a desire for the subject to survive meant the abandonment of its intellectual principles. As Butt (1997) comments:

> This pragmatism clearly sits somewhat uneasily with the need for any subject to achieve intellectual respectability and to prove its academic merit (p. 1)

The problematic link between school and university geography clearly endures. Achieving a close, symbiotic relationship between university-based academics—whose primary aim is to push back the frontiers of their discipline through research, and geography teachers in schools—who must prioritise knowledge transfer over creation, is obviously difficult. Many teachers are undoubtedly concerned about losing touch with their parent discipline: a consequence of the diminishing relevance of what they previously learned as undergraduates, and a reflection of their inability to keep pace with the expanding frontiers of the discipline. These tensions are ever present for teachers, although their responses vary considerably—those who are frustrated by their inability to 'keep up' may claim, defensively, that academic geography has strayed into irrelevance, expansionism and interdisciplinary enquiry which they believe to be stereotypical, even facile, representations of the real world. Such teachers may find recent developments within their discipline to be either incomprehensible, unpalatable or unacceptable. Fisher (1970), commenting on these issues half a century ago, referred to this situation as geography 'over extending its periphery at the expense of its base' (p. 374). A more optimistic appreciation, expressed by Bailey (1992), recognises that:

> The reforms in academic and school geography (which) began in the mid-1960s and which have continued ever since enable modern geography to make distinctive and substantial contributions to the education of young people, which it certainly could not have made in its unreformed state (p. 65)

2.9 Can We 'Bridge the Divide'?

It has been regularly stated that a 'chasm', 'gap', 'border' or 'discontinuity' persists between geography education in schools and universities (see Goudie 1993; Machon and Ranger 1996; Bradford 1996; Marsden 1997; Bonnett 2003; Butt 2008a; Johnston 2009; Hill and Jones 2010, Butt and Collins 2018; Butt 2019). At times this has threatened the future of both the academic discipline and the school subject:

As geography in higher education on the one hand segments into more and more 'adjectival geographies', and, on the other, as disciplinary boundaries break down, the fossilisa-
tion of school geography in archaic conceptual and pedagogical moulds looks increasingly
likely. Geography in higher education is becoming increasingly separated from the subject
in schools (Bale 1996, p. 5)

It is, however, important to consider the ways in which this 'divide' might be
bridged—although it is doubtful that the gap will ever be closed completely. A
symbiosis clearly must exist—with both 'sides' understanding that they need connections to function effectively. There should be no intention to create uniformity,
nor any attempt to fashion a smooth continuity and progression of geographical
themes—better to strive for a mutual, coherent and agreed understanding of the subject, recognisable by both sectors. But what is the role of geography educators, and
geography education researchers, in bridging he divide?

There are a number of agents, mediators and ambassadors who should be involved
in bringing universities and schools closer together. These include academic geographers, initial teacher educators, professional associations, awarding bodies and geography teachers—all of whom can influence the content of school geography and the
themes for geography education research (see Table 2.1). Central to this bridge are
the efforts of new entrants to the teaching profession—commonly these are recent
graduates who bring fresh learning from the academy into the classroom, enriched
by the latest advances in disciplinary research. For many 'beginning teachers' this
is a difficult transition—they not only have to deal with the enormous commitment expected of new practitioners, but also teach a form of geography in schools
that may be 20 years out of date! Each new cohort of geography teachers—prepared through programmes of initial teacher education (ITE) in universities or, with
increasing likelihood, in schools—must make their own bridge between university
and school geography; even though many find that the geography syllabuses taught
in schools reflect very little of the themes and content they have recently studied
within the academy. Transference of contemporary geographical knowledge, understanding and skills by new practitioners into the classroom can prove unsettling (see
Marsden 1997, Barratt Hacking 1996; Walford 1996b; Brooks 2010b, c): they often
feel inadequately supported by mentors in school—who may themselves understand
little of recent disciplinary advances in their subject, or fear that examiners may not
recognise (or reward) students who attain new disciplinary knowledge.

2.10   The 'Knowledge Turn'

Since 2010, an aspect of education policy change that has helped to draw together
geography education in schools and universities has been the intention of Coalition and Conservative governments, in England, to promote subject knowledge in
schools. The Schools White Paper, *The Importance of Teaching* (DfE 2010), encouraged schools to consider more closely the subject content they taught to children,
raising the question of what constituted 'essential knowledge'. Both Lambert (2011)

Table 2.1 Bridging the divide—updating the content of school geography, from Lambert and Jones (2017), reprinted with permission of the publishers, Routledge

Activity	Agents
Professional development conferences and events	Professional associations (e.g. GA and GA branches, RGS-IBG and GA conferences)
Academic conferences and events	Academic geographers and initial teacher educators (with some geography teachers) (e.g. COBRIG, Association of American Geographers Conference, IGU, ESRC 'Engaging Geographies' seminar series, RGS-IBG and GA conferences)
Producing textbooks/journal articles for school students/geography teachers	Geography teachers, academic geographers and/or initial teacher educators (in schools and universities) (e.g. *Teaching Geography, Geography Review, Geography*)
Producing scholarly/research texts	Academic geographers and/or initial teacher educators (in schools and universities) (e.g. GEReCo, Rawling and Daugherty 1996; Kent 2000'; Butt 2011)
Research projects	Geography teachers in association with academic geographers and/or initial teacher educators (in schools and universities) (e.g. Young People's Geographies Project)
Curriculum Development Projects	Notably subject associations (e.g. see under 'projects' on geography.org.uk)
'Mediation'	'Mediators' and 'Ambassadors' working in/with geographers in schools (e.g. GA Chief Executive/Professor of Geography Education; RGS-IBG subject officers; key geography academics; initial teacher educators in geography; geography undergraduates in schools; A level geography students attending day 'outreach/widening participation' courses in university geography departments)
Special Interest Groups	As represented in professional associations (IGU, GA, RGS-IBG, etc.)
Political lobbying for government funded initiatives	Professional associations (GA, RGS-IBG) (e.g. Action Plan for Geography); 'mediators'
Award bearing courses/CPD (Masters, EdD, PhD in geography education)	University Schools of Education
Initial teacher education	New geography teachers, with geography educators (e.g. PGCE and PGDipEd courses)
Development and review of examination specifications[a]	Awarding bodies in association with academic geographers, teacher educators and geography teachers (Butt and Collins 2018)

[a]We cannot escape the sustained influence of the awarding bodies on the content of geography taught in schools. Many geography teachers express a desire to include up-to-date research in their teaching and may be encouraged to do so by their choice of syllabus. However, most teachers will only teach content which they believe will be credited by the examiners, who may favour 'traditional' (and possibly outdated) answers

and Mitchell (2011) welcomed the stimulus this gave geography educators to now (re)connect with their subject discipline, and to move on from what they saw as an unhelpful period when teachers and teacher educators focussed too closely on pedagogy and generic 'thinking skills'. Indeed, Mitchell (2011, 2017) characterises this as an intellectual impoverishment of geography education, caused by an 'emptying of subject knowledge'. The subsequent 'knowledge turn' therefore pushed geography education forward: leaving behind a phase dominated by irrelevant 'moral' and 'ethical' themes, according to Standish (2007, 2009, 2012), whose influence Kathryn Ecclestone and Dennis Hayes recognised as stretching across the curriculum, heralding a 'dangerous rise in therapeutic education' (Ecclestone and Hayes 2009). Lambert (2011) concurs, noting that a tight focus on pedagogy and adjectival geographies had occurred at the expense of the contribution of subject knowledge and the acceptance of geography teachers as 'knowledge workers' (see also, Mitchell 2011, 2017). Geography educationists and researchers argued that the acquisition of a more theorised and sophisticated understanding of knowledge by teachers—incorporating an appreciation of Young's conceptions of 'powerful knowledge' (Young 2008), of Hirsch's notions of core knowledge and cultural literacy (Hirsch 1987, 2007), and of the possibly of embracing a 'capability approach' to geography (Lambert and Morgan 2010)—would ensure that geography curriculum development in schools would advance.

2.11 Reform of Public Examinations

Reforms of school geography curricula, and of public examinations, have had an unmistakable impact upon geography education, both in schools and in the academy. This is particularly the case with respect to adjustments in the expected geography subject knowledge of school students, some of whom will progress to study geography in universities. Conservative governments in the UK have, since the first decade of the new millennium, focused on the reintroduction of a more traditional style of public examinations. This has meant that General Certificate of Secondary Education (GCSE), Advanced Supplementary and Advanced Level ('AS' and 'A' level) geography examinations in England became increasingly 'knowledge-led'—implying a closer association with, and bigger contribution from, academics in university geography departments. The creation of an A-level Content Advisory Board (ALCAB) by the Coalition government in 2014 was designed to significantly alter the geography content of 'A' and 'AS' level syllabuses—academic geographers on the Board, alongside GA and RGS-IBG members and a practicing teacher, were expected to be the main drivers of change. This process led to 'a degree of reconnection of HE with A level content' (Evans 2015)—although one view of such developments was that it heralded a return to the somewhat paternalistic relationship between teachers and academics previously witnessed in the Madingley era. Butt and Collins (2018) are slightly more sanguine about the reform of public examinations in geography, asserting:

> There must always be strong connections between [geography in schools and universities]
> if the discipline of geography is to remain healthy, for one important purpose of schools is
> to introduce disciplinary knowledge to young people

2.12 Teachers as Researchers

This chapter has principally concentrated on the contribution of academics, either working in geography departments or departments of education in higher education institutions, to geography education and research. However, this neglects the considerable input to geography education research made by those classroom-based practitioners who are also researchers. Recognition of the role of the 'teacher as researcher', or as co-workers alongside academic researchers, goes back to Lawrence Stenhouse (1975) and the influential Humanities Curriculum Project in England in the 1970s. Stenhouse made the assumption that if teachers were not engaged with, and had no ownership of, education research they could not be expected to use it to help shift their own practice. For Stenhouse, educational research was key to supporting school improvement—where curriculum development, and the research that underpinned it, belonged to the teacher; here professional researchers would support teachers in testing research ideas and considering provisional findings. He also saw curriculum development and the advancement of teaching skills as being closely linked to teacher self-development. The connections with the professional development of teachers is interesting, going back to the work of Schon (1987) (see also Butt and Macnab 2013; Brooks 2010a). However, we should also be aware of wide-ranging criticisms of teacher researchers and of their research—including the role that practitioner research plays in professional development—even from among those who have traditionally promoted practitioner enquiry and action research (see Elliott 2009; Carr and Kemmis 1986; McNiff 1988; Naish 1996).

The 'teacher as researcher', and the status of the research methods often used by such practitioners, has attracted considerable attention. The main issue concerns the *outcomes* of practitioner research—specifically, the types of knowledge generated and whether this is of practical or theoretical value. Additionally, the methods used by practitioner researchers—which tend to be qualitative and interpretivist, often applied in small scale studies—have attracted critical comment. This links us back to the problems associated with adopting a 'what works', or 'how to', research focus—forms of research that are particularly popular with practitioners for understandable reasons. The dangers are that non generalisable research may neither add much to our knowledge, nor help us to solve more persistent educational problems. What is unacceptable is that many teacher researchers assume that they have discovered something new and significant though their, clearly well-intentioned, research studies—even though similar findings are already known, or would have been revealed simply by undertaking a more substantial review of extant research literature. Research findings only rarely present neat solutions to problems. Unfortunately, many teacher-researchers (perhaps by necessity?) appear to accept their own 'answers' too quickly, often on the basis of limited empirical evidence. Practitioner–research may be 'the best it can be'—in terms of design, research question formation, methods, methodology, and overall quality—but we should recognise that its impact may be of most value in the professional transformation of teachers, rather than in pushing back research frontiers (Furlong et al. 2003). This is *not* to casually denigrate the efforts of teacher-researchers, but to understand the contexts in which their research is produced—we need to consider that much practitioner research typically represents the start, rather than the culmination, of interesting research enquiries. These issues go to the root of many of the criticisms of geography education research since the mid-1990s; research which has been perceived as:

> generally disconnected from educational research in other disciplines and characterized by studies that are mostly descriptive, incidental and anecdotal (Solem et al. 2013, p. 220)

2.13 Research and Initial Teacher Education

Despite the sustained pressures which have affected initial teacher education in the UK in recent years (Hartley 2000), those working in the sector have always sought to generate significant research findings in geography education. However, we should also recognise that ITE students have a significant contribution to make in connecting academic geography, school geography and geography education research; indeed, it is possible that their impact on geography education in schools may be greater than that initiated by geography education research and researchers. The fact that 'beginning teachers'—having progressed through different geography, or geography-related, undergraduate degrees—arrive at their pre-service training with a range of subject knowledge, understanding and skills is often perceived as an issue. Nevertheless, aside from some inherent difficulties this may cause with respect to individual trainees, new teachers act as an important conduit through which research ideas from the academy are regularly brought into schools; this is a contention held by many school geography mentor teachers and academics involved in teacher preparation (Butt 2018). These thoughts are echoed by Bradford (1996), who noted that:

> Newly and recently trained teachers can form a very useful resource and agent of change. They may be well qualified to prepare packages of new material, addressing emerging areas of the subject, for publication through such organisations as the GA (p. 287)

Walford (1996a) concurs, but from a different perspective-referring to schools as supplying the 'seed-corn' of students who will study for geography degrees, amongst whom some will eventually train to teach geography. This functional, mutually dependent, circular relationship—from school student, to university under-graduate, to postgraduate trainee teacher, to school teacher—has worked reasonably well for decades. Importantly, geography teachers in schools, academic geographers in universities, geography education researchers, geography educators in education departments, and the students they 'produce' must each work to ensure that this relationship continues to function effectively. Disturbance in any one element of this system inevitably affects each of the others.[8] It is therefore salutary to reflect on the changes that have shaped initial teacher education in the modern era. For over 20 years governments in the UK, of every political stripe, have sought to reform initial teacher education—seemingly often for ideological, rather than practical, reasons. The damage inflicted on what was once considered to be a world-leading model of teacher preparation in England—which nurtured a genuine partnership between schools and higher education institutions—has been immense. Initial teacher education has dramatically shifted its location: physically, in the sense that 'partnership' training now occurs almost entirely in schools; and intellectually, in that the expertise of university-based teacher educators and researchers is seemingly no longer trusted, or required—school-based mentors having become the sole source of intellectual development and challenge for beginning teachers. There is empirical evidence that this is having a significant, negative impact on the preparation of new entrants to the profession, on geography curriculum development in schools, and on research in geography education (Butt 2015; Tapsfield et al. 2015). For the foreseeable future, teacher education in England will reside in schools—indeed, it is doubtful that even if universities were given an opportunity to 'take back' the pre-service preparation of teachers they would consider doing so, given the risk of future political interference. The effect on research in geography education is palpable. As Butt and Collins (2018) lament:

> The role of the specialist geography educator in ITE is therefore sadly disappearing. Given the significant decline in numbers of trainee geography teachers based in HEIs, and the increasing placement of trainee teachers on school-based routes, this is likely to lead to narrower forms of teacher preparation and the loss of subject specialisms. This will inevitably have far reaching effects on research and scholarship in geography education and ultimately on the quality of geographical knowledge experienced by students in schools.

2.14 Diversity of Research Activity

Furlong (2013) visualises the diversity of research activity in education as 'challenging': a consequence of the marriage of different educational research traditions—each with their own processes, methods and methodologies. This diversity is also a function of the different aims, audiences (either academic or public) and expectations associated with the various strands of education research:

> diversity is not just in terms of topic or educational sector, but in terms of method (… design experiments, randomised control trials, longitudinal cohort studies, large-scale sur-

---

[8]A pertinent example of this relationship beginning to fail was provided by Davidson and Mottershead's (1996) observations on the development of physical geography. 'A' level geography courses have not traditionally prepared students well for the demands of studying physical geography in university degree courses—which have a more scientific and positivistic, but less interactive and enquiry based, approach than students are used to in schools.

veys, interview studies and detailed qualitative case studies); in terms of theoretical framing
(from atheoretical empiricism, to labour economics, to child development); and in terms
of fundamental purpose (academic and applied research and practitioner enquiry) (Furlong
2013, p. 101)

Such diversity can be viewed both as a strength, and as a weakness. One observer
might celebrate this miscellany, while another might view the constant subdivision
and fragmentation of educational research into ever smaller fields of enquiry as
damaging and wasteful of resources. The problem of research endeavours fracturing
into sub divisions—creating discrete, themed, sub communities—is that it diminishes
the contribution of each research group. Carr (2006) notes that although education
researchers often claim to be part of a large and coherent research community, any
prospect of achieving greater homogeneity in the field is essentially fanciful. With
its wide range of different intellectual traditions, approaches to research, research
questions and theoretical positions, the world of education research and researchers
cannot convincingly declare any sense of unity. Indeed, a less charitable interpretation
might view the tendency to divide into ever smaller factions as simply a ruse by
emergent groups to gain individual status, funding, recognition and influence. As a
consequence, sub fields of education research and researchers often have weak claims
to internal intellectual coherence—for some this approaches dishonesty, given that
groups may essentially be 'held together' by their own specialist themed journals,
publications, institutions, professional associations, conferences and seminars. As
such, education researchers—not least in geography education—can have difficulty
in firmly establishing their truth claims: these being mostly practice-based, suffering
from historical and paradigmatic variations, and revealing an unwelcome tendency
to introspection.

2.15 Conclusions

We have seen how post war changes in England's politics, economy and society
have impacted on schools and universities and, more prosaically, on research in
geography education. During this period, it is fair to say that the development of
geography teaching in schools witnessed a number of 'false dawns'. For example, in
1979, the editor of *Teaching Geography,* Patrick Bailey,[9] proclaimed a 'new maturity'
in geography education in schools—such that 'the shape of the new geography in
schools is clear, as are the appropriate methods of teaching it. All we have to do now
is make the best practice general' (p. 31). Each of the seven justifications advanced
by Bailey for such an optimistic evaluation was subsequently challenged by Kirby
and Lambert (1981)—the first most strongly: that sound theoretical foundations were

---

[9]Bailey's optimistic view of the future of geography education in schools continued into the late
1980s. In 1988 he confidently stated that geography had gained a 'place in the sun' (Bailey 1988),
something that others were much less sanguine about (see Hall 1990; Lambert 1991; Rawling 1991;
Roberts 1991; Butt 1992).

apparent in the geography taught in schools. As they correctly observed, the process of subject evolution is both lengthy and complicated—a range of intervening factors affecting whether subjects will survive and prosper in schools. Additionally, the persistence of the gap between the research frontier of academic geography and the types of geography commonly taught in schools remains a concern.

Arguably, the most tangible advances in geography education in the UK were associated with the curriculum development projects sponsored by the Schools Council[10] in the 1970s—the Geography 16–19 Project being perhaps the most noteworthy. Significantly, this project started by considering the *educational* needs of students, before deliberating on the necessary contributions from academic geography—a shift from the traditional focus on the subject, towards the ways in which studying geography could support the learner (Naish and Rawling 1990). The Schools Council projects had a profound impact on the teaching of geography in many schools, and on the nature of research in geography education, but only a somewhat tangential relationship with academic geography. Interestingly, the changing knowledge and skill sets exhibited by incoming undergraduate geography students—many of whom had recently studied 16–19 A level Geography—forced the tertiary sector (perhaps for the first time) to take greater note of their students' subject preparation and educational backgrounds. Many in the academy disliked what they considered to be a decline in standards of geographical knowledge. Binns (1996) commented ruefully:

> Remarkably few higher education teachers have a detailed understanding of geography teaching at A level, let alone of the pre-16 age group. Some would cynically suggest that interest in school geography among higher education teachers only increases when their student applications are falling! (Binns 1996, p. 44)

At the start of this chapter I repeated Haubrich's (1996a, b) assertions that the period from the mid-1970s to the mid-1990s saw research in geographical and environmental education make substantial progress in terms of its scale, coverage and methods. Haubrich also claimed that research publications had shown a significant upturn, both in their quality and quantity. He was not alone. In the same year Gerber and Williams (1996)—perhaps not quite so confidently, but certainly with a positive spin—reflected on the same period, commenting:

> twenty years ago the main researchers amongst geographical educators were university lecturers in a few centres in a few countries such as the United Kingdom, Germany, France and the USA. Currently the researchers consist of university academics, teachers, postgraduate and undergraduate students, community members, professional associations and school students. This diverse band of researchers are distributed widely across the countries of the world. They are not only working with colleagues in their own country, but with colleagues from other countries in similar and contrasting contexts (p. 5)

While there are some self-evident truths in these statements, my concern is that the progress they report was either illusory or has not been sustained in the period from the mid-1990s to the current day.

---

[10]Two other projects, both of which were initiated in the 1970s were Geography for the Young School Leaver (GYSL or Avery Hill Project) and Geography 14–16 (Bristol Project). The locations stated in the parenthesis reveals the higher education institutions in which the projects were based.

In the following chapter our attention turns to more contemporary developments in geography education research and theory.

References

Bailey, P. (1988). A place in the sun: The role of the Geographical Association in establishing Geography in the National Curriculum of England and Wales, 1975–1989. *Journal of Geography in Higher Education, 13*(2), 144–157.

Bailey, P. (1992). A case hardly won: Geography in the National Curriculum of English and Welsh Schools. *Geographical Journal*, pp. 63–74.

Bale, J. (1996). Mediated knowledge: Things we never read in the 70s. Paper delivered to *UDE Tutors' Conference*, Matlock Bath, March 2.

Barratt Hacking, E. (1996). Novice teachers and their geographical persuasions. *International Research in Geographical and Environmental Education, 5,* 77–86.

Binns, T. (1996). School geography: They key questions for discussion. In E. Rawling & R. Daugherty (Eds.), *Geography into the twenty-first century* (pp. 37–56). London: Wiley.

Boardman, D. (1988). *The impact of a curriculum: Project geography and the young school leaver.* Birmingham: Educational Review Publications.

Bonnett, A. (2003). Geography as the world discipline: Connecting popular and academic geographical imaginations. *Area, 35*(1), 56–63.

Bradford, M. (1996). Geography at the secondary/higher education interface: Change through diversity. In E. Rawling & R. Daugherty (Eds.), *Geography into the twenty-first century.* Chichester: Wiley.

Bradford, M., & Kent, A. (1977). *Human geography: Theories and their application.* Oxford: Oxford University Press.

Brock, C. (Ed.). (2015). *Education in the United Kingdom.* London: Bloomsbury.

Brooks, C. (Ed.). (2010a). *Studying PGCE Geography at M Level.* London: Routledge.

Brooks, C. (2010b). Developing and reflecting on subject expertise. In C. Brooks (Ed.), *Studying PGCE Geography at M Level: Reflection, research and writing for professional development.* London: Routledge.

Brooks, C. (2010c). How does one become a researcher in geography education? *International Research in Geographical and Environmental Education, 19*(2), 115–118.

Brooks, C., Butt, G., & Fargher, M. (Eds.). (2017). *The Power of geographical thinking.* Dordrecht: Springer.

Butt, G. (1992). Geography. In P. Ribbins (Ed.), *Delivering the national curriculum: Subjects for secondary schooling* (pp. 157–175). Longmans: Harlow.

Butt, G. (1997). An investigation into the Dynamics of the National Curriculum Geography Working Group (1989–1990). Unpublished Ph.D., University of Birmingham, Birmingham.

Butt, G. (2008). Is the future secure for geography education? *Geography, 93*(3), 158–165.

Butt, G. (Ed.). (2011). *Geography, education and the future.* London: Continuum.

Butt, G. (2015). What impact will changes in teacher education have on the geography curriculum in schools? Presentation at *RGS-IBG Annual International Conference*, University of Exeter, 3 September 2015.

Butt, G. (2018). *What is the future for subject-based education research?* Public seminar delivered at the Oxford University Department of Education. 22 October (www.podcasts.ox.ac.uk).

Butt, G. (2019). Bridging the divide between school and university geography—'Mind the Gap!'. In S. Dyer, H. Walkington, & J. Hill (Eds.), *Handbook for learning and teaching geography* (pp. 1–10). London: Elgar.

Butt, G., & Collins, G. (2013). Can geography cross 'the divide'? In D. Lambert & M. Jones (Eds.), *Debates in geography education* (pp. 291–301). RoutledgeFalmer: Abingdon.

Butt, G., & Collins, G. (2018). Understanding the gap between schools and universities. In M. Jones & D. Lambert (Eds.), Debates in geography education (2nd ed., pp. 263–274). London: Routledge.

Butt, G., & MacNab, N. (2013). Making connections between the appraisal, performance management and professional development of dentists and teachers: 'We sat down and we talked, and we've never formally sat down and talked before'. Professional Development in Education, 39(5), 841–861.

Carr, W. (2006). Education without theory. British Journal of Educational Studies, 54(2), 136–159.

Carr, W., & Kemmis, S. (1986). Becoming critical: Education, knowledge and action research. London: Falmer.

Castree, N., Fuller, D., & Lambert, D. (2007). Geography without borders'. Transactions of the Institute of British Geographers, 32, 129–132.

Catling, S. (2017). Reflections: An introduction and an auto(geo) biography. In S. Catling (Ed.), Reflections on primary geography: Conference participants' perspectives on aspects of primary geography (pp. 13–28). Sheffield: Register of Research in Primary Geography/GA.

Chorley, R., & Haggett, P. (Eds.). (1965). Frontiers in geographical teaching. London: Methuen.

Chorley, R., & Haggett, P. (Eds.). (1967). Models in geography. London: Methuen.

Clifford, N. (2002). The future of geography: When the whole is less than the sum of its parts. Geoforum, 33, 431–436.

Daugherty, R., & Rawling, E. (1996). New perspectives for geography: An agenda for action. In E. Rawling & R. Daugherty (Eds.), Geography into the twenty-first century (pp. 1–15). Chichester: Wiley.

Davidson, J., & Mottershead, D. (1996). The experience of physical geography in schools and higher education. In E. Rawling & R. Daugherty (Eds.), Geography into the twenty-first century (pp. 307–322). London: Wiley.

Department for Education (DfE). (2010). The importance of teaching: The Schools White Paper. London: The Stationery Office.

Department for Education and Science (DES). (1988). Education reform act. London: HMSO.

Department for Education and Science (DES). (1991). Geography in the national curriculum (England). London: HMSO.

Ecclestone, K., & Hayes, D. (2009). The dangerous rise of therapeutic education. London: Routledge.

Elliott, J. (2009). Research based teaching. In S. Gewirtz, P. Mahoney, I. Hextall, & A. Cribb (Eds.), Changing teacher professionalism. London: Routledge.

Evans, M. (2015). Reconsidering geography at the schools-HE boundary; the ALCAB experience. Presentation at RGS-IBG Annual International Conference, University of Exeter, 3 September 2015.

Finn, M. (2018). Why reformed A levels are not preparing undergraduates for university study. The Conversation. Accessed 5 October.

Firth, R. (2011). Debates about knowledge and the curriculum: Some implications for geography education. In G. Butt (Ed.), Geography, education and the future (pp. 141–164). London: Continuum.

Fisher, C. (1970). Wither regional geography? Geography, 55(4), 373–389.

Furlong, J. (2013). Education: An anatomy of the discipline. London: Routledge.

Furlong, J., Salisbury, J., & Coombs, J. (2003). The best practice research scholarship scheme. An evaluation and final report to the DfES. University of Cardiff School of Social Science.

Gerber, R., & Williams, M. (Eds.). (1996). Qualitative research in geographical education. Armidale: University of New England.

Goodson, I. (1983). School subjects and curriculum change. London: Croom Helm.

Goudie, A. (1993). Schools and universities: The great divide. Geography, 78(4), 338–339.

Graves, N. (1975). Geography in education. London: Heinemann.

Gregory, D. (1978). Ideology, science and human geography. London: Hutchinson.

Haggett, P. (1996). Geography into the next century: Personal reflections. In E. Rawling & R. Daugherty (Eds.), *Geography into the twenty-first century* (pp. 11–18). London: Wiley.

Haggett, P. (2015). Madingley: Half-century reflections on a geographical experiment. *Geography, 100*(1), 5–11.

Haggett, P., & Chorley, R. (1989). From Madingley to Oxford: A foreword to remodelling geography. In W. Macmillan (Ed.), *Remodelling geography* (pp. xv–xx). Oxford: Blackwell.

Hall, D. (1990). The national curriculum and the two cultures: Towards a humanistic perspective. *Geography, 75*(4), 313–324.

Hartley, D. (2000). Shoring up the pillars of modernity. Teacher education and the quest for certainty. *International Studies in the Sociology of Education, 10*(2), 113–131.

Harvey, D. (1969). *Explanation in geography*. London: Edward Arnold.

Harvey, D. (1973). *Social justice and the city*. Baltimore: John Hopkins University Press.

Haubrich, H. (1992). Geographical education research 2000: Some personal views. *International Research in Geographical and Environmental Education, 1*(1), 52–56.

Haubrich, H. (1996a). Foreword. In M. Williams (Ed.), *Understanding geographical and environmental education. The role of research* (pp. xi–xii). London: Cassells.

Haubrich, H. (1996b). *Geographical education 1996: Results of a survey in 38 countries*. Mimeo.

High School Geography Project (HSGP). (1971). *American High School Geography Project. Geography in an urban age*. Toronto, ON, Canada: Collier-Macmillan.

Hill, J., & Jones, M. (2010). 'Joined-up geography": Connecting school-level and university-level geographies. *Geography, 95*(1), 22–32.

Hirsch, E. (1987). *Cultural literacy: What every American needs to know*. New York: Houghton Mifflin.

Hirsch, E. (2007). *The knowledge deficit: Closing the shocking gap for American children*. New York: Houghton Mifflin.

Holloway, S., & Valentine, G. (Eds.). (2004). *Children's geographies: Playing, living, learning*. London: Routledge.

Hore, P. (1973). A teacher looks at the new geography. In R. Walford (Ed.), *New directions in geography teaching* (pp. 132–137). Longmans: Harlow.

Jackson, P. (2018). 125 years of the Geographical Association. *Geography, 103*(3), 116–121.

Johnston, R. (2009). On geography, geography and geographical magazines. *Geography, 94*(3), 207–214.

Kent, A. (Ed.). (2000). *Reflective practice in geography teaching*. London: Paul Chapman Publishing.

Kirby, A., & Lambert, D. (1981). Seven reasons to be cheerful? … or school geography in youth, maturity and old age. In R. Walford, (Ed.), *Signposts for geography teaching* (pp. 113–119). Harlow: Longmans.

Lambert, D. (1991). Too late for debate? *Times Educational Supplement*, 29 March.

Lambert, D. (2011). Reviewing the case for geography, and the 'knowledge turn' in the English National Curriculum. *The Curriculum Journal, 22*(2), 243–264.

Lambert, D. (2015). Research in geography education. In G. Butt (Ed.), *MasterClass in geography education* (pp. 15–30). London: Bloomsbury.

Lambert, D., & Jones, M. (Eds.). (2017). *Debates in geography education* (2nd ed.). London: Routledge.

Lambert, D., & Morgan, J. (2010). *Teaching geography 11–18: A conceptual approach*. Maidenhead: McGraw Hill/Open UP.

Livingstone, D. (1992). *The geographical tradition*. Oxford: Blackwell.

Long, M., & Roberson, B. (1966). *Teaching geography*. London: Heinemann.

Machon, P., & Ranger, G. (1996). Change in school geography. In P. Bailey & P. Fox (Eds.), *Geography teacher's handbook*. Sheffield: Geographical Association.

Marsden, W. (1997). On taking the geography out of geographical education—Some historical pointers on geography. *Geography, 82*(3), 241–252.

McNiff, J. (1988). *Action research. Principles and practice*. London: Macmillan.

Mitchell, D. (2011). A "knowledge turn"—Implications for geography initial teacher education (ITE). Paper presented at *the IGU-CGE Conference*, Institute of Education, University of London, April 2011.

Mitchell, D. (2017). *Geography curriculum making in changing times* (Unpublished Ph.D.). Institute of Education, University College London.

Morgan, J. (2008). Curriculum development in 'New times'. *Geography, 93*(1), 17–24.

Morgan, J. (2014). Foreword. In M. Young, D. Lambert, C. Roberts, & M. Roberts (Eds.), *Knowledge and the future school: Curriculum and social justice* (pp. ix–ix243). London: Bloomsbury.

Morgan, J., & Firth, R. (2010). 'By our theories shall you know us': The role of theory in geographical education. *International Research in Geographical and Environmental Education., 19*(2), 87–90.

Naish, M. (1987). Geography in British Schools: A medium for education. In K. Husa, C. Vielhaber, & H. Wohlschagel (Eds.), *Festschrift Zum 60 Geburtstag van Ernest Troger* (pp. 99–109). Hirt, Vienna: Beitrage zur Didaktic der Geographie.

Naish, M. (1996). Action research for a new professionalism in geography education. In A. Kent, D. Lambert, M. Naish, & F. Slater (Eds.), *Geography in education: Viewpoints on teaching and learning* (pp. 321–343). Cambridge: Cambridge University Press.

Naish, M. (2000). The geography curriculum of England and Wales from 1965: A personal view. In D. Lambert & D. Balderstone (Eds.), *Learning to teach geography in the secondary school*. London: RoutledgeFalmer.

Naish, M., & Rawling, E. (1990). Geography 16–19: Some implications for higher education. *Journal of Geography in Higher Education., 14*(1), 55–75.

Naish, M., Rawling, E., & Hart, C. (1987). *Geography 16–19: The contribution of a curriculum project to 16–19 education*. London: Longman.

Newby, P. (1980). The benefits and costs of the quantitative revolution. *Geography, 65,* 13–18.

Peet, R. (Ed.). (1977). *Radical geography: Alternative viewpoints on contemporary social issues.* Chicago: Maaroufa Press.

Rawding, C. (2013). How does geography adapt to changing times? In D. Lambert & M. Jones (Eds.), *Debates in geography education* (2nd ed., pp. 282–290). London: Routledge.

Rawling, E. (1991). Making the most of the national curriculum: The implications for secondary school geography. *Teaching Geography, 16*(3), 130–131.

Rawling, E. (1996). Madingley revisited? In E. Rawling & R. Daugherty (Eds.), *Geography into the twenty-first century* (pp. 3–8). London: Wiley.

Rawling, E. (2001). *Changing the subject. The impact of national policy on school geography 1980–2000.* Sheffield: Geographical Association.

Rawling, E., & Daugherty, R. (Eds.). (1996). *Geography into the twenty-first century*. Chichester: Wiley.

Robbins, L. (1963). *Higher education*. Committee on Higher Education (The Robbins Report). London: HMSO.

Roberts, M. (1991). On the eve of the geography national curriculum: The implications for secondary school geography. *Geography*, pp. 331–342.

Robinson, R. (1981). Quantification and school geography—A clarification. In R. Walford (Ed.), *Signposts for geography teaching* (pp. 94–96). Longmans: Harlow.

Ross, A. (2000). *Curriculum: Construction and critique*. London: Taylor and Francis.

Schon, D. (1987). *Education the reflective practitioner*. San Francisco: Jossey-Bass.

Slater, F. (1982). *Learning through geography*. London: Heinemann.

Smith, D. (1973). Alternative 'relevant' professional roles. *Area, 5,* 1–4.

Solem, M., Lambert, D., & Tani, S. (2013). Geocapabilities: Toward an international framework for researching the purposes and values of geography education. *Review of International Geographical Education Online (RIGEO), 3,* 214–229.

Standish, A. (2007). Geography used to be about maps. In R. Whelan (Ed.), *The corruption of the curriculum*. London: Civitas.

Standish, A. (2009). *Global perspectives in the geography curriculum: Reviewing the moral case for geography*. London: Routledge.

Standish, A. (2012). *The false promise of global learning. Why education needs boundaries*. London: Continuum.

Stenhouse, L. (1975). *Introduction to curriculum research and development*. London: Heinemann.

Stoddart, D. (1986). *On geography*. Oxford: Blackwell.

Tapsfield, A., Roberts, M., & Kinder, A. (2015). *Geography initial teacher education and teacher supply in England*. Sheffield: Geographical Association.

Tate, S., & Swords, J. (2013). Please mind the gap: Students' perspectives of the transition in academic skills between A-level and degree-level geography. *Journal of Geography in Higher Education, 37*(2).

Thrift, N. (2002). The future of geography. *Geoforum, 33,* 291–298.

Tidswell, V. (1977). *Pattern and process in human geography*. London: University Tutorial Press.

Unwin, T. (1996). Academic geography: The key questions for discussion. In E. Rawling & R. Daugherty (eds.) Geography into the twenty-first century (pp. 19–36). London: Wiley.

Walford, R. (1973). Introduction. In R. Walford (Ed.), *New directions in geography teaching* (pp. 1–6). Harlow: Longmans.

Walford, R. (Ed.). (1981). *Signposts for geography teaching*. Harlow: Longmans.

Walford, R. (1991). National curriculum: Burden or opportunity? *Teaching Geography, 16*(1), 32.

Walford, R. (1996a). Geography 5–19: Retrospect and prospect. In E. Rawling & R. Daugherty (Eds.), *Geography into the twenty-first century*. London: Wiley.

Walford, R. (1996b). What is geography? An analysis of definitions provided by prospective teachers of the subject. *International Research in Geographical and Environmental Education, 5,* 69–76.

Walford, R. (2001). *Geography in British Schools 1850–2000*. London: Woburn Press.

Young, M. (2008). *Bringing knowledge back in: From social constructionism to social realism in the sociology of education*. London: Routledge.

Young, M. (2013). Powerful knowledge: An analytically useful concept or just a 'sexy sounding term'? A response to John Beck's 'Powerful knowledge, esoteric knowledge, curriculum knowledge'. *Cambridge Journal of Education, 43*(2), 195–198.

Young, M., & Lambert, D., with Roberts, C., & Roberts, M. (2014). *Knowledge and the future school: Curriculum and social justice*. London: Bloomsbury.

Young, M., & Muller, J. (2010). Three educational scenarios for the future: Lessons for the sociology of knowledge. *European Journal of Education, 45*(1), 11–27.

Chapter 3
Contemporary Developments in Geography Education Research and Theory

3.1 Context

This chapter focuses closely on the range of ideas considered significant in contemporary geography education research and theory. It foregrounds certain themes that have proved attractive to researchers in geography education (such as, *inter alia*, progression, capabilities and geocapablities, powerful knowledge and powerful disciplinary knowledge, Future 3 curricula, Young People's Geographies, 'curriculum making', Geographic Information Systems (GIS) and fieldwork), but inevitably cannot claim to offer a comprehensive appreciation of *all* the avenues of research that geography educators pursue.[1] Nonetheless, the research themes selected have some conceptual links—many of their key concepts overlapping in ways which reveal synergies between the research projects that have engaged with them. This chapter also highlights the growth in outlets for reporting research in geography education, noting where geography education researchers choose to publish their work, the establishment of 'new' journals in the field (such as *International Research in Geographical and Environmental Education* (IRGEE) in 1992, and the re-launch of the Geographical Association's main academic journal *Geography* to more fully reflect educational content in 2007), the increase of online and open access journals (such as *Review of International Geographical Education online* (RIGEO) in 2011, and *J Reading: Journal of Research and Didactics in Geography* in 2012), and the use of blogs and social media.

One contention offered—in agreement with Morgan and Firth (2010)—is that geography education research is currently characterized by an over adoption of theories, methods and questions closely related to classroom-based issues. This builds on points raised earlier in the book, which argue that much contemporary research activity in geography education is dominated by teacher-researchers who carry out small

---

[1] A different version of this chapter, for example, might devote space to ethnogeographies, inferentialism, complexity theory, etc.

© Springer Nature Switzerland AG 2020
G. Butt, *Geography Education Research in the UK: Retrospect and Prospect*, International Perspectives on Geographical Education,
https://doi.org/10.1007/978-3-030-25954-9_3

scale, action research projects. This tendency, Morgan and Firth (2010) have asserted, has restricted 'outreach' into other areas of theory to such an extent that since the early 1990s geography education researchers have witnessed 'the end of theory'. The trend is further exacerbated by the drive to relocate initial teacher education entirely in schools in England—which has involved the removal of many research-active geography education specialists in higher education institutions (HEIs) from the process of teacher education, with both themselves and their institutions stigmatized as purveyors of 'useless theory'.

It is perhaps helpful to start with an overview of the themes for research in geography education that have been largely ignored and conversely those which have attracted attention. At the start of the new millennium Roberts (2000) highlighted some areas which she felt had been neglected with regard to research into the teaching and learning of geography. These were as follows:

- children's prior understanding of concepts in human geography
- resources other than textbooks, with attention to the implications of accessibility of new and varied sources of data
- learners' uses and understanding of resources other than maps
- processes of teaching and learning in real-life classroom situations

While any list such as this can be criticised for its breadth and lack of specificity, the attempt Roberts makes to define themes for future attention among geography education researchers is to be applauded. The act of compilation shows how difficult, indeed dangerous, it is to attempt to outline topics for future research. Lambert (2015) bravely attempts a similar 'illustrative' overview of themes researched in geography education early in the twenty-first century,[2] mindful that this list is neither exhaustive, nor references all the researchers who have worked in these areas. He states that these themes draw on perspectives from:

> philosophy (Firth 2011; Winter 2006, 2010, 2013a, b), cultural studies (Morgan 2007, 2008b), childhood studies (Biddulph 2011a; Martin and Catling 2011), and cognitive and conceptual development (Bennetts 2005a, b; Taylor 2011)… curriculum studies (Rawling 2001; Roberts 1996; Morgan 2008a), learning styles and pedagogy (Leat and Higgins 2002; Roberts 2013), technologies, including GIS (Fargher 2013a, b), personalization (Jones 2013), fieldwork (Rickinson et al. 2004; Dunphy and Spellman 2009), development education (Lambert and Morgan 2011), assessment and evaluation (Butt et al. 2011)

3.2 Introduction

Education as a discipline is not solely defined by its epistemological roots or by the arguments it advances to substantiate its existence—which tend to focus on aims, theories, research questions and methods, the nature of evidence and ontology. It is

---

[2]For a more complete overview of the research themes of geography educationists up to 1997, see Foskett and Marsden (1998).

also delineated by a range of sociological factors which explain how the discipline has become established and currently functions (see Barnett 2009; Furlong 2013). As such, it might be claimed that education has a strong *institutional* reality—that is, it continues to be represented in many universities—but nevertheless exists despite its general lack of *epistemological* coherence. There are inherent dangers in such a situation, for when pressures bear on education as a discipline, and on its subject-based research and researchers, this lack of coherence make it a target for disapprobation.

It is worthwhile considering the main types of theories that researchers in geography education employ. Hirst (1996) discusses how teachers have traditionally developed their own theories based on classroom practice—indeed, he believes that teachers commonly theorise about their work 'in their own way'. This echoes Morgan and Firth's (2010) concerns outlined above. The issue to be addressed, in Hirst's opinion, is whether such theories are credible and useful—they may have been forged in the heat of teachers' critical reflection on their practice, but they also need to be exposed to the scrutiny of others. Their 'fit' with established and commonly applied theories of education is crucial—for although teachers' 'theories' may be helpful in hypothesising their day-to-day work, they may not withstand intense testing by those more fully versed in theory-building and application. Reflection and judgement *within* a professional community is important, but should not be seen as the end of the process. It can be argued that by the very nature of their transition from being teachers in schools to working as academics in higher education many education researchers began with (and may still apply) a large degree of practical wisdom—and theorising—based on their previous school experiences. Often subject-based researchers in education do not stray too far from this 'practical theorising', as schools tend to remain at the centre of their research enquiries. However, all researchers must understand, and engage with, theories from beyond their immediate domain.

In the rather selective examples of contemporary research themes in geography education outlined below, I may be guilty of personal bias in choosing which to represent and which to ignore. No systematic method has been applied to select these themes. They are chosen because of the interest they have generated among geography education researchers—which has resulted in a body of work developing, rather than 'one off' pieces of research which have not stimulated further research activity, or interest. Whilst I understand that this unscientific approach lays the text open to criticisms of partiality, preference and prejudice, I nevertheless ask that areas of omission are, on this occasion, at least tolerated. I have tried to include research themes that have not only attracted the attention of geography educationists, but have also opened up enquiries that link to the parent discipline of Geography as well as to areas of theory beyond the somewhat narrow bounds of geography education. Others have done so before. An example of a publication in which, arguably, the link between school and academic geography in a *research* context is strong is John Huckle's edited work *Geography Education—reflection and action* (Huckle 1983). Here, as Rawding (2013) comments, the chapter headings reveal a strong connection with the (then) recent changes in academic geography; for example, in the 'New perspectives' section there are chapters on: Behavioural Geography; Humanistic Geography; Geography through Art; Welfare approaches to geography; Radical geography; Political education; Development education; Environmental education; and Urban studies. There is a realisation that education theory must be grounded in practice, but also connect with the academic discipline—with the themes offered in Huckle's book being applicable to (attractive to?) school teachers, as well as geographers and geography educationists in the academy. The 1980s, while not perhaps having the qualities of a 'golden age' of connection between academic geography and school geography in terms of research, application of theory and publication, did witness the production of a variety of work—often politically left of centre—that attempted to engage with issues of social justice, educational change and shifts in the parent academic discipline. The series *Contemporary Issues in Geography and Education* (1984–1987), for example, included themes for teachers relating to new movements in academic geography. Indeed:

> it was not seen as unusual for texts of the time to discuss geographical education from the perspective of academic paradigms within the discipline (Rawding, p. 284)

Much of this chapter discusses teachers' relationships with their subject and its parent discipline. Here it is important to recognise the work of Brooks (2007, 2016) who has focussed on the significance both of the discipline and subject of geography in teacher preparation and development, even though this connection may not always be recognised as central to practice.

3.3   Progression

Geography educators and researchers have for many years struggled with notions of how children 'progress' in their attainment of geographical concepts and ideas. As Taylor (2013) reminds us:

> Learning involves change in someone's knowledge, understanding, skills, or attitudes … but a neutral idea of 'change' is not enough. The change must be seen as valuable, as moving in a positive direction, as progress (p. 302).

However, as Taylor subsequently explores, this leads us into contentious, indeed political, territory—who decides what is 'valuable' to learn (the government, academics, employers, teachers, assessors, students?), and how does this change given that geography as a subject is continually under construction ('in progress')? The nature of any discipline, and of the school subjects that are their products, reveal that the educational content that is thought to be of value, and therefore promoted, shifts over time.

During the construction of the geography national curriculum (GNC) in England members of the Geography Working Group (GWG) were well aware of the dearth of research evidence they could draw on concerning continuity and progression[3] in

---

[3]I have adopted Bennetts (1995) brief definitions of two key terms: continuity as 'the persistence of significant features of geographical education as pupils move through their school system'; and progression as 'the focus on how pupils' learning advances' (Bennetts 1995, p. 75).

geography education. This was obviously an issue, for the ways in which children progress in geography would subsequently prove central to debates about national curriculum levels, assessment and conceptual understandings of the subject. At the time, the Chair of the working group stated:

> There are always difficulties of progression in a content-rich subject like geography, it is more difficult than in a reasoning subject like Mathematics (Fielding 1990)

This encapsulated what many believed was an attempt by the Chair to side step time consuming debate about progression in geography, in the knowledge that supporting research evidence in this area was lacking. Unfortunately, the consequence of this was to label geography as a subject in which content is simply amassed, reasoning is downplayed, and progression is based solely on content accumulation—important questions about which geographical concepts were 'harder' to learn, and indeed whether progression could ever be tied to assessed levels, were largely avoided. As Lambert (1996) noted, the research evidence for linking geographical content, concepts and understanding to assessed levels was simply non-existent:

> without sufficient evidence to suggest whether or not it is possible, let alone desirable ... (we are) forced to describe progression in distinctive levels; what concentrates the mind further is the assumption that this should acquire national acceptance and agreement (p. 269)

One of the GWG members, Michael Storm, admitted at the time:

> progression has always been a great weakness intellectually with geography in schools (Storm 1995)

This is a situation where a lack of existing research evidence greatly hampered the delivery of a policy imperative—the creation of the first geography national curriculum in England—impeding the GWG from making confident statements about progression in geography. However, once the curriculum was enshrined in a statutory order the problem was simply assumed to have gone away—there was no time to open up a larger conceptual debate during the life of the working group, nor to undertake research that could immediately feed into policy making. As such, some aspects of progression were stated with assurance in the GNC (for example, the need to progress from studying oblique images to vertical images in air photographs), whilst others were not (say, on progression of area studies). This revealed how ideas about progression held among working group members were:

> largely based on personal experience, expectations and beliefs rather than on harder edged research findings[4] (Butt 1997, p. 239)

Trevor Bennetts has stood at the forefront of geography educators and researchers who have wrestled with the concepts of progression and geographical understanding (Bennetts 1981, 1995, 1996, 2005a, b). His work takes us well beyond the aspects

---

[4]Members of the GWG claim that the research evidence for progression was discussed during the creation of the GNC—particularly the Brunerian concept of a spiral curriculum involving the revisiting of concepts as the child progresses, but without the direct repetition of previous content studied.

of progression foregrounded in the GNC—as detailed in an appendix to a scheme of work which targeted aspects of vocabulary, knowledge of places, patterns and processes, geographical thinking, geographical explanation, geographical investigation and map skills (QCA 2000). Here the 'official' conception of progression was captured in hierarchical level descriptions which were brief, bland and often somewhat obvious. These offered the teacher a simple notion of moving the learner 'from' and 'to': for example, in 'vocabulary' students should move from 'using a limited geographical vocabulary' to 'precise use of a wider range of vocabulary'; while in 'knowledge of places' they should progress from 'geographical knowledge of some places', to an 'understanding of a wider range of areas and links between them'. These are, one could argue, self-evident steps that might be expressed similarly at any key stage, or even at university level.

Bennetts' (2005a) more carefully considered analysis, however, acknowledges that progression of understanding firstly depends on the gap between what the learner needs to understand and what they already know, or have experienced. He focuses on seven aspects of progression:

(i)   complexity—of experience, information, ideas, cognitive tasks,
(ii)   abstraction—of ideas about processes, relationships, values, but also forms of presentation
(iii) precision—being more exact, and knowing when that is appropriate and useful
(iv)   making connections and developing structures—from applying simple ideas to experience and making simple links, to use of sophisticated conceptual models and theories
(v)   breadth of context—in which explanations occur, especially spatial and temporal contexts
(vi)   association of understanding with cognitive abilities and skills
(vii) association of understanding with affective elements—attitudes and values, value laden nature of some ideas (after Bennetts 2005a)

Although only some of these aspects are *explicitly* geographical—for example, the reference to *spatial* contexts in (v) above—they represent an attempt to describe progression in a wider educational sense in terms of breadth, depth, movement from concrete to abstract concepts, and use of a range of techniques (SCAA 1994; QCA 2000; Marsden 1995; Bennetts 1995, 2005a, b; Taylor 2013). These have obvious connections with more generic work on progression in education (see Bloom 1956; Krathwohl 2002), but arguably do not take us much further in terms of progression of *geographical* thinking. What is clear is that 'making progress' in geography should be more than the mere acquisition of an increasing number of facts (Biddulph 2018). As Firth (2013) comments, this represents a particular view of knowledge that positions it as inert, objective and given—and, as a consequence, remarkably easy to assess— but which is essentially inadequate, in educational terms, as a representation of progression in learning. We also have evidence that geographical *concepts*, rather than content, provide us with the essential building blocks for curriculum and progression planning (Marsden 1997; Brooks 2017; Rawling 2016, 2018; Biddulph 2018).

More contemporary work has been completed on progression in geography education, at the international scale. For example, *Learning Progressions in Geography Education: International Perspectives* (Muniz Solari et al. 2017), offers a more global appreciation of the concept—it includes a chapter by Biddulph and Lambert on the English context which looks at the historical development of ideas about geographical progression against the backdrop of policy and statutory definitions. This is important because, as we have seen above, the GNC with its level descriptions became the de facto definition of student progress in geography—a technocratic national system to 'ensure' progress without recourse to research evidence, or indeed professional judgement (James 2014). Here we must acknowledge that our current understanding of progression in geography involves an element of inference, often based on assessment evidence, for doubt still exists as to the precise nature of progress made by students in geography (Bennett 2011).

3.4 Capability and GeoCapability

David Lambert, in his inaugural professorial lecture at the [then] University of London, Institute of Education, offered a powerful argument for the adoption of a capability approach to geography education (Lambert 2009). Making his case for finding new ways of appreciating geography as a subject in the school curriculum, Lambert articulated how geography—alongside other subjects—had begun to get lost in a 'post disciplinary orthodoxy' (notwithstanding the fact that it *does* exist in a subject-based national curriculum in England). He argued that geography education could enhance 'human capability' and aid the development of human potential, such that learners become 'self–fulfilled and competent individuals, informed and aware citizens, and critical and creative 'knowledge workers'' (Lambert 2009, p. 17). Disciplinary knowledge remains important in this process, which Lambert recognises in his description of a capabilities approach that 'may be very productive in enabling a fresh understanding of subject disciplines serving broad educational goals' (p. 16). This also provides a link to contemporaneous work on powerful knowledge and the Future 3 curriculum, as advanced by Young and Muller (2010) (see Sects. 3.4 and 3.5).

The volume of research in geography education associated with the growth of interest in capability, and newer conceptions of 'geocapabilities', is now considerable. Originally devised by the economist, Sen (1980, 1985), and the philosopher, Nussbaum (2000; Nussbaum and Sen 1993), the concept of capability draws upon ideas from the fields of human welfare and the economics of development. The capability approach offers a conceptual framework to help understand the components of a 'good life'—the things people are capable of doing, thinking and achieving and the freedoms these afford to live life in the way one chooses. Interestingly, Sen rejected calls to provide a clear articulation of what human capabilities are—in part to avoid their collection into any form of 'tick list' while also recognising that these may change according to social, cultural and economic settings. However, Nussbaum (2000, 2013), has listed a set of ten individual, universal human capabilities based on the opportunities people have to 'do and be'—a list that is expansive and therefore difficult to capture concisely in the school curriculum. On the face of it, the capabilities approach may initially appear tangential to educational matters— however, geography educationists have risen to the challenge of applying its theories and concepts to the geography curriculum. The concept removes us from a simple calculation of wealth, well-being, access to services, employment, health care and education, towards a nuanced appreciation of what it means to live well—where the quality of human functioning (or, technically, 'substantial freedoms') is key. Here is the link to education—without the enrichment of experiencing geography education, one might argue, learners face 'capability deprivation': the education of young learners is deficient or deprived if geography education is removed. Extending this idea one might also pursue the extent to which a capability approach helps us to define our educational aims, purposes and goals in geography. Of the ten capabilities outlined by Nussbaum, the following have most relevance to the geographical education of young people:

– senses, imagination and thought. That is, being capable of reasoning, imagining and thinking that accepts certain principles of 'humanity' and which moves the learner beyond the basic provision of education towards literacy, numeracy and scientific understanding. Here the potential for humanities education, including geography education, appears considerable.
– practical reasoning. Including a concept of what is 'good' and applying this in personal life planning, and reflection on one's actions.
– affiliation. Living and working successfully with others, showing concern and empathy for the situations others face, and being able to interact positively in social situations.
– control over one's environment. Being able to participate in political action and make choices that affect one's life, including the protection of free speech and association (after Lambert 2009).

Nussbaum considers 'capability deprivation' as a concept which has applicability to education in the sense that learners may be deprived of aspects of knowledge, understanding and skills (or, more broadly, 'functionings') that would enable them to flourish as well educated individuals (see Saito 2003; Walker 2005; Hinchliffe 2007a, b; Hart 2009). As Lambert (2009) states, this is in essence a human welfare approach:

> with its stress on freedom and choice, diversity and human possibility or potential, (it) clearly has an educational dimension …. educationists, working in both richer and poorer countries, can apply a capability perspective to the education process itself, in terms of capability-building of both individuals and societies. (Lambert 2009, p. 17)

Importantly, connections to the epistemological roots of the academic discipline of geography, and how these relate to subject learning, are strengthened within the capabilities approach—'what is learnt, and how it is learnt, matters' (p. 18). Lambert (2009) argues that official conceptions of subjects (typically provided by examination syllabuses, specifications, national curriculum programmes of study, etc.) restrict our potential understandings and hold them open to easy dismissal by policy makers as 'outdated nineteenth century constructs'. Indeed, Lambert and Morgan (2010) assert that the capability approach provides an opportunity for geography educationists to articulate an underlying *purpose* for their subject, as 'capability provides a framework for clarifying educational goals' (p. 64) (cited in Bustin 2016). Without an appreciation of the capabilities of geography education, it is posited, the subject of geography becomes easy to side-line in a post disciplinary world.

The notion of capabilities leads us to a consideration of 'geocapabilities'—those capabilities which have a direct geographical expression, or which support the aims of geography education. For Lambert (2009) these include capabilities that are concerned with:

– enhancing individual freedoms (autonomy and rights)
– choices about how to live (citizenship and responsibilities)
– being creative and productive in a knowledge-economy (economy and culture)

These each have a particular geographical expression in the form of place and world knowledge, relational understanding of people and places, empowerment, and the consideration of alternative social, economic and environmental futures. Although much of the original work on GeoCapabilities as a conceptual framework began with geography educationists in England (see Lambert 2011b, 2016) the capabilities approach has gained support globally. Two internationally funded research projects—GeoCapabilities 1 (2012–2013) and GeoCapabilities 2 (2013–2016) have been completed. The first, discussed in Solem et al. (2013), was funded by the US National Science Foundation's Geography and Spatial Science Program and involved the Association of American Geographers (AAG). It compared the aims of geography education in the US, England and Finland to see whether geography could contribute to the capabilities identified by Nussbaum (2000); the second, discussed in Lambert et al. (2015), was funded by the EU's Comenius[5] funds, involving nine project partners, subject associations, universities and schools from across the globe. This more substantial and ambitious project sought to promote the concept of powerful knowledge of geography in a capability framework, positioning teachers as curriculum leaders with responsibility for devising the knowledge content of their geography curricula (Bustin 2016).

In establishing the pilot GeoCapabilities project three of Nussbaum's capabilities were chosen (see above), modified to suit the aims of geography education. Specifi-

---

[5]The future prospects for the modest numbers of geography education researchers who have previously benefited from being engaged with EU funded research projects are uncertain. The UK electorate, having narrowly voted in a national referendum in 2016 to come out of the EU, now face the eventual probability of 'Brexit'. This will make researcher involvement in EU projects less likely, more costly and, in organisational terms, additionally complex. As McKie (2019) states: 'Britain has a first-class record in attracting research funds from the European Union; the money it pulls in greatly exceeds the cash it invests in research budgets for the EU'. For example, in the period 2007–13 Britain paid a total of Euros 5.4bn towards research, development and innovation activities in the EU. In return, it received Euros 8.8bn in EU grants for research projects 'carried out at universities and other scientific centres around the world' (p. 8).

cally these questioned whether geography education, with a view to promoting social justice and human welfare, could:

1. Promote individual autonomy and freedom and the ability of children to use their imagination and to be able to think and reason?
2. Help young people identify and exercise their choices in how to live, based on worthwhile distinctions with regard to their citizenship and to sustainability?
3. Contribute to understanding one's potential as a creative and productive citizen in the context of the global economy and culture? (Lambert et al. 2013)

Lambert et al. (2013) had previously provided the theoretical underpinnings for the international collaborative projects using the term 'GeoCapabilities'. In turn the projects claimed to 'respond in new ways to enduring challenges facing geography teachers in schools' (p. 723)—specifically by defining geography's contribution to the education of all young people, addressing its 'divergence' in educational settings, and explaining its highly disparate expression as a research discipline in university departments. The first project was demonstrably international: in attempting to create a framework for communicating the aims and purposes of geography in schools *across* jurisdictions, it noted the various ways in which geography curricula were described (as national standards, national curriculum, regional schemes of work) and conveyed (as a discrete subject, in social studies, in the humanities, or through other cross curricular allegiances). The stated goal was eventually to establish 'a secure platform for the international development of teachers' capacities as creative and disciplined innovators' (p. 723)—not least as geography 'curriculum makers'. At the start of the project team members discussed the educational aims commonly expressed in other countries. There was some agreement that the application of human capital theory, albeit with its narrow conception of education serving to develop workers to increase international competitiveness, was apposite. In many ways neoliberal concerns for global competitiveness have led to the development of school curricula which valorise 'learning to learn' and the promotion of skill development above the gaining of disciplinary knowledge—leading to regressive and over socialised curricula (see Sect. 3.4). The adoption of geocapabilities thinking helped shift curricular aims away from the development of human capital and towards human flourishing. GeoCapabilities also offered the prospect for students to gain greater access to disciplinary knowledge, through enabling them to achieve "epistemic ascent" (Winch 2013).

The GeoCapabilities projects' approach to understanding disciplinary knowledge (the "whatness" of geography) is broadly based on the concepts of social realism (Young et al. 2014)—itself a recent focus for debate in geography education circles in the United Kingdom (Firth 2011, 2013; Major 2013). The projects helped to connect the capabilities approach to wider conceptual work on curriculum content, structure and contemporary debates on the sociology of knowledge, introducing and developing Bernstein's ideas of pedagogic rights and Young's notion of powerful knowledge. As Bustin (2016) explains, in educational terms:

> Being told what to think, rather than how to think, would be an example of capability deprivation and the capability approach is able to articulate this difference (p. 112).

He concludes that the project teams, in their development of an original model of capability with relation to curriculum thinking in geography, revealed that:

> If the discussions of pedagogy in subjects is about 'how' to teach that subject, then powerful knowledge articulates 'what' to teach; capability captures 'why' teach it, questioning how the powerful knowledge of that subject will benefit the educated person (p. 275).

Uhlenwinkel et al. (2016) describe the findings of the second international Geo-Capabilities project[6] which, combined with the concept of 'powerful disciplinary knowledge' (Young 2008, 2013), investigated the purposes and values of geography education in four European countries. This is not an inconsiderable challenge, given the different conceptions of geography, the various ways in which it is taught, and the diversity of educational aims and structures that exist across different national jurisdictions. The development of human capabilities among young learners was pursued through geography education to see whether learning geography helped young people to prepare themselves to lead valuable lives (using the question 'What is the role of geography in helping students reach their full potential and enhance human well-being'?). In Finland, Germany, the Netherlands and Sweden geography teachers and teacher educators were asked what role they thought geography played in developing their students' 'human potential'. One important justification for teaching geography as a school subject—a common finding across jurisdictions—was 'to increase the capability of young people as (responsible) citizens' (Uhlenwinkel et al. 2016, p. 10). Many respondents also linked geography education to notions of sustainability and diversity (or tolerance and respect for differences); although it was acknowledged that some qualities that geography education encourages can also be learned in other subjects (such as critical, creative and responsible thinking). As Uhlenwinkel et al. (2016) conclude:

> powerful disciplinary knowledge in all four countries is described in terms of world knowledge and understanding the world using geographical perspectives such as looking at human and nature interactions, using the concepts of scale and of local-global relationships, studying geographical issues (e.g. climate change) and linking these to personal (or individual or communal) choices. These ideas encompass some of the aspects of what in the Anglophone community has become known as geographical thinking (Uhlenwinkel et al. 2016, p. 10).

Despite the varying contexts in which geography is taught in schools, the common features of geography education shared across borders suggest that the research findings of the geocapabilities projects have widespread relevance. Geography was under pressure in all the countries surveyed, threatened either by the rise of integrated courses or by other intervening subjects. Where subject status was not threatened by cuts in curriculum time, geography was often marginalised because it was taught

---

[6] A three-year long GeoCapabilities project was funded by the EU-Comenius programme in late 2013. The European partners were England, Belgium, Finland, Greece and Turkey, with USA as a 'third country' partner. This international base was described by Lambert as 'essential', serving to attract 'associate partners' from the Netherlands, Germany, Czech Republic, Portugal, Sweden, Singapore and China. Although international in scope the project was discussed as being 'paradoxically, or counter intuitively, small scale, interpretive and collaborative' (Lambert et al. 2013, p. 4).

by non-specialist teachers, affecting the quality of geography teaching. Geography, like all subjects, found difficulties in unilaterally expressing its educational aims and goals, whilst a common outcome of the project was that despite pre-service training being provided for geography teachers, subject content was often neglected. Uhlenwinkel et al. (2016) therefore welcomed the Geocapabilities approach with its:

> ability to build a bridge between powerful disciplinary knowledge, general educational goals and the development of the capabilities of individual students. If this link was elaborated in the realm of theoretical thinking in the context of teacher training it may help to support the curriculum thinking of geography teachers (p. 11)

In conclusion, the geocapabilities projects offer a model for how geography education research and curriculum making—which utilises theories and concepts from beyond the discipline—could also create a forum for discussion about educational research at the international scale. However, the project team realised that:

> Ideas generated from within other cultural settings may themselves pose challenges for the Anglophone self-conception, not least the Nordic tradition of subject didactic, which on the surface looks similar to 'curriculum making'. The term 'didactics' has a negative connotation in English unlike in Dutch, German, Finnish or Swedish (Gundem and Hopmann 1998). If the British reject 'didactics', but promote "curriculum making" there is a huge potential for misunderstandings (p. 11)

Importantly, geography educationists and researchers involved in the projects drew on a breadth of literature from educational studies (see Walker 2005), teacher development (Page 2004; Raynor 2004); school leadership (Bates 2004); special educational needs (Terzi 2004); and higher education (Deprez and Butler 2001; Watts and Bridges 2003; Hinchcliffe 2006).

3.5   Powerful Knowledge and Powerful Disciplinary Knowledge (PDK)

The recent focus in geography education research on knowledge and curriculum matters follows a period when other areas of educational research concentrated mostly on education policies and practices, identity and marginalisation. In England the development of a discourse about the importance of teachers' subject knowledge, in part encouraged by two Government White Papers (2010, 2016) and the Carter review of initial teacher education (ITE) (Carter 2015), has had a profound impact on geography education research. Policy reforms highlighted inconsistencies in the ways in which subject knowledge was being addressed in schools, as well as tacitly acknowledging the work of subject-based education researchers. Deng (2018) has noted how the question of content—that is, knowledge in the curriculum—had, until recently, 'all but disappeared from global policy and academic discourses concerning teaching and teachers' (p. 371). He describes how:

> Across the globe there has been a shift in curriculum policy from a concern with content selection and organization to a preoccupation with academic standards, learner outcomes

and high-stakes testing (Yates and Collins, 2010; Young 2009). Accompanying that shift is a move to depict teaching as focused on promoting students' academic outcomes measured by high stakes tests, and teachers as accountable for students' learning outcomes, through the employment of evidence-based practices (Hopmann 2008) (Deng 2018, p. 371)

Contributions from researchers such as Biesta (2012, 2017)—who argues that the needs of learners have been over-emphasised compared to the needs of teachers and subjects, effectively pushing teachers to the margins of the education process— have further stimulated debate about the importance of developing teachers' subject knowledge and the dangers of 'learnification'.

As Beck (2013) states, with the agreement of Slater and Graves (2016), the term 'powerful knowledge' is attractive and appealing to educators, after all 'who would *not* want … children to have access to such a thing?' (Beck 2013, p. 184 emphasis in original). But the educational purposes of gaining powerful knowledge—itself a slippery concept—requires further isolation and explanation: how does the concept of powerful knowledge play out differently in the sciences and humanities? what is the place of everyday knowledge? how does knowledge relate to pragmatic frames of reference? and what is 'emancipatory knowledge'? Three tensions recognised by Beck (2013) are significant: (i) that academic knowledge is self-referential, esoteric and difficult to access by disadvantaged groups, (ii) that education suffers from its attempts to balance breadth and specialisation, particularly in a globally competitive world that demands academic success and vocational competence, and (iii) that esoteric knowledge is a part of 'high culture', which itself perpetuates social and cultural hierarchies and exclusion. Additionally, Young (2013) acknowledges the need for the 'curriculum question' to be addressed with respect to the practicalities of teaching with a view to developing powerful knowledge among learners—to refocus the debate on epistemic, rather than social, aspects of powerful knowledge.

As Young (2013) concludes:

> The 'powerful knowledge' argument is that, in devising a curriculum, it is the knowledge structures where we have to start, and that the analytical distinction between curriculum and pedagogy is crucial. Any attempt to develop a pedagogy that imagines it can avoid, rather than work with, the 'epistemic constraints' of a subject will be doomed to fail (p. 198)

Interestingly, many educationists have found it hard to discuss the contribution of disciplinary knowledge to school subjects without being deemed reactionary, conservative and 'backwards-looking' in terms of their views on curriculum development. Young uses the concept of powerful knowledge to argue for a subject-based form of state education, but he is at pains to remind us that he did not endorse the Conservative-led championing of traditional subjects in England by the Secretary of State for Education, Michael Gove (who served from 2010–2014).

Wheelahan (2007) reveals how gaining access to powerful knowledge is restricted for children from working class backgrounds, limiting their intellectual and social horizons:

> Unless students have access to the generative principles of disciplinary knowledge, they are not able to transcend the particular context. Students need to know how these complex bodies of knowledge fit together if they are to decide what knowledge is relevant for a particular

purpose, and if they are to have the capacity to transcend the present to imagine the future. Knowledge is not under their control. This simultaneously denies them epistemic access to the structure of knowledge relevant in their field and social access to (what Bernstein called) 'the unthinkable' (Wheelahan 2007, p. 648 cited in Beck 2013, p. 182)

Here the ideas of Bernstein, and other sociologists in education, consider knowledge structures by drawing on the concepts of social and critical realism for their epistemological base. Taking Young's conception of powerful knowledge, and knowledge of the powerful, Beck identifies 'tensions' which restrict the ways in which powerful knowledge can be extended to leaners from socially and economically disadvantaged groups. However, this is a further spur to research into the educational contribution of formal disciplinary knowledge, which Furlong and Lawn (2010) assert still remains an extremely important element of educational research. Although Jackson (2006) reminds us that geography, as a discipline, is concept rather than content led—with geographers using sets of concepts 'that help us see the connections between places and scales that others frequently miss' (p. 203)—he too recognises the importance of disciplinary knowledge.

The concept of 'powerful knowledge', as promoted by Michael Young,[7] has rapidly achieved a central place in subject and disciplinary thinking in geography education. Indeed, as Lambert et al. (2013) explain:

Young argued that as a matter of social equity, all young people have the right to be introduced to powerful (disciplinary) knowledge. This is a social realist position, usefully discussed by Firth in the context of the geography curriculum in English schools (Firth 2011, 2013), which counters both the extreme relativist positioning of much "progressive" skills-led thought in education and those who propose "traditionalist" knowledge-led perspectives who see the contents of the school curriculum as a fairly fixed selection of the canon of "core knowledge" (Hirsch 1987, 2007) (p. 730)

Roberts (2014a, b) critiques Young, stating that to achieve powerful knowledge pedagogy has to be similarly 'powerful'—students do not just acquire knowledge by being in schools, they need skilled teachers to create the bridge between the discipline of geography and their understanding, comprehension and application of the subject. This, of course, has strong connections to notions of 'pedagogic content knowledge' (Shulman 1987) and the work of Brooks (2007, 2016). Roberts also acknowledges that children bring their 'everyday' geographical knowledge to the classroom and recognises the importance of their own ability to make meaning.

Stoltman and Lidstone (2015), in an editorial for the journal *International Research in Geographic and Environmental Education,* recount an 'interview' with David Lambert—actually taken from written responses to previously submitted questions—which reflects the rise in discussion about powerful knowledge at 'national and international conferences'. Recognising Lambert's central involvement with the concept of powerful disciplinary knowledge in geography, this discursive piece helpfully reiterates Young's distinction between 'powerful knowledge' and 'knowledge of the powerful'—but also illustrates many of the problems associated with the 'inward

---

[7]Although note that the term is present in Wheelahan's (2007) paper, which pre-dates much of Young's development of the concept.

looking' perspectives held amongst many geography education researchers. Lambert reveals how geography educationists at the Institute of Education (including Norman Graves, Michael Naish and Frances Slater) either failed to involve themselves with the concepts of powerful knowledge, and subject framing, which other colleagues[8] were developing—or were too engaged in geography education research of a different stripe. The idea, from Bernstein (1999), that scientific subjects are 'strongly framed and vertical', whilst geography is 'weakly framed and horizontal' as a discipline, encouraged little response from geographers. In a rejoinder to Lambert's interview two of his former colleagues, Slater and Graves (2016), subsequently challenge his statement about the separation of learning geography from learning skills as a 'false opposition'. The belief that powerful geography needed clearer articulation before teachers could use it as a formulation for curriculum making, and for selecting 'what to teach', is stated elsewhere (White 2018), but is nonetheless pertinent. Interestingly, Lambert's (2016) response pushes the question of the ways in which geography can be considered to represent powerful knowledge back to teachers, who he characterises as formulating 'geography lessons that have become, in the most extreme cases, free from any meaningful connection to geography as a discipline or system of thought' (p. 192).

Lambert, in the Stoltman and Lidstone (2015) editorial, captures the theoretical underpinnings of powerful knowledge as being:

1. evidence based
2. abstract and theoretical (conceptual)
3. part of a system of thought
4. dynamic, evolving, changing—but reliable
5. testable and open to challenge
6. sometimes counter-intuitive
7. outside the direct experience of the teacher and the learner
8. discipline based (in domains that are not arbitrary or transient) (after Young 2010 p. 4)

Additionally, Lambert and Biddulph argue that teachers need to readily engage with:

> specialist theoretical knowledge that has been created by the subject disciplines, which are not entirely arbitrary. The knowledge exists outside the direct experience of the student and is powerful, *because* it is derived in part by the specialist communities and the disciplined procedures that produce and verify it (Lambert and Biddulph 2015, p. 216).

The connections between powerful knowledge as a theoretical construct, and its representation in school curricula—which test the applicability of the concept to geography education—were debated by Margaret Roberts and Michael Young at a Geography Education Research Collective (GEReCo) research seminar at the Institute of Education in London in 2013. Despite Young's insistence that powerful knowledge should eschew 'everyday' forms of knowledge—which have

---

[8] Both Young and Bernstein worked at the Institute of Education at the time.

recently dominated the curriculum, obscuring the epistemic foundations for school subjects—Roberts (2014a, b) considers that everyday knowledge is important, and closely related, to many themes in both academic and school geography. The influence of powerful knowledge, as a concept, has now spread worldwide in geography education. It has been debated in professional, as well as academic, journals of geography education and is therefore becoming more accessible to classroom teachers—for example, Roberts' (2017) article in *Teaching Geography*: 'Geography education is powerful if…'. Here Roberts (2017) argues that geographical knowledge is particularly powerful if it enables students: to make connections between their everyday knowledge and school geography; to transform the ways in which they understand the world; to be aware of the values dimension of decisions that affect local, national and world geography; to develop the skills needed to deal with the complexity of geographical knowledge and to develop understanding; and to take an active part in learning'.

3.6 Future 3 Curriculum

Work by Muller and Young (2008) has further explored questions of the ways in which knowledge is powerful, the aspects of subject knowledge we want learners to acquire, and how related pedagogy should be organised in the school curriculum. The concept of Future 1, 2 and 3 curricula have generated much interest among geography education researchers worldwide, particular concerning its links with powerful knowledge—helping to further articulate how geography is best represented in the curriculum (see Brooks et al. 2017). Debates about the 'Futures curricula' are enhanced by Muller and Young (2008) who question the non-arbitrariness of knowledge domains, the connections between school subjects and disciplines, and the importance of subject knowledge within the curriculum. Essentially, this helps us consider how knowledge is defined in school subjects, including geography, and the boundaries within which knowledge is contained.

In the Future 1 curriculum, subject boundaries are given and fixed: this Future is associated with an *under-socialised* conception of knowledge. Here curricula promote traditional knowledge—which Morgan (2014) refers to as the 'time honoured collection of ideas, theories, 'Great Books', and facts'—and view geography education in schools as having been 'emptied out' of geographical knowledge (Mitchell 2017). The essential problem is that subject knowledge is perceived as 'given' and contained in curricula which can be used to help select high achievers, from particular social groups, for progression into the academy. The dominant knowledge traditions are highly prized in the Future 1 curriculum—knowledge that is static and socially conservative—which help an elite system of education that favours the few to continue. This under-socialised view of knowledge does not sufficiently recognise the social, historical and cultural conditions of its production and encourages a transmission style of pedagogy (Morgan 2014; Hammond 2015; Butt 2017). The main drawback of this curriculum is that it:

> Treats access to knowledge as the core purpose of the curriculum and assumes that the range
> of subjects and the boundaries that define knowledge are largely given. It tends towards … a
> 'curriculum for compliance' and in extreme cases encourages little more than memorization
> and rote learning (Young 2011, p. 182)

The Future 2 curriculum has few subject boundaries and is associated with an *over-socialised* conception of knowledge. Here we see the weakening of knowledge boundaries, the integration of some school subjects (such as the integration of geography, history and religious studies into 'Humanities'), and curriculum content being understood in terms of its educational outcomes and generic skills. The everyday knowledge of students is favoured in this curriculum—'knowledge is no longer treated as given… but seen as constructed in response to particular needs and interests' (Young et al. 2014, p. 59)—with pedagogic styles being child centred and facilitative, stimulating 'knowledge building'. This form of curriculum tends to be more social inclusive, encouraging higher 'staying on' rates into senior school, and possibly the academy. The overall result is that:

> In its most extreme forms Future 2 argues that because we have no objective way of making
> knowledge claims, the curriculum should be based on the learner's experiences and interests
> and that somehow these can be equated with the interest of society (Young et al. 2014)

Future 3 curricula offer the prospects for Future 1 and 2 conceptions of subject curricula to be 'brought together', to enable the creation and acquisition of new knowledge. Essentially, this curriculum enables 'boundary maintenance to be observed, prior to boundary crossing' (Butt 2017, p. 20). Here knowledge is viewed as a social product, but sanctioned by scholarly communities within their established traditions, conventions and rules. Academic communities therefore safeguard the development of school subjects in accordance with epistemic, rather than arbitrary, rules—boundary maintenance occurs ('Is this geography?') and boundary crossing is permitted, but must be acknowledged (Morgan 2014). Worthwhile knowledge is therefore determined by disciplinary norms, and has a status *beyond those who produce it*. Knowledge is viewed as dynamic, contested and changing, leading to a Future 3 curriculum that:

> Treats subjects as the most reliable tools we have for enabling students to acquire knowledge
> and make sense of the world… it implies that the curriculum must stipulate the concepts
> associated with different subjects and how these are related. It is this link between concepts,
> contents and activities that distinguishes a Future 3 curriculum from Hirsch's lists of 'what
> every child should know' (Young 2014, p. 67)

As Lambert (2018) reminds us, there are three recurring keywords in education—knowledge, teaching and curriculum—which represent the defining categories for the multiplicity of roles and functions that are thrust onto schools. But what *must* remain at the core of what schools and teachers do is knowledge, encouraging us to consider closely:

> how teachers can make specialist, often abstract, knowledge available in a way that motivates
> and engages the interests of all students—as a 'pedagogic right' … so as to enable new and
> powerful ways to understand the world and how it works (Lambert 2018, p. 357)

3.7   Young People's Geographies (YPG) Project

Biddulph (2016) describes how she, with her colleague Roger Firth, encouraged their student teachers to draw on their geographical 'passions' to re-contextualise their disciplinary knowledge in collaboration with local schools. This gave rise to what became known as the promotion of 'Fantastic Geographies'. Against a backdrop of government education policy which appeared to adopt a deficit view of the knowledge that new entrants to the profession possessed (DfE 2016), the over surveillance of ITT (Furlong and Smith 2013), and the widening of a 'curriculum chasm' between universities and schools (Butt 2019), Biddulph (2016) sought to celebrate how recent entrants to teaching could:

bring new insights and perspectives *from* the discipline *to* the school subject (p. 9)

'Fantastic Geographies' demonstrated for Biddulph how school geography could achieve stronger academic integrity, reconnect with its disciplinary roots at curriculum level, and utilise beginning teachers as agents of curriculum change. Young geography graduates entering the profession are especially important to this process 'because of their more recent engagements with developments in the discipline' (p. 10). Inspired by a presentation by Anoop Nayak at the Association of American Geographers conference in Chicago in 2006—which articulated his ethnographic research with young people in Newcastle around themes of economic reform, employment insecurity and new urban cultures (Nayak 2003, 2006)—Firth and Biddulph developed ideas from 'Fantastic Geographies' to launch the 'Young People's Geographies' (YPG) project. This project was conceived not only to engage with the pedagogical and geographical knowledge of teachers and academics, but also with the 'personal geographies' of school students, to inform the curriculum making process.

Drawing on research from beyond the disciplines of geography and education, YPG incorporated concepts associated with the development of young people's lives. It therefore sought to address issues concerning power relationships, local-global interconnections, and social and cultural participation, while being grounded in conceptions of the discipline of geography (Biddulph 2016). It was important to the project founders that secondary school teachers[9] remained connected to their subject discipline and brought, in a re-contextualised form, research-led developments in academic geography into the classroom. The project exemplified how research from other disciplinary areas—from subject-based education beyond geography and from areas of educational theory underused by geography educationists—could combine to create important new understandings for geography teachers. As such, YPG engaged with research into students' perceptions of learning both history and geography (Adey and Biddulph 2002; Biddulph and Adey 2004), student voice[10] (Fielding

---

[9]Research exists into the preferences of younger children (see Catling and Martin 2011), but YPG was essentially a secondary school project.

[10]Bernstein (2000) refers to the 'acoustics' of research projects—specifically, whose voices get heard, whose do not, and why.

2006; Arnot and Reay 2007) and the personal geographies of young people, often drawing on their social and political actions and the geographies this produced (Hopkins 2008; Hopkins and Alexander 2010; Skelton 2010; Valentine 2010). The notion of 'conversations' between students, teachers and academic geographers was important to the YPG project, with intentions to challenge the power structures that tended to dominate curriculum making:

> Young people's geographies emphasise the agency of young people, give voice to their experience and, most importantly, make a statement about the rights of young people to have their experiences recognised and valued (Firth and Biddulph 2009, p. 33).

With its own website—which initially introduced the project, and subsequently presented outcomes of its work—this was a research project that took seriously the need to disseminate its findings (Biddulph 2011a, b, 2012). The website detailed the project's aims and principles, its school curriculum development plans, teaching resources, students' comments and evaluation reports. Details of conference presentations and project publications were shared nationally and internationally. The project leaders recognised the need to engage with high quality geography education research, but also realised the expectations to deliver publications that detailed practical ideas and resources for teaching (see Firth and Biddulph 2009).

The independent project evaluations of the YPG project endorsed the 'bridge' it had created between pedagogic and curriculum processes, and disciplinary knowledge—while also contributing in a major way to the Geographical Association's (GA's) manifesto and emergent conceptions of curriculum making (Hopwood 2007, 2008; Davidson 2009; GA 2009). What is interesting is the ways in which the project, building on work by Bereiter and Scardamalia (2003), placed great emphasis on disciplinary knowledge, the creation of knowledge, and how knowledge could be democratised through the application of inclusive pedagogic processes. As Biddulph (2016) explains, YPG was about the relationship between developments in the discipline, dialogic pedagogies and the potential of drawing young people into curriculum making in a way that reflected their rich and multi layered geographies:

> The notion of subject authority is important if YPG is to avoid becoming an 'anything goes geography' or some vague attempt to 'personalise' learning. This would be too permissive an approach to the curriculum and could lead to careless geographical learning by students: to be reliant just on young people's personal experiences is in danger of leading to a curriculum lacking in challenge, promoting introspection and failing to take students beyond what they already know (Biddulph 2012, p. 161).

In research terms the project foregrounded attempts to democratise the curriculum for both teachers and students, whilst keeping the discipline of geography at the very heart of the curriculum making process. The inclusivity of its curriculum thinking meant that teachers could not sidestep their responsibilities for curriculum design and implementation, but had to use (and update) their subject knowledge and expertise to ensure that their contributions maintained momentum. As such, subject knowledge was brought to the fore, geographical content became important again, and disciplinary concepts remained central to curriculum making. The similarities to the aims of promoting powerful knowledge in geography education are not coincidental (see Young and Muller 2010; Young et al. 2014).

3.8 Curriculum Making

The concept of 'curriculum making', as it has been determined in England, has recently attracted attention among a number of researchers in geography education (Mitchell 2017; Lambert and Biddulph 2015; Biddulph 2012; Brooks 2007; Lambert and Morgan 2009; GA 2009). Although it exists as a discrete set of ideas, the thinking associated with the concept of 'curriculum making' has been central to many contemporary research projects in geography education—notably GeoCapabilities, Young People's Geographies and developments surrounding the application of powerful knowledge—and has also been called upon to support the initial teacher education and professional development of geography teachers. At its core 'curriculum making' emphasises the responsibility of geography teachers to construct their own geography curricula, and to take ownership and agency over the process of curriculum design. This is important even in the context of countries who have nationally defined curricula, and educational policies and frameworks which dictate the overall aims and content of school subjects. To fully deliver on their professional responsibilities every teacher must at least adapt, or more fundamentally 'make', the curricula they teach—their recognition of this responsibility is obviously important (Lambert et al. 2013). Curriculum making in geography draws on the interaction of three elements—the teachers' knowledge, expertise and skills; the students' interests and needs; and the disciplinary roots of the subject of geography. Although in its modern form the concept of curriculum making is relatively new to the field of geography education, it has historical origins in early iterations of curriculum theory (see Bobbit 1924). More recently it has been used as a way to support teachers' professional development (GA 2009; Lambert and Morgan 2010) with the Geographical Association creating a contemporary interpretation of curriculum making for geography education—referring to geography teachers as the 'enactors of the curriculum' and 'curriculum makers' (GA 2009, p. 30). This contrasts sharply with the bureaucratisation and centralisation of the curriculum widely reported in recent educational research (see Ball 2008).

Biddulph (2016), drawing on the work of Smith (2000) and Kelly (2009), contrasts between four models of curriculum making:

	Model	Key ideas	Key proponent
1	Cultural transmission model	The curriculum as a body of knowledge to be transmitted from the teacher to the student	No particular proponent, but associated with educational ideologies which focus on learning content

(continued)

(continued)

	Model	Key ideas	Key proponent
2	Objectives-led model	The curriculum as a predetermined set of outcomes, the successful achievement of which can be measured	Tyler (1949)
3	Process model	The curriculum is shaped by teacher-student interaction and the focus is on students' meaning making	Stenhouse (1975)
4	Praxis model	A development of the process model, except it is committed to broader ideals, especially those that are emancipatory	Freire (1996), Giroux (1992)

(After Biddulph 2016, p. 58)

She contends that curriculum making can be juxtaposed with the process model of the curriculum proposed by Stenhouse, or the praxis models of Friere and Giroux, arguing that these are more fluid, less fixed, more tentative and provisional accounts—allowing for the dynamic and contested nature of geographical knowledge to be recognised. The predetermined outcomes of the objectives-led curricula, and the prescribed content of cultural transmission, may be less messy, less challenging and easier to teach—but ultimately less 'rewarding' in terms of knowledge creation and understanding. Here is the notion of 'boundary working,' as teachers and students grapple with complex, theoretical ideas and develop a critical understanding of key concepts in geography. Importantly, teachers and students—as well as the subject itself—exert equal influences on the enacted curriculum.

There are big questions about where power and control reside in the construction of the school curriculum. If schools are in the business of transacting knowledge what does the core of this knowledge look like, and what should be the statutory framework employed to deliver it—a national curriculum, national standards, or something else? Does this 'authorized knowledge' (see Lambert and Hopkin 2014) get taught, and what accompanies it in the form of education aims, preferred pedagogies and expectations of performance? In any educational system there are precious few guarantees about what actually gets taught and learnt, despite the use of high stakes inspection and examination regimes. In any country the geography curriculum experienced by students may not match the intentions of the curriculum makers, be they stated in the form of standards, a national curriculum, or statutes.

The movement away from a curriculum of delivery (Pring 2012) towards one that engages with subjects such as geography, recognises the need for teachers to introduce their students to notions of complexity, theoretical knowledge and knowledge that may appear counterintuitive, or which contradicts with common sense, everyday notions (Lambert and Biddulph 2015). Here:

Curriculum making

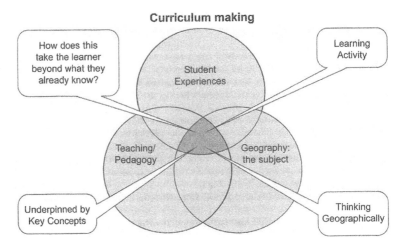

Fig. 3.1 The three pillars of curriculum making in geography (from Action Plan for Geography. www.geographytecahingtoday.org.uk)

the teacher-subject relationship is more than a mere gesture to 'keeping up' with academic developments, or a superficial searching for connections with student experiences. It involves a continuing relationship between teachers and geography that is sustained throughout their career (Biddulph 2016, pp. 55–56)

In geography education there is now a popular and often reproduced model of curriculum making which encapsulates the concept (Fig. 3.1).

The notion of curriculum making has now created widespread interest among geography educators and geography education researchers in many countries and jurisdictions.

3.9   Geographic Information Systems (GIS)

van der Schee (2003), at the start of the twenty-first century, indicated an expectation that the growth of new media and technologies would 'accelerate the renewal of geographic education' (p. 205) for geography students, educators and researchers alike. While noting the need for systematic research and development, internationally, into the educational impacts of the digital revolution—not least, its potential to enhance the quality of modern geography teaching—van der Schee outlined from the Dutch perspective the opportunities that rapid growth in the use of new technologies would bring, whilst regretting that 'empirical research in this field is scarce' (p. 206). The lack of dedicated research in geography education, at this stage, was particularly galling given that computers had already been used in many schools, and in initial teacher education, in developed countries for almost 20 years. A body of largely practical knowledge and experience concerning new technologies had grown in geography education in the UK, and beyond, but this was not commensurate with the scale of the digital revolution that was occurring (see, for example, Bednarz and Audet 1999; Fitzpatrick 1990, 1993; Fisher 2000; Hill and Solem 1999; Houtsonen 2003; Kent 1983; Seong 1996; Leask 1999; Martin 2000; Nellis 1994; Tapsfield 1991; van der Schee et al. 1996; Watson 2000; Wiegand and Tait 2000; Wong Yuk Yong 1996). Here we turn to a consideration of the educational use of GIS in schools, and to the response to its deployment made by geography education researchers.

Developments in geospatial technologies have moved rapidly in recent years, but school geography and geography education research still needs to realise the educational potential that Geographic Information Systems (GIS[11]) can offer:

> The digitisation of spatial data has transformed the ways in which geographical information can now be represented through information technology. New geo-technologies[12] have made place 'virtually accessible' to the individual internet user in ways that were not feasible until relatively recently (Fargher 2013a, p. 13)

The educational application of GIS has the power to help integrate geographical knowledge into schools—with its ability to collect, select, present and analyse geographic data and to represent and interpret place in various ways (Schuurman 2004). Bednarz and Bednarz (2004), in the US context, supported the incorporation of GIS into schools based on three 'justifications': the workplace justification (specifically, that using GIS is an essential skill, making students who possess such skills more employable); the education justification (that GIS enhances geographical learning); and the place-based justification (that GIS supports the study of local environments, often through fieldwork). GIS interfaces with GIScience—a new field which encapsulated both cartography and geography, growing as information technology began to be used increasingly to process geographical information from the 1960s onwards. Advocates argued that GIS tools could be utilised to ask questions about spatial relationships that would not be possible outside the application of GIScience (Fargher 2013b; Schuurman 2004). However, it must be remembered that the representation and communication of geographical information in any cartographic form, including those produced by GIS, involves cartographical abstraction, generalisation, visuality and criticality which all combine to influence our sense of place (Boardman 1983; Harley 1989). This not only leads to choices about what is represented, how it is shown, and whether this depiction is simplified or exaggerated, but also exemplifies the need for geography and geography education to help direct and justify such choices.

---

[11]GIS used in schools tends to be adapted from commercial versions used in industry including: ArcGIS, MapInfo, Idrisi and geobrowsers such as Google Earth, Nasa Worldwind and Bing Maps (Fargher 2013a). We must recognise that commercially produced GIS software packages, both in US and European contexts, were not initially developed with education in mind (Bednarz and van der Schee 2006).

[12]Geo-technologies describes the range of geographically-oriented technologies including GIS, Global Positioning Systems (GPS), earthviewers and hand-held digital devices capable of mediating geographical information.

Despite the power of GIS to support the learning of geographical concepts, many commentators have recognised that its potential has not yet been fully realised in schools. As Fargher (2013b) notes, even as late as 2011 the inspection agency OfSTED reported that the use of GIS in English schools was limited, even in geography classrooms (OfSTED 2011). She goes on to comment:

> Other school-based research on GIS has often highlighted the barriers to its use (Audet and Paris 1997; Kerski 2003; Bednarz and van der Schee 2006). More recently, whilst more traditional and costly GIS remains difficult to access for many, the proliferation of free online GIS on a range of platforms (PC, laptop, network, mobile) has started to make digital geographical information readily available (Elwood 2008). Teachers and pupils are beginning to gain more access to the use of these technologies in the formal curriculum (Fargher 2013b, p. 15)

Research in geography education continues to produce evidence that despite the initially slow uptake of GIS and related technologies in primary and secondary schools, the pedagogical advantages of using these applications can be noteworthy (see Kerski 2001; Bednarz and Bednarz 2004; Favier and van der Schee 2012). The period when teachers associated the educational use of 'new' technologies with difficulties in their adoption, application and the adequacy of professional training—as reported, among others, by Fargher and Rayner (2012), Fargher (2013a), and Baker et al. (2009)—must surely now be over, as teachers and the wider population become far more proficient in using similar technologies in their personal lives.

Alongside supplementing fieldwork activities (Favier and van der Schee 2009; Buchanan-Dunlop 2008), supporting enquiry learning (Scheepers 2005), developing spatial orientation skills (Hemmer et al. 2013, 2015), understanding spatial citizenship (Jekel et al. 2015), and visualising geographical phenomena (Lei et al. 2009), it must also be recognised that GIS is not necessarily only applicable to studying geography—it has cross curricular applications, including supporting data capture, analysis, problem solving, modelling, evaluation of environmental impact and planning. Fargher (2013a) reports that the use of GIS within and beyond the realms of geography education is most common in schools in the US, given their tradition of broader social science education. This, of course, strengthens the argument for sound geographical education to support the use, analysis and 'meaning making'—indeed knowledge production—associated with the consumption of spatial data through GIS technology.

As Fargher (2013b) concludes:

> Teaching and learning in twenty-first century school geography should encompass critical and expedient use of spatial thinking through geographic information systems (GIS). Young people's meaningful learning about pertinent issues such as climate change, geopolitical shifts and patterns of economic uncertainty require careful consideration of the significance of geographical context and both the specificity and interconnectedness of place(s) (p. 19)

3.10   Issues with GIS

Despite some legitimate criticism of GIS use in school—specifically, that it is deterministic in the ways in which it narrows the geographical perspective of young learners to 'think geographically' (Sui 1994), that its positivistic approaches to spatial scientific thinking restrict students' abilities to qualify as well as quantify (Pickles 1995; Sui 2004), and that it is often applied without appropriate attention to underlying geographical concepts and principles (Butt 2018)—the reluctance of geography education researchers, in some jurisdictions, to engage more fully with research into the educational uses of GIS is perhaps surprising. Certain concerns about GIS are difficult to overcome. For example, the veracity of much spatial data, and the ways in which it is represented, are significant worries—particularly with respect to easily accessible 'volunteered' geographies available via the internet. These offer specific geographical visions of the world, but their quality is questionable and untested. Pickles (2002) commented almost twenty years ago about the apparent 'malleability' of electronic images, and the fact that geographical theory had remained surprisingly silent about the ways in which GIS was affecting the discipline, its data sources and the uses to which spatial imaging could be put. This point goes to the very heart of how geography, as a synthesising discipline, interprets and uses spatial scientific information—a contention that has not yet been answered by those who strongly support the use of GIS (Sinton and Lund 2007), particularly with respect to the development of spatial literacy (Kerski 2003; Bednarz 2001; Lei et al. 2009; Sinton and Bednarz 2007; National Research Council 2006). Where geography education research has an important role to play is in the investigation of ways in which data science interfaces with GIS—helping geographers and educationists further understand the critical application of GIS in gathering and processing spatial, digital data. Here geographical thinking and established disciplinary principles may assist in making sense of complex data. A tangible example of this, with distinct educational applications, is the Worldmapper project. Established by the Social and Spatial Inequalities Group—a global research network dedicated to the reduction of social inequalities worldwide—Worldmapper was developed as a website in 2006, featuring 700 maps which indicated global variations in equality using socio-economic data (Barford 2006). This website was relaunched in 2018 with ten categories—Connectivity, Economy, Education, Environment, Habitation, Health, Identity, People, Resources and Society—each, again, featuring cartograms, 'maps whose areas are re-sized according to the topic of interest', that can be utilised in geography education (Henning 2018, p. 66).

Walshe (2018) notes that 'despite the apparently compelling evidence of the relevance of GIS for geographers from industry and its focus in policy, the uptake of GIS in schools has been piecemeal' (p. 46). She comments that current debate on how to engage geography teachers more fully with GIS applications in the classroom has focused on issues such as overcoming costs, improving the availability of equipment, and enhancing technological knowhow—but that even where training and software has been provided for teachers, take-up is still not convincing (Walshe

2017). Walshe concludes that the problem may centre on the difficulty that teachers often have in 'finding the geography' within many aspects of GIS application, resulting in their disinclination to engage with it pedagogically. US research suggests that GIS can be used actively to develop spatial skills—to ask spatial questions, visualise spatial data, and perform spatial analysis—although in some respects the educational outcomes and benefits, particularly in terms of mapping, are equivocal (see Kerski 2000). Fargher (2017) also reminds us that spatial thinking is but one aspect of *geographical* thinking, in its broadest sense—indeed, as Walshe (2018) concludes, 'rather than viewing GIS as a skill to be ticked off, we should see it as a means to develop a greater capacity for geographical thinking' (p. 48).

3.11   Research into the Educational Application of GIS

Research into the application of GIS in schools, and beyond, is currently dominated by work from the US (Walshe 2018)—but with significant, and growing, contributions from researchers elsewhere (see, for example, Lam et al. 2009; Demirci 2011; Demirci et al. 2013a, b; Gyrl and Jekel 2012; Gryl et al. 2010; Lidstone 2010; Lidstone and Stoltman 2006; Muñiz Solari et al. 2015; Roulston 2013; Roulston and Young 2013; van der Schee and Scholten 2009). The difference in the ways in which geography education is delivered in schools in the US and UK is worthy of note, given that this affects the impact of GIS-related education and related research. The individual state-based education systems, and curricula, witnessed in each of the US states contrasts appreciably with the national curriculum employed in English, Welsh and Northern Irish schools—where geography education is delivered within a separate, dedicated subject. Nonetheless, in US schools a strong history of social science, earth science and environmental education connects strongly with GIS use—particularly with respect to fieldwork, enquiry-based learning and aspects of citizenship (Kerski et al. 2013). Fargher (2013a) identifies the impact of educational funding from the Environmental Systems Research Institute (ESRI), since 2000, which supports the professional development of teachers with respect to GIS use in schools. This funding has been beneficial for training social sciences teachers, as well as those in science and STEM subjects, who have developed enquiry-based learning techniques using GIS.

Fargher (2013a) notes three common foci around which discussion of GIS use in schools occurs:

i.   The nature of spatial thinking with GIS in school geography education
ii.   The role of enquiry-learning in GIS-based lessons
iii. The technical challenges that teachers face in developing both their own ICT-related GIS skills and pedagogies around GIS.

In the US context the links between GIS and the promotion of enquiry-based learning has been investigated (see Bednarz 2001; Baker 2000; Keiper 1999; Scheepers 2005), as well as its role in promoting spatial skills (Gershmehl 2008). By contrast, in the UK, although enquiry-learning has traditionally been strongly promoted in geography education (see Roberts 2003), its connection with GIS-related education have been less apparent (although see Newcombe 1999; Bednarz 2001; Van Joolingen et al. 2007). Given the rather fragmented nature of geography education in US schools, much research into the educational application of GIS has occurred in cross curricular contexts, or in relation to the public understanding of GIS in education (Sinton and Lund 2007; Sinton and Bednarz 2007). However, Kerski (2008) has explored how GIS and geography education have combined to promote enquiry learning, dismissing notions that geographical learning comprises narrowly of the accumulation of geographical, or place related, facts.

The challenges of using GIS successfully in the classroom are therefore notable (Demirci 2011). They require teachers not only to apply relevant subject content knowledge, but also to achieve technical understanding and to adopt new pedagogic practices which embrace the ways in which young people learn with technology (see Kerski 2003; Mishra and Koehler 2006; Shulman 1987). With rapid advances in technology, teachers' knowledge of different packages and applications quickly date, affecting their pedagogical content knowledge. The factors that influence teachers' decisions to become involved in GIS related curriculum innovation, in US and European schools, have been researched by Bednarz and van der Schee (2006). Both highlight the influence of power, authority and manageability of technology, as well as considering aspects of educational consistency—that is, whether an innovation fits with current curricula, practices and systems. Kerski et al. (2013) have extended this work, exploring the global landscape of GIS use in secondary education in 33 countries; while in the context of case studies in Australia, US and Europe, Kirschner and Davis (2006) and Baker et al. (2009) have highlighted how the availability of professional development has influenced teachers becoming competent users of technology. The lack of guidance for teachers wishing to develop their pedagogic practice with respect to deploying GIS in the classroom—particularly for those teachers who are *not* geographers—is noted in many studies.

In conclusion, as Fargher's (2013a) doctoral research into the role of GIS in constructing relational place knowledge through school geography education discovered:

- Teachers use GIS to visualise spatial patterns in teaching and learning
- Teachers use GIS to make connections between spatially-referenced data sets
- Teachers employ mainly closed enquiry strategies in teaching and learning with conventional GIS
- Teachers using hybrid GIS employ more open-enquiry strategies with GIS
- Teachers' GIS experience and expertise has a major influence on their GIS practice (p. 208).

She also notes that:

Though highly efficient, speedy and well-suited to gathering, processing and displaying large amounts of digital geographical information, [using GIS] could lead to rather deterministic and formulaic learning unless the class teacher is aware of these limitations and is prepared to supplement these types of methods with more open-ended learning about place (p. 226)

As such,

> The mass digitization of data has actually made our need to be conscious and critical of the language and meanings behind the signs and symbols present in GIS. As with any semiotic medium, it is important to be aware of the different discourses which are being represented within it (p. 238)

Without doubt, huge amounts of digital information and data relevant to geography education is now readily, easily and instantly available through the use of various technologies. What must be guarded against—in what is sometimes referred to pejoratively as the 'post-truth era'—is the easy acceptance of such information on face value. Also, as Parkinson (2013) argues, from the perspective of practitioners who are considering curriculum making and professional development in geography education using 'new' technology: 'we should continually ask the question: 'is technology always the best tool to use?'' (p. 202).

3.12 Fieldwork

The educational importance of fieldwork—commonly understood to refer to students leaving the classroom to learn through first-hand experience outdoors (Kent et al. 1997; Boyle et al. 2007)—and of research into approaches to fieldwork commonly used in geography and outdoor education, cannot be denied (Dillon et al. 2006; Remmen and Frøyland 2014). Foskett (1999) details how for most geographers[13] the significance of fieldwork 'needs no reinforcement', being embedded within their personal biographies and experiences associated with learning the subject (see also, McPartland 1996; Catling et al. 2010). In many countries, the requirements for geography candidates to be examined with respect to their fieldwork skills, and to report on previous field-based enquiries in geography—in addition to working in the field being a statutory element of national curricula and standards—provide a helpful lever for schools to offer residential, or day, fieldwork experiences for children. However, the importance placed on gaining fieldwork experience, as revealed in different geography curricula, is highly variable (Foskett 1999)[14]—in UK schools the value of students engaging in fieldwork has traditionally been stressed (Cook 2011), with residential experiences in geography education for sixth formers in the 1970s often being extended into the lower years by the 1980s. This was championed, in part, by the requirements that geography candidates produce fieldwork reports to be assessed in public examinations, and facilitated through the growth of fieldstudy centres across Local Authorities and by the Field Studies Council. The

---

[13]Other subjects also value field experience highly, of course. See, for example, accounts of the importance of fieldwork in earth science (Hawley 2012), and the natural sciences (Lambert and Reiss 2014).

[14]Foskett's (1999) paper *Fieldwork in the Geography Curriculum—International Perspectives and Research Issues* introduces a *Forum* which contains contributions from geography education researchers based in the US, Holland, England, China and Australia.

subsequent removal, or watering down, of formal assessment requirements relating to fieldwork—alongside growing concerns among teachers and parents about safety, financal, and organisational pressures—has led to a decline in field activity. There are parallels elsewhere—in Australia and New Zealand geography students similarly enjoyed a variety of forms of fieldwork, until the growth of integrated social studies and humanities courses heralded a falling-off in fieldstudy activity (Lidstone 1988). In other countries the tradition of fieldwork in geography education is more limited (Bednarz 1999), or less established, for example:

> In the USA, geography fieldwork has largely been ignored in curriculum development, and in many states in Europe the role of fieldwork has been marginal. In most less developed countries, resource constraints have meant fieldwork developments have been a very low priority (Foskett 1999, p. 159).

Fieldwork is still considered by many geography educationists to be integral to the 'all round' development of the young geographer, essential in the provision of a range of practical, organisational and academic skills, knowledge and understanding of the discipline. Essentially fieldwork 'enables the 'theory' of the classroom to engage with the 'practice' of the real world, bringing both together in a unique and fulfilling way' (Butt 2002, p. 64). Additionally, alongside other forms of outdoor education, fieldwork regularly provides the most motivational, memorable and distinctive elements of learning for young people—developing their social and personal skills (such as team working and decision making) (Lai 1999); bringing together cognitive and affective education (Caton 2006a), promoting aesthetic, values and skills-based education; and expanding the 'personal geography' of the individual (Gair 1997; Foskett 1997, 2000; Caton 2006b). Students, not surprisingly, regularly rate the social aspects of fieldwork very highly (Dunphy and Spellman 2009; Powell 2002).

The range of possible approaches to fieldwork, and the skills these involve, have been the focus for previous research in geography education and elsewhere (Lambert and Reiss 2014). Job (1996) details how fieldwork activities and skills lie along a continuum—from teacher-led activities with pre-determined outcomes at one end, to heuristic pupil-centred approaches at the other—either emphasising quantification (which may be rather reductionist in nature), or affective learning (which may be more holistic). Within this framework lies fieldwork that is deterministic, being structured around hypothesis testing, as well as more loosely regulated field excursions (sometimes called 'look, see' fieldwork). Both may be teacher-led, but stress differently elements of quantification and affective learning. 'Enquiry' and 'discovery' fieldwork usually includes more pupil-centred approaches, whilst fieldwork that is described as 'earth education' is almost wholly affective in purpose (Bland et al. 1996). Research has suggested that enquiry-led fieldwork, where students are encouraged to devise their own questions to investigate and are supported with respect to appropriate methods of data collection to apply, often within a frame of structured key questions, has often been favoured in UK schools—particularly in response to the demands of public examinations (Butt 2002; Foskett 1997; Roberts 2013; Hart and Thomas 1986). With the intervention of new technologies, fieldwork has even become virtual (Buchanan-Dunlop 2008).

Unfortunately, in many countries, fieldwork activity—particularly that which is residential in nature—has come under severe pressure, often being deemed too expensive, too demanding, or too time consuming to organise. Fieldwork is now less highly regarded, or considered more difficult to assess, by many public examination bodies—it is often believed to be disruptive to other teaching in schools, and may lack the support of certain religious and cultural groups (Herrick 2010). In higher education, a renewed interest in researching fieldwork may be linked to threats to its future viability (Kinder 2013; Herrick 2010). It appears that difficulties abound. The demands on teachers to draw upon their planning skills, subject knowledge, locational and site knowledge, pedagogical expertise and health and safety awareness have also been noted (Kinder 2013; May and Richardson 2005; Holmes and walker 2006). Educationally, there are also dangers that teachers may assume too much of students with respect to their being able to 'read' landscapes and environments, or to interpret what they see (Job 1996; Rynne 1998). Indeed, Kinder (2013) argues that the benefits of fieldwork may be built on assumptions, rather than research evidence, and that the evidence that does exist points to affective rather than cognitive gain:

> Gold et al. (1991) have suggested that 'there is no clear evidence of the general value of fieldwork' (p. 27), a view echoed by Kent and Foskett (2002) who note that 'we often struggle to provide evidence to support our beliefs about the benefits of fieldwork' (p. 177). Whilst relatively little research has been conducted to investigate the cognitive gains of selecting first-hand investigation over classroom-based strategies, those who have ventured into this field tend to emphasise affective benefits (Fuller et al. 2000; Herrick 2010) (p. 188).

Nonetheless, the *prima facie* case for supporting fieldwork in geography education remains strong. Research into outdoor learning by Rickinson et al. (2004), which reviewed 150 pieces of research published over a decade, suggests that effective fieldwork has a number of benefits in promoting both affective and conceptual learning, but requires skilled planning, execution and follow up by educators. Further research may develop a case for fieldwork entitlement based on arguments about its benefits with respect to well-being, fitness, health and the challenges it provides to the increasingly sedentary, online existences of many young people (Moss 2012; Parkinson 2009).

3.13   Nature of Geography Education Research Publications

Lambert (2015) reminds us of the various outlets available for the publication of research evidence. Much of the research and scholarship conducted in the field of geography education is published in textbooks and teachers' manuals which support the professional practice of geography teachers—such as publications from the GA in its 'Learning to Teach' series (see Lambert and Balderstone 2010). Publications of this kind are laudable and much needed—professional handbooks offer important, direct support to classroom practitioners—but may not qualify as containing research that could claim to be truly rigorous and impactful. In essence these are practical texts, often covering similar topics and drawing on *established* research in the field (or introducing small scale case study research evidence to back up previous findings). The scholarship on which such publications are based has its place, but does not necessarily aspire to push back research frontiers—it is often categorised as 'what works' enquiry, or is an extension of good sense, practice and experience. The markets for such publications tends to be among trainee teachers, those in the early stages of their careers, or teachers who seek professional support for their day-to-day classroom practice.

Different from such handbooks—in content, structure and tone—are texts that are more research-informed (or even research-driven), such as those specifically designed to support geography teachers in conducting their own research. These may have generic sections on education theory or research methods, but will almost certainly draw more heavily on theoretical and conceptual information—for example, texts that are designed specifically for those teachers who wish to gain Masters credits (either as part of their initial teacher education, or for an award bearing Masters course) through undertaking some small scale, but rigorous, research project (see, for example, Butt 2015). Similar books may also be aimed at mentor teachers in schools, or postgraduate researchers (possibly at doctoral level, including both EdD and PhD students), or initial teacher educators in universities and schools. Beyond these texts—which fall somewhere between a handbook and a research tome—are those that are entirely about research, possibly written by an international collection of academics and scholars. These publications may pursue just one theme, eschewing more generalist approaches and content (see, for example, Brooks et al. 2017, or Lambert and Jones 2017).

Much research in education, and elsewhere, results in some form of publication in journals—these journals tend to be classified as either professional or academic, subject-based or generic. Journals like *Teaching Geography* and *Primary Geography*, two of the journals produced by the GA, would be classed as professional journals— these are practice-based and composed of short articles (up to 1500 words). The articles might review and reference the research of others in the field, but would often only include limited conceptual, theoretical or original material. They are demonstrably subject-based and designed for teachers, specialising in the pursuit of popular issues in geography education. Academic journals, such as *International Research in Geographical and Environmental Education* are much more research driven, with the focus of articles (up to 6000 to 7000 words in length) not necessarily being based in classroom practice, but often of a theoretical, philosophical or conceptual nature. Articles tend to include a clearly articulated research methodology as well as a substantial literature review (Chang and Kidman 2018). Online, open access, academic journals, which are also peer reviewed—such as *Review of International Geographical Education Online* (RIGEO) and *J Reading*—have opened up the possible outlets for publication by geography education researchers, potentially with wider audiences globally.[15]

Geography education researchers can also choose to publish in journals whose content attracts attention from a wider, more generic, audience beyond that solely made up of geography educators—international, peer reviewed, academic education journals such as (say) *Educational Review*, or *Oxford Review of Education,* would fall into this category. Lambert (2015) rightly asserts that it is important for researchers in geography education to write for (and read) such journals, which may be less 'safe' that those which contain only work from geography educationists. This helps geography education researchers avoid the issue of their discourse becoming:

'internalist' and full of insider orthodoxies and unchallenged assumptions (p. 19)

It is unrealistic to expect that geography teachers will regularly engage with a plethora of academic journals as part of their day-to-day practice; this may only occur if they are involved in a research project of their own, or are tasked to solve a particular educational problem in their school. Members of senior management teams in schools, and Heads of Department, might also be drawn into reading this research at specific times, for particular practical purposes. Academics must recognise this. It is foolhardy to expect the situation to change—the 'middle way' may be for 'digests' of pertinent research to be made more available to busy teachers. By doing so, academics would acknowledge the characteristics of their audience and show a greater understanding of the time-limited way in which many teachers have to work.

According to some there are wider, and inherently more serious, implications if teachers do not regularly engage with research material. Grossman (2008) and Abbott (1998) suggest that as part of a professional body, with professional responsibilities, teachers should be expected to 'problem solve' within their own practice. Research evidence may help them to do so. They warn that if teachers do not strive to solve their own problems, someone else will be directed to take the initiative—often with less palatable results. Many would argue that this situation already exists: as politicians, consultants, education 'experts', awarding bodies and academics all seem keen to involve themselves in advising teachers, and directing what they do. A more research-informed teaching workforce might be able to counter some of the more foolish, fashionable and fantastic 'solutions' to the issues they face in the classroom and beyond.

---

[15]In addition, the IGU CGE website (www.igu-cge.org) is a source of information on upcoming conferences, projects, news and publications in geography education—as well as providing background information on the aims and history of the IGU CGE. The website was improved after 2012, featuring a newsletter and access to information about research publications, doctoral theses, reports, etc. The website serves as a free resource for all those involved in geography education and geography education research.

3.14   Conclusions

In what is, self-evidently, a chapter that is highly selective of its content I have nonetheless explored some research themes that have recently attracted the attention of researchers in geography education, both nationally and internationally. The chapter started by questioning the reasons why some potential areas of research are pursued, even championed, by geography education researchers, and others largely ignored. The motivations to carry out research into one area, or another, may be complex: some research will be driven by specific problems that a number of geography educators face—for example, a body of research into the use of maps, the language of maps and graphicacy grew in the 1970s and 1980s, not only from particular issues in geography classrooms but also from links with areas of cognitive development research (see Blades and Spencer 1986; Boardman 1983). Some research is driven by policy—at various times governments have focussed their education policy on (say) improving assessment, on lesson planning, or on curriculum construction, which has attracted the attention of geography education researchers, and others. What is also apparent is that the scale at which researchers can afford to undertake their work is a very strong influence on research outcomes and subsequent publications. If researchers come together only in small numbers to undertake research, then the ambitions of their research projects, the research questions that can be answered, and the publications that result must surely be modest. Here is the incentive for geography education researchers to work collectively across national boundaries to pursue their common research goals.

References

Abbott, A. (1998). *A system of professions*. Chicago: Chicago University Press.

Adey, K., & Biddulph, M. (2002). 'What's the point?' Pupils' perceptions of the relevance of studying history and geography to their future employability. *Career Research and Development, 6*, 30–32.

Arnot, M., & Reay, D. (2007). A sociology of pedagogic voice: Power, inequality and pupil consultation. *Discourse Studies in the Cultural Politics of Education, 28*(3), 311–325.

Audet, R., & Paris, J. (1997). GIS implementation model for schools: Assessing the critical concerns. *Journal of Geography, 96*(6), 282–300.

Baker, T. (2000). *Applications of GIS in the K-12 science classroom*. Paper presented at the 20th Annual ESRI User Conference. San Diego Conference Centre, July, 2000.

Baker, T., Palmer, A., & Kerski, J. (2009). A national survey to examine teacher professional development and implementation of desktop GIS. *Journal of Geography, 108*(4–5), 174–185.

Ball, S. (2008). *The education debate: Policy and politics in the twenty-first century*. London: Policy Press.

Barford, A. (2006). Worldmapper: The world as you've never seen it before. *Teaching Geography, 31*(2), 68–75.

Barnett, R. (2009). Knowing and becoming in the higher education curriculum. *Studies in Higher Education, 34*(4), 429–440.

Bates, R. (2004). *Developing capabilities and the management of trust: Where administration went wrong*. Australian Association for Research in Education Annual conference, Melbourne, December.

Beck, J. (2013). Powerful knowledge, esoteric knowledge, curriculum knowledge. *Cambridge Journal of Education, 43*(2), 177–193.

Bednarz, S. (1999). Fieldwork in K-12 geography in the United States. *International Research in Geographical and Environmental Education, 8*(2), 164–170.

Bednarz, S. (2001). Thinking spatially: Incorporating geographic information science in pre and post-secondary education. In L. Houtsonen, & M. Tammilehto (Eds.), *Innovative practices in geographical education. Proceedings of the symposium of the IGU commission on geographical education*. Helsinki, August 6–10.

Bednarz, S., & Audet, R. (1999). The status of GIS technology in teacher preparation programs. *Journal of Geography, 98*, 60–67.

Bednarz, R., & Bednarz, S. (2004). Geography education: The glass is half full and it's getting fuller. *The Professional Geographer, 56*(1), 22–27.

Bednarz, S., & van der Schee, J. (2006). Europe and the United States: The implementation of geographic information systems in secondary education in two contexts. *Technology, Pedagogy and Education, 15*(2), 191–205.

Bennetts, T. (1981). Progression in the geography curriculum. In R. Walford (Ed.), *Signposts for geography teaching* (pp. 165–185). Longmans: Harlow.

Bennetts, T. (1995). Continuity and progression. *Teaching Geography, 20*(2), 75–79.

Bennetts, T. (1996). Progression and differentiation. In P. Bailey & P. Fox (Eds.), *Geography teachers' handbook* (pp. 81–93). Sheffield: Geographical Association.

Bennetts, T. (2005a). Progression in geographical understanding. *International Research in Geographical and Environmental Education, 14*(2), 112–132.

Bennetts, T. (2005b). The links between understanding, progression and assessment in the secondary geography curriculum. *Geography, 90*(2), 152–170.

Bennett, R. (2011). Formative assessment: A critical review. *Assessment in Education: Principles, Policy and Practice, 18*(1), 5–25.

Bereiter, C., & Scardamalia, M. (2003). Learning to work creatively with knowledge. In E. De Corte, L. Verschaffel, N. Entwistle, & J. van Merrienboer (Eds.), *Unravelling basic components and dimensions of powerful learning environments* (pp. 55–68). EARLI Advances in Learning and Instruction Series.

Bernstein, B. (1999). Vertical and horizontal discourse. *British Journal of Sociology of Education, 20*(2), 157–173.

Bernstein, B. (2000). *Pedagogy, symbolic control and identity: Theory, research, critique*. Oxford: Rowman and Littlefield Publishers.

Biddulph, M. (2011a). Young people's geographies: Implications for school geography. In G. Butt (Ed.), *Geography, education and the future* (pp. 44–59). London: Continuum.

Biddulph, M. (2011b). Articulating student voice and facilitating student agency. *The Curriculum Journal, 22*(3), 381–401.

Biddulph, M. (2012). 'Spotlight on… young people's geographies and the school curriculum. *Geography, 97*(3), 155–162.

Biddulph, M. (2016). *What does it mean to be a teacher of geography? Investigating the teacher's relationship with the curriculum*. Unpublished Ph.D., University of Nottingham.

Biddulph, M. (2018). Primary and secondary geography: Common ground and some shared dilemmas. *Teaching Geography, 71*(3), 101–104.

Biddulph, M., & Adey, K. (2004). Pupil perceptions of effective teaching and subject relevance in history and geography at KS3. *Research in Education, 71*, 1–8.

Biesta, G. (2012). Giving teaching back to education: Responding to the disappearance of the teacher. *Phenomenology and Practice, 6*(2), 35–49.

Biesta, G. (2017). *The rediscovery of teaching*. London: Routledge.

Blades, M., & Spencer, C. (1986). Map use by young children. *Geography, 71*(1), 47–52.

Bland, K., Chambers, D., Donert, K., & Thomas, T. (1996). Fieldwork. In P. Bailey & P. Fox (Eds.), *Geography teachers' handbook* (pp. 165–175). Sheffield: Geographical Association.

Bloom, B. (1956). *Taxonomy of educational objectives: The classification of educational goals.* New York: McKay.

Boardman, D. (1983). *Graphicacy and geography teaching.* London: Croom Helm.

Bobbit, F. (1924). The new technique of curriculum making. *The Elementary School Journal., 25*(1), 45–54.

Boyle, A., Maguire, S., Martin, S., Milsom, C., Nash, R., Rawlinson, S., et al. (2007). Fieldwork is good: The student perception and the affective domain. *Journal of Geography in Higher Education, 31*(2), 299–317.

Brooks, C. (2007). *Towards understanding the influence of subject knowledge in the practice of 'expert' geography teachers.* Unpublished Ph.D. thesis. Institute of Education, University of London.

Brooks, C. (2016). *Teacher subject identity in professional practice: Teaching with a professional compass.* London: Routledge.

Brooks, C. (2017). Becoming a geography teacher in the primary school. In S. Catling (Ed.), *Reflections on primary geography* (pp. 171–173). Sheffield: Geographical Association.

Brooks, C., Butt, G., & Fargher, M. (Eds.). (2017). *The power of geographical thinking.* Dordrecht: Springer.

Buchanan-Dunlop, J. (2008). Virtual fieldwork. In D. Mitchell (Ed.), *ICT in secondary geography: A short guide for teachers.* Sheffield: Geographical Association.

Bustin, R. (2016). *An investigation into geocapability and future 3 curriculum thinking in geography.* Unpublished Ph.D. UCL Institute of Education.

Butt, G. (1997). *An investigation into the dynamics of the national curriculum geography working group (1989–1990).* Unpublished Ph.D., University of Birmingham, Birmingham.

Butt, G. (2002). *Reflective teaching of geography 11-18: Meeting standards and applying research.* London: Continuum.

Butt, G. (Ed.). (2015). *Masterclass in geography education.* London: Bloomsbury.

Butt, G. (2017). Debating the place of knowledge within geography education: reinstatement, reclamation or recovery? In C. Brooks, G. Butt, & M. Fargher (Eds.), *The power of geographical thinking* (pp. 13–26). Dordrecht: Springer.

Butt, G. (2018). *What is the future for subject-based education research?* Public seminar delivered at the Oxford University Department of Education. 22 October www.podcasts.ox.ac.uk.

Butt, G. (2019). Bridging the divide between school and university geography—'Mind the Gap!'. In S. Dyer, H. Walkington, & J. Hill (Eds.), *Handbook for learning and teaching geography* (pp. 1–10). London: Elgar.

Butt, G., Weeden, P., Chubb, S., & Srokosz, A. (2011). The state of geography in English secondary schools: An insight into practice and performance in assessment. *International Research in Geographical and Environmental Education, 15*(2), 134–148.

Carter, A. (2015). *The carter review of initial teacher training (ITT).* London: HMSO.

Catling, S., Greenwood, R., Martin, F., & Owens, P. (2010). Formative experiences of primary geography educators. *International Research in Geographical and Environmental Education, 19*(4), 341–350.

Catling, S., & Martin, F. (2011). Constructing powerful knowledge: The primary geography curriculum as an articulation between academic and children's (ethno-) geographies. *The Curriculum Journal, 22*(3), 317–335.

Caton, D. (2006a). *Theory into practice: New approaches to fieldwork.* Sheffield: Geographical Association.

Caton, D. (2006b). Real world learning through geographical fieldwork. In D. Balderstone (Ed.), *Secondary geography handbook.* Sheffield: Geographical Association.

Chang, C.-H., & Kidman, G. (2018). Reflecting on recent geographical and environmental education issues. *International Research in Geographical and Environmental Education, 27*(1), 1–4.

Cook, V. (2011). The origins and development of geography fieldwork in British Schools. *Geography, 96*(2), 69–74.

Davidson, G. (2009). *Young people's geographies: Evaluation report*. 2008/9. Sheffield: Geographical Association.

Demirci, A. (2011). Using Geographic Information systems (GIS) at schools without a computer laboratory. *Journal of Geography, 110*(2), 49–59.

Demirci, A., Karaburun, A., & Kalar, H. (2013a). Using Google Earth as an educational tool in secondary school geography lessons. *International Research in Geographical and Environmental Education, 22*(4), 277–290.

Demirci, A., Karaburun, A., & Mehmet, U. (2013b). Implementation and effectiveness of GIS-based projects in secondary schools. *Journal of Geography, 112*(5), 214–228.

Deng, Z. (2018). Rethinking teaching and teachers: Bringing content back into conversation. *London Review of Education, 16*(3), 371–383.

Department for Education (DfE). (2016). *Educational excellence everywhere*. London: HMSO.

Deprez, L., & Butler, S. (September, 2001). *The capabilities approach and economic security for low income women in the US: Securing access to higher education under welfare reform*. Conference on Justice and Poverty, Examining Sen's Capability Approach. St. Edmunds College, Cambridge.

Dillon, J., Rickinson, M., Teamey, K., Morris, M., Choi, M. Y., Sanders, D., et al. (2006). The value of outdoor learning: Evidence from research in the UK and elsewhere. *School Science Review, 87*(320), 107–111.

Dunphy, A., & Spellman, G. (2009). Geography fieldwork, fieldwork value and learning styles. *International Research in Geographical and Environmental Education, 18*(1), 19–28.

Elwood, S. (2008). Geographic information science: New geovisualization technologies—Emerging questions and linkages with GiScience research. *Progress in Human Geography, 33*(2), 256–263.

Fargher, M. (2013a). *A study of the role of GIS in constructing relational place knowledge through school geography education*. Unpublished Ph.D. thesis. Institute of Education, University of London.

Fargher, M. (2013b). Geographic information (GI); How could it be used. In D. Lambert & M. Jones (Eds.), *Debates in geography education* (pp. 206–218). London: Routledge.

Fargher, M. (2017). GIS and the power of geographical thinking. In C. Brooks, G. Butt, & M. Fargher (Eds.), *The power of geographical thinking* (pp. 151–164). Dordrecht: Springer.

Fargher, M., & Rayner, D. (2012). United Kingdom: Realizing the potential for GIS in the school geography curriculum. In A. Milson, A. Demirci, & J. Kerski (Eds.), *International perspectives on teaching and learning with GIS in secondary schools*. London: Springer.

Favier, T., & van der Schee, J. (2009). Learning geography by combining fieldwork with GIS. *International Research in Geographical and Environmental Education, 18*(4), 261–274.

Favier, T., & van der Schee, J. (2012). Exploring the characteristics of an optimal design for inquiry-based geography education with geographic information systems. *Computers & Education, 58*, 666–677.

Fielding, M. (2006). *Leadership, collegiality and the necessity of Person-Centred education. Teaching and learning research programme thematic seminar series: Culture, values, identities and power*. University of Exeter.

Fielding, L. (9 June, 1990). *Geography in the national curriculum*. Times Educational Supplement.

Firth, R. (2011). Debates about knowledge and the curriculum: Some implications for geography education. In G. Butt (Ed.), *Geography, education and the future* (pp. 141–164). London: Continuum.

Firth, R. (2013). What constitutes knowledge in geography? In D. Lambert, & M. Jones (Eds.), *Debates in geography education* (pp. 59–74). London and New York: Routledge.

Firth, R., & Biddulph, M. (2009). Young people's geographies. *Teaching Geography, 34*(1), 32–34.

Fisher, T. (2000). Developing the educational use of information and communications technology: Implications for the education of geography teachers. In C. Fisher & T. Binns (Eds.), *Issues in geography teaching* (pp. 50–65). London: RoutledgeFalmer.

Fitzpatrick, C. (1990). Computers in geography instruction. *Journal of Geography, 89,* 148–149.
Fitzpatrick, C. (1993). Teaching geography with computers. *Journal of Geography, 92,* 4.
Foskett, N. (1997). Teaching and learning through fieldwork. In D. Tilbury & M. Williams (Eds.), *Teaching and learning geography* (pp. 189–201). London: Routledge.
Foskett, N. (1999). Forum: Fieldwork in the geography curriculum—International perspectives and research issues. *International Research in Geographical and Environmental Education, 8*(2), 159–163.
Foskett, N. (2000). Fieldwork and the development of thinking skills. *Teaching Geography, 25*(3), 126–129.
Foskett, N., & Marsden, B. (Eds.). (1998). *A bibliography of geographical education 1970–1997.* Sheffield: Geographical Association.
Freire, P. (1996). *Pedagogy of the oppressed* (2nd ed.). London: Penguin Books.
Fuller, I., Rawlinson, S., & Bevan, R. (2000). Evaluation of student learning experiences in physical geography fieldwork: Paddling or pedagogy? *Journal of Geography in Higher Education, 24*(2), 199–215.
Furlong, J. (2013). *Education: An anatomy of the discipline.* London: Routledge.
Furlong, J., & Lawn, M. (2010). *The disciplines of education: Their role in the future of education research.* London: Routledge.
Furlong, J., & Smith, R. (2013). Introduction. In Furlong & Smith (Eds.), *The role of higher education in initial teacher training* (pp. 1–7). London: Routledge.
Gair, N. (1997). *Outdoor education: Theory and practice.* London: Cassell.
Geographical Association (GA). (2009). *A different view: A manifesto from the geographical association.* Sheffield: Geographical Association.
Gershmehl, P. (2008). *Teaching geography.* New York: Guilford Press.
Giroux, H. (1992). *Border crossings: Cultural workers and the politics of education.* London: Routledge.
Gold, J., Jenkins, A., Lee, R., Monk, J., Riley, J., Shepherd, I., et al. (1991). *Teaching geography in higher education: A manual of good practice.* Oxford: Blackwell.
Grossman, P. (2008). Responding to our critics: From crisis to opportunity in research on teacher education. *Journal of Teacher Education, 59*(1), 10–23.
Gryl, I., & Jekel, T. (2012). Re-centring geoinformation in secondary education: Toward a spatial citizenship approach. *Cartographica, 47*(1), 18–28.
Gryl, I., Jekel, T., & Donert, K. (2010). GI and spatial citizenship. *Learning with GI, V,* 2–11.
Gunder, C., & Hopmann, D. (1998). On the evaluation of curriculum reforms. *Journal of Curriculum Studies, 35,* 459–478.
Hammond, L. (2015). *Can powerful pedagogies be used to support powerful knowledge in geography education?* Oxford: presentation to GTE conference 31 January.
Harley, J. (1989). Deconstructing the map. *Cartographica: The International Journal for Geographic Information and Geovisualization, 26*(2), 1–20.
Hart, C. (2009). Quo vadis? The capability space and new directions for the philosophy of education research. *Studies in Philosophy and Education, 28,* 391–402.
Hart, C., & Thomas, T. (1986). Framework fieldwork. In D. Boardman (Ed.), *Handbook for geography teachers* (pp. 205–218). Sheffield: Geographical Association.
Hawley, D. (2012). The 'real deal' of earth science: Why, where and how to include fieldwork in teaching. *School Science Review, 94*(347), 87–100.
Hemmer, I., Hemmer, M., Kruschel, K., Neidhardt, E., Obermaier, G., & Uphues, R. (2013). Which children can find a way through a strange town using a streetmap? Results of an empirical study on children's orientation competence. *International Research in Geographical and Environmental Education, 22*(1), 23–40.
Hemmer, I., Hemmer, M., Neidhardt, E., Obermaier, G., Uphues, R., & Wrenger, K. (2015). The influence of children's prior knowledge and previous experience on their spatial orientation skills in an urban environment. *Education 3-13, 43*(2), 184–196.
Henning, B. (2018). Worldmapper: Rediscovering the world. *Teaching Geography, 43*(2), 66–68.

Herrick, C. (2010). Lost in the field: Ensuring student learning in the 'threatened' geography field-trip. *Area, 42*(1), 108–116.

Hill, D., & Solem, M. (1999). Geography on the Web-changing the learning paradigm. *Journal of Geography, 98*, 100–107.

Hinchliffe, G. (2006). *Beyond key skills: Exploring capabilities*. Networking Day for Humanities Careers Advisers in London. 16 June.

Hinchliffe, G. (2007a). Special issue on the concept of capability and its application to questions of equity, access and the aims of education. *Prospero, 13*(3), 1–4.

Hincliffe, G. (2007). Beyond key skills: The capability approach to personal development. *Prospero, 13*(3), 5–12.

Hirsch, E. (1987). *Cultural literacy: What every American needs to know*. New York: Houghton Mifflin.

Hirsch, E. (2007). *The knowledge deficit: Closing the shocking gap for American children*. New York: Houghton Mifflin.

Hirst, P (1996). The demands of professional practice and preparation for teachers. In J. Furlong, & R. Smith (Eds.), *The role of higher education in initial teacher training* (pp. 23–34). London, Kogan page.

Holmes, D., & Walker, M. (2006). Planning geographical fieldwork. In D. Balderstone (Ed.), *Secondary geography handbook* (pp. 47–54). Sheffield: Geographical Association.

Hopkins, P. (2008). Young, male, Scottish and Muslim: A portrait of Kabir. In C. Jeffery & J. Dyson (Eds.), *Telling young lives: Portraits of global youth* (pp. 485–502). Philadelphia: Temple.

Hopkins, P., & Alexander, C. (2010). Politics, mobility and nationhood: Upscaling young people's geographies. *Introduction to Special Section, Area 42*(2), 142–144.

Hopmann, S. (2008). No child, no school, no state left behind: schooling in the age of accountability. *Journal of Curriculum Studies, 40*(4), 417–456.

Hopwood, N. (2007). *Young people's geographies: Evaluator's report*. Sheffield: Geographical Association.

Hopwood, N. (2008). *Young people's geographies: Evaluator's report*. Year 2. Sheffield: Geographical Association.

Houtsonen, L. (2003). Maximising the use of communications technologies in geographical education. In R. Gerber (Ed.), *International handbook on geographical education* (pp. 47–63). Kluwer.

Huckle, J. (Ed.). (1983). *Geographical education: Reflection and action*. Oxford: Oxford University Press.

Jackson, P. (2006). Thinking geographically. *Geography, 91*(3), 199–204.

James, M. (2014). *Assessing without levels: From measurement to judgement*. School Leaders Conference: Nottingham.

Jekel, T., Gryl, I., & Schulze, U. (2015). Education for spatial citizenship. In O. Muñiz Solari, A. Demirci, & J. van den Schee (Eds.), *Geospatial technologies and geography education in a changing world. Advances in geographical and environmental sciences*. Tokyo: Springer.

Job, D. (1996). Geography and environmental education: An exploration of perspectives and strategies. In A. Kent, D. Lambert, M. Naish, & F. Slater (Eds.), *Geography in education: Viewpoints on teaching and learning*. Cambridge: CUP.

Jones, M. (2013). What is personalized learning in geography? In D. Lambert & M. Jones (Eds.), *Debates in geography education* (pp. 116–128). London: Routledge.

Keiper, A. (1999). GIS for elementary students: An inquiry into a new approach to learning geography. *Journal of Geography, 98*(2), 47–59.

Kelly, A. (2009). *The curriculum: Theory and practice* (6th ed.). London: Sage.

Kent, A. (1983). *Geography teachers and the Micro*. York: Longman.

Kent, A., & Foskett, N. (2002). Fieldwork in the school geography curriculum; pedagogical issues and developments. In M. Smith (Ed.), *Teaching geography in secondary schools*. London: Routledge Falmer.

Kent, M., Gilbertson, D., & Hunt, C. (1997). Fieldwork in geography teaching: A critical review of the literature and approaches. *Journal of Geography in Higher Education, 21*(3), 313–332.

Kerski, J. (2000). *The implementation and effectiveness of geographic information systems technology and methods in secondary education*. Unpublished dissertation. Department of Geography, University of Colorado.

Kerski, J. (2003). The implementation and effectiveness of geographic information systems technology and methods in secondary education. *Journal of Geography, 102*(3), 128–137.

Kerski, J. (2008). The role of GIS in digital earth education. *International Journal of Digital Earth, 1*(4), 326–346.

Kerski, J., Demirici, A., & Milson, A. (2013). The global landscape of GIS in secondary education. *Journal of Geography, 112*(6), 232–247.

Kesrki, J. (2001). A national assessment of GIS in American high schools. *International Research in Geographical and Environmental Education, 10*(1), 72–84.

Kinder, A. (2013). What is the contribution of fieldwork to school geography? In D. Lambert & M. Jones (Eds.), *Debates in geography education* (pp.180–192). London: Routledge.

Kirschner, P., & Davis, N. (2006). Pedagogic benchmarks for information and communications technology in teacher education. *Technology, Pedagogy and Education, 12*(1), 125–147.

Krathwohl, D. (2002). A revision of bloom's taxonomy: An overview. *Theory into Practice, 41*(4), 212–218.

Lai, K.-C. (1999). Freedom to learn: A study of the experiences of secondary school teachers and students in a geography field trip. *International Research in Geographical and Environmental Education, 8*(3), 239–255.

Lam, C.-C., Lai, E., & Wong, J. (2009). Implementation of geographic information system (GIS) in secondary geography curriculum in Hong Kong: Current situations and future directions. *International Research in Geographical and Environmental Education, 18*(1), 57–74.

Lambert, D. (1996). Assessing pupils' attainment and supporting learning. In A. Kent, et al. (Eds.), *Geography in education: Viewpoints on teaching and learning* (pp. 260–286). Cambridge: Cambridge University Press.

Lambert, D. (2009). *Geography in education: Lost in the post? Inaugural lecture*. Institute of Education, University of London.

Lambert, D. (2011). Reframing school geography: A capability approach. In G. Butt (Ed.), *Geography, Education and the Future* (pp. 127–140). London: Continuum.

Lambert, D. (2015). Research in geography education. In G. Butt (Ed.), *MasterClass in geography education* (pp. 15–30). London: Bloomsbury.

Lambert, D. (2016). Editorial. *International Research in Geographical and Environmental Education, 25*(3), 189–194.

Lambert, D. (2018). *Editorial: Teaching as a research-engaged profession. Uncovering a blind spot and revealing new possibilities. London Review of Education, 16*(3), 357–370.

Lambert, D., & Balderstone, D. (2010). *Learning to teach geography*. London: Routledge.

Lambert, D., & Biddulph, M. (2015). The dialogic space offered by curriculum-making in the process of learning to teach and the creation of a progressive knowledge-led curriculum. *Asia Pacific Journal of Teacher Education, 43*(3), 210–224.

Lambert, D., & Hopkin, J. (2014). A possibilist analysis of the geography national curriculum in England. *International Research in Geographical and Environmental Education, 23*(1), 64–78.

Lambert, D., & Jones, M. (Eds.). (2017). *Debates in geography education* (2nd ed.). London: Routledge.

Lambert, D., & Morgan, J. (2009). Corrupting the curriculum: The case of geography. *London Review of Education, 7*(2), 147–157.

Lambert, D., & Morgan, J. (2010). *Teaching geography 11-18: A conceptual approach*. Maidenhead: McGraw Hill/Open UP.

Lambert, D., & Morgan, J. (2011). *Geography and development: Development education in schools and the part played by geography teachers*. DERC Research Paper 3. London: Institute of Education.

Lambert, D., & Reiss, M. (2014). *The place of fieldwork in geography and science qualifications*. London: UCL/Institute of Education.

Lambert, D., Solem, M., & Tani, S. (2013). Achieving human potential through geography education: A capabilities approach to curriculum making in schools. *Annals of the Association of American Geographers, 105,* 723–735.

Leask, M. (1999). Teaching and learning with ICT: An introduction. In S. Capel, M. Leask, & T. Turner (Eds.), *Learning to teach in the secondary school* (pp. 2–12). London: Routledge.

Leat, D., & Higgins, S. (2002). The role of powerful pedagogical strategies in curriculum development. *The Curriculum Journal, 13*(1), 77–85.

Lei, P., Kao, G., Lin, S., & Sun, C. (2009). Impacts of geographical knowledge, spatial ability and environmental cognition on image searches supported by GIS software. *Computers in Human Behaviour, 25*(6), 1270–1279.

Lidstone, J. (1988). Teaching and learning geography through fieldwork. In R. Gerber, & J. Lidstone (Eds.), *Developing skills in geographical education.* Brisbane: IGUCGE.

Lidstone, J. (2010). Forum: The geospatial web and geographical education. *International Research in Geographical and Environmental Education, 19*(1), 51.

Lidstone, J., & Stoltman, J. (2006). Editorial: Searching for, or creating, knowledge: The roles of google and gis in geographical education. *International Research in Geographical and Environmental Education, 15*(3), 205–209.

Major, B. (2013). *Geography: A powerful knowledge.* http://www.academia.edu/1327261/Geography_-_A_Powerful_Knowledge. Last accessed March 28, 2015.

Marsden, W. (1995). *Geography 11-16: Rekindling good practice.* London: David Fulton.

Marsden, W. (1997). On taking the geography out of geographical education—Some historical pointers on geography. *Geography, 82*(3), 241–252.

Martin, F. (2000). Models for ICT training in geography. In A. Kent (Ed.), *Research forum 2. Information and Communications Technology* (pp. 65–70). London: IGU CGE and Institute of Education, University of London.

Martin, F., & Catling, S. (2011). Contesting powerful knowledge: The primary geography curriculum as an articulation between academic and children's (ethno-) geographies. *Curriculum Journal, 22*(3), 317–15.

May, S., & Richardson, P. (2005). *Managing safe and successful fieldwork.* Sheffield: Geographical Association.

McKie, R. (2019). *After we leave, bright young scientists won't come to join us here in Britain* (p. 8). Observer. 20.1.19.

McPartland, M. (1996). Walking in our own footsteps: Autobiographical memories and the teaching of geography. *International Research in Geographical and Environmental Education, 5*(1), 57–62.

Mishra, P., & Koehler, M. (2006). Technological pedagogical content knowledge: A framework for teacher knowledge. *Teachers' College Record, 108*(6), 1017–1054.

Mitchell, D. (2017). *Geography curriculum making in changing times.* Unpublished Ph.D. Institute of Education, University College London.

Morgan, J. (2007). School geography and the politics of culture. *Geography Compass, 1*(3), 373–388.

Morgan, J. (2008a). Curriculum development in 'New times'. *Geography, 93*(1), 17–24.

Morgan, J. (2008b). Contesting Europe: Representations of space in English school geography. *Globalisation, Societies, Education, 6*(3), 281–290.

Morgan, J. (2014). Foreword. In M. Young, D. Lambert, C. Roberts, & M. Roberts (Eds.), *Knowledge and the future school: Curriculum and social justice* (pp. ix–xii). London: Bloomsbury.

Morgan, J., & Firth, R. (2010). 'By our theories shall you know us': The role of theory in geographical education. *International Research in Geographical and Environmental Education, 19*(2), 87–90.

Moss, S. (2012). *Natural childhood.* Available from www.nationaltrust.org.uk.

Muller, J., & Young, M. (2008). *Three scenarios for the future: Lessons from the sociology of knowledge.* London: DCSF and Futurelab.

Muñiz Solari, O., Demirci, A., & van der Schee, J. (Eds.). (2015). *Geospatial technologies and geography education in a changing world.* International Geographical Union-Commission on Geographical Education (IGU-CGE), (IGU). Japan: Springer.

Muñiz Solari, O., Solem, M., & Boehm, R. (Eds.). (2017). *Learning progressions in geography education. International perspectives.* National Center for Research in Geography Education. Berlin: Springer.

National Research Council (NRC). (2006). *Learning to think spatially: GIS as a support system in the K–12 Curriculum.* Washington, D.C.: The National Academies Press.

Nayak, A. (2003). *Race, place and globalization: Youth cultures in a changing world.* London: Berg.

Nayak, A. (2006). Displaced masculinities: Chavs, youth and class in the post-industrial city. *Sociology, 40*(5), 813–831.

Nellis, M. (1994). Technology in geographic education: Reflections and future directions. In R. Bednarz & J. Peterson (Eds.), *A Decade of reform in geographical education: Inventory and prospect* (pp. 51–57). Indiana, PA: National Council of Geographical Education.

Newcombe, L. (1999). Developing novice teacher ICT competence. *Teaching Geography, 24*(3), 128–132.

Nussbaum, M. (2000). *Women and human development.* Cambridge, MA: Cambridge University Press.

Nussbaum, M. (2013). *The therapy of desire: Theory and practice in hellenistic ethics.* New Jersey: Princeton University Press.

Nussbaum, M., & Sen, A. (Eds.). (1993). *The quality of life.* Oxford: Clarendon Press.

OfSTED (Office for Standards in Education). (2011). *Geography: Learning to make a world of difference.* London: HMSO.

Page, E. (1 June, 2004). *Teacher and pupil development and capabilities.* Work in Progress. Capability and Education Network seminar, St. Edmunds College, Cambridge.

Parkinson, A. (2009). *Fieldwork-an essential part of a geographical education.* Available from www.geography.org.uk.

Parkinson, A. (2013). How has technology impacted on the teaching of geography and geography teachers? In D. Lambert & M. Jones (Eds.), *Debates in geography education* (pp. 193–205). London: Routledge.

Pickles, J. (1995). Representations in an electronic age: Geography, GIS, and democracy. In J. Pickles (Ed.), *Ground truth: The social implications of geographic information systems.* New York: Guilford Press.

Pickles, J. (2002). *A history of spaces: Cartographic reason, mapping and the geo-coded world.* London: Routledge.

Powell, R. (2002). The Sirens' voices? Field practices and dialogue in geography. *Area, 34*(3), 261–272.

Pring, R. (2012). Importance of philosophy in the conduct of educational research. *Journal of International and Comparative Education, 1*(1), 23–30.

QCA (Qualifications and Curriculum Authority). (2000). *A scheme of work for key stage 3: Teacher's guide.* London: QCA/DfEE.

Rawding, C. (2013). How does geography adapt to changing times? In D. Lambert & M. Jones (Eds.), *Debates in geography education* (2nd ed., pp. 282–290). London: Routledge.

Rawling, E. (2001). *Changing the subject. The impact of national policy on school geography 1980–2000.* Sheffield: Geographical Association.

Rawling, E. (2016). The geography curriculum 5-19: What does it all mean? *Teaching Geography, 41*(1), 6–9.

Rawling, E. (2018). Reflections on progression in learning about place. *Journal of Geography, 117*(3), 128–132.

Raynor, J. (1 June, 2004). *Girls' development and capabilities in Bangladesh.* Capability and Education Network seminar, St. Edmunds College, Cambridge.

Remmen, K.-B., & Frøyland, M. (2014). Implementation of guidelines for effective fieldwork designs: Exploring learning activities, learning processes, and student engagement in the classroom and the field. *International Research in Geographical and Environmental Education, 23*(2), 103–125.

Rickinson, M., Dillon, J., Tearney, K., Morris, M., Choi, M., Sanders, D., et al. (2004). *A review of research on outdoor learning*. London: NFER and Kings College, London.

Roberts, M. (1996). Interpretations of the geography national curriculum: A common curriculum for all? *Journal of Curriculum Studies, 27*(2), 187–205.

Roberts, M. (2000). The role of research in supporting teaching and learning. In A. Kent (Ed.), *Reflective practice in geography teaching* (pp. 287–295). London: Paul Chapman.

Roberts, M. (2003). *Learning through enquiry: Making sense of geography in the key stage 3 classroom*. Sheffield: Geographical Association.

Roberts, M. (2013). *Geography through enquiry*. Sheffield: Geographical Association.

Roberts, M. (2014a). Curriculum leadership and the knowledge-led school. In M. Young, D. Lambert, C. Roberts, & M. Roberts (Eds.), *Knowledge and the future school: Curriculum and social justice* (pp. 139–158). London: Bloomsbury.

Roberts, M. (2014b). Powerful knowledge and geographical education. *The Curriculum Journal, 25*(2), 187–209.

Roberts, M. (2017). Geographical education is powerful if …. In *Teaching geography* (pp. 6–9). Sheffield: Geographical Association.

Roulston, S. (2013). GIS in Northern Ireland secondary schools: Mapping where we are now. *International Research in Geographical and Environmental Education, 22*(1), 41–56.

Roulston, S., & Young, O. (2013). GPS tracking of some Northern Ireland students—Patterns of shared and separated space: Divided we stand? *International Research in Geographical and Environmental Education, 22*(3), 241–258.

Rynne, E. (1998). Utilitarian approaches to fieldwork: A critique. *Geography, 83*(3), 205–213.

Saito, M. (2003). Amartya Sen's capability approach to education: A critical exploration. *Journal of Philosophy of Education, 37,* 17–34.

SCAA (School Curriculum Assessment Authority). (1994). *The review of the national curriculum: A report on the 1994 consultation*. London: SCAA.

Scheepers, D. (2005). GIS in the geography curriculum. Position IT (pp. 40–45).

Schuurman, N. (2004). *GIS: A short introduction*. London: Blackwell.

Sen, A. (1980). Equality of what? The Tanner lecture on Human Values, delivered at Stanford University, 22 May 1979.

Sen, A. (1985). *Commodities and capabilities*. Amsterdam: North Holland Press.

Seong, K. T. (1996). Interactive multimedia and GIS applications for teaching school geography. *International Research in Geographical and Environmental Education, 5*(3), 205–212.

Shulman, L. (1987). Knowledge and teaching: Foundations of the new reform. *Harvard Educational Review, 57*(1), 1–23.

Sinton, D., & Bednarz, S. (2007). Putting the 'G' in GIS. In D. Sinton & J. Lund (Eds.), *Understanding place: GIS and mapping across the curriculum*. Redlands: ESRI Press.

Sinton, D., & Lund, J. (2007). *Understanding place: GIS and mapping across the curriculum*. Redlands: ESRI Press.

Skelton, T. (2010). Taking young people as political actors seriously: Opening the borders of political geography. *Area, 42*(2), 145–151.

Slater, F., & Graves, N. (2016). Editorial. *International research in geographical and environmental education, 25*(3), 189–194.

Smith, M. (2000). *Curriculum theory and practice*. The Encyclopedia of Informal Education. www.infed.org/biblio/b-curric.htm. Accessed January 21, 2019.

Solem, M., Lambert, D., & Tani, S. (2013). Geocapabilities: Toward an international framework for researching the purposes and values of geography education. *Review of International Geographical Education Online (RIGEO), 3,* 214–229.

Stenhouse, L. (1975). *Introduction to curriculum research and development*. London: Heinemann.

Stoltman, J., & Lidstone, J. (2015). Editorial—Powerful knowledge in geography: IRGEE editors interview Professor David Lambert, London Institute of Education, October 2014 *International Research in Geographical and Environmental Education, 24*(1), 1–5.

Storm, M. (1995). Geography in schools: The state of the art. *Geography, 74*(4), 289–298.

Sui, D. (1994). GIS and urban studies: Positivism, post-positivism, and beyond. *Urban Geography, 15*(3), 258–278.

Sui, D. Z. (2004). GIS, cartography, and the "third culture": Geographic imaginations in the computer age. *The Professional Geographer 56*(1), 62–72.

Tapsfield, A. (1991). From computer-assisted learning to information technology. In R. Walford (Ed.), *Viewpoints on geography teaching: The Charney Manor conference papers*. London: Longman.

Taylor, L. (2011). Investigating change in young people's understandings of Japan: A study in learning about distant place. *British Educational Research Journal, 37*(6), 1033–1054.

Taylor, L. (2013). What do we know about concept formation and making progress in learning geography? In D. Lambert & M. Jones (Eds.), *Debates in geography education* (pp. 302–313).

Terzi, L. (September, 2004). On education as a basic capability. In *4th International Conference on the Capability Approach*. University of Pavia.

Tyler, R. (1949). *Basic principles of curriculum and instruction*. Chicago: University of Chicago Press.

Uhlenwinkel, A., BénekerT, Bladh G, Tani, S., & Lambert, D. (2016). GeoCapabilities and curriculum leadership: Balancing the priorities of aims-based and knowledge-led curriculum thinking in schools. *International Research in Geographical and Environmental Education, 26*(4), 327–341.

Valentine, G. (2010). Boundary crossings: Transitions from childhood to adulthood. *Children's Geographies, 1*(1), 37–52.

van der Schee, J. (2003). New media will accelerate the renewal of geographic education. In R. Gerber (Ed.), *International handbook on geographical education* (pp. 205–213). Dordrecht: Kluwer.

van der Schee, J., & Scholten, H. (2009). Geographical information systems and geography teaching. In H. Scholten, R. van de Velde & N. van Manen (Eds.), *Geospatial technology and the role of location in science*. GeoJournal Library, 96. Dordrecht: Springer.

van der Schee, J., Schoenmaker, G., Trimp, H., & van Westrhenen, H. (Eds.). (1996). *Innovation in geographical education*. Utrecht/Amsterdam: IGU and Centre for Geographical Education of the Free University of Amsterdam.

Van Joolingen, W., De Jong, T., & Dimitrakopoulou, A. (2007). Issues in computer supported inquiry learning in science. *Journal of Computer Assisted Learning, 23*, 111–119.

Walker, M. (2005). Amartya Sen's capability approach and education. *Educational Action Research, 13*(1), 103–110.

Walshe, N. (2017). Developing trainee teacher practice with geographical information systems (GIS). *Journal of Geography in Higher Education*, pp. 608–628.

Walshe, N. (2018). Geographical information systems for school geography. *Geography, 103*(1), 46–49.

Watson, D. (2000). Issues raised by research into ICT and geography education. In A. Kent (Ed.) *Research forum 2. Information and communications technology* (pp. 21–32). London: IGU CGE and Institute of Education, University of London.

Watts, M., & Bridges, D. (June, 2003). *Accessing higher education: What injustice does the current widening participation agenda seek to reform?* St. Edmunds College, Cambridge.

Wheelahan, L. (2007). How competency-based training locks the working class out of powerful knowledge: A modified Bernsteinian perspective. *British Journal of Sociology of Education, 28*, 637–651.

White, J. (2018). The weakness of 'powerful knowledge'. *London Review of Education, 16*(2), 325–335.

Wiegand, P., & Tait, K. (2000). The developing and piloting of a software cartographic tool giving user control over small scale thematic maps. In A. Kent (Ed.), *Research forum 2. Information and communications technology* (pp. 1–20). London: IGU CGE and Institute of Education, University of London.

Winch, C. (2013). Curriculum design and epistemic ascent. *Journal of Philosophy of Education, 47*(1), 128–146.

Winter, C. (2006). Doing justice to geography in the secondary school: Deconstruction, invention and the national curriculum. *British Educational Research Journal, 54*(2), 212–229.

Winter, C. (2010). Places, spaces, holes for knowing and writing the earth: The geography curriculum and Derrida's Khora. *Ethics and Education, 4*(1), 57–68.

Winter, C. (2013a). Enframing geography: Subject, curriculum, knowledge, responsibility. *Ethics and Education, 7*(3), 277–290.

Winter, C. (2013b). 'Derrida applied': Derrida meets Dracula in the geography classroom. In M. Murphy (Ed.), *Social theory and education research: Understanding Foucault, Habermas, Bourdieu and Derrida*. London: Taylor and Francis.

Wong Yuk Yong, J. (1996). Geographic information systems education in the Asia-Pacific region. *International Research in Geographical and Environmental Education, 5*(3), 196–198.

Yates, L., & Collins, C. (2010). The absence of knowledge in Australian curriculum reforms. *European Journal of Education, 45*(91), 89–102.

Young, M. (2008). *Bringing knowledge back in: From social constructionism to social realism in the sociology of education*. London: Routledge.

Young, M. (2009). Education, globalisation and the 'voice of knowledge'. *Journal of Education and Work, 22*(3), 193–204.

Young, M. (2010). Alternative educational futures for a knowledge society. *European Educational Research Journal, 9*(1), 1–12.

Young, M. (2011). Discussion to Part 3. In G. Butt (Ed.), *Geography, education and the future* (pp. 181–183). London: Continuum.

Young, M. (2013). Powerful knowledge: An analytically useful concept or just a 'sexy sounding term'? A response to John Beck's 'powerful knowledge, esoteric knowledge, curriculum knowledge'. *Cambridge Journal of Education, 43*(2), 195–198.

Young, M. (2014). Curriculum theory and the question of knowledge: A response to the six papers. *Journal of Curriculum Studies, 47*(6), 820–837.

Young, M., Lambert, D., Roberts, C., & Roberts, M. (2014). *Knowledge and the future school: Curriculum and social justice*. London: Bloomsbury.

Young, M., & Muller, J. (2010). Three educational scenarios for the future: Lessons for the sociology of knowledge. *European Journal of Education, 45*(1), 11–27.

Chapter 4
Geography Education Research Methods

4.1 Context

This chapter explores the influences on the researcher's choice of research methods in geography education. In doing so it inevitably touches on the types of research themes that tend to be chosen by geography education researchers, and on the scale and focus of their research. This brings us back to an over-riding consideration about how geography education brings together its two 'big ideas'—geography and education—and what this implies for the selection of suitable research questions and methods. Research in geography education arguably centres on how geography can best contribute to the education of young people, however this chapter recognizes that such research can be pulled in various directions by educational fashions and fads and by the launch of new educational policies and ideas. Indeed, the thoughts and actions of influential individuals in the field of geography education research may also help to determine what appears to be important for others to research, both nationally and internationally.

The consequences of the predominance of qualitative enquiries, and of their related methods and methodologies, in geography education research—not least the favouring of action research approaches among many practitioner researchers—are briefly explored. This leads to a discussion about the suitability of commonly applied research methods, with a consideration of their impact on the validity and reliability of research findings. The pertinent question of whether the entire range of research methods are realistically at the disposal of the majority of researchers in geography education—given issues of the scale and cost of their research (both in terms of time and money)—is reflected on. This chapter also gives space to a discussion about who geography education research is ultimately 'for'.

© Springer Nature Switzerland AG 2020

G. Butt, *Geography Education Research in the UK: Retrospect and Prospect*, International Perspectives on Geographical Education, https://doi.org/10.1007/978-3-030-25954-9_4

4.2   Introduction

Morgan (2015) provides us with a characteristically thought-provoking perspective on the current use of research methods in geography education. Specifically referencing three chapters written on 'Researching' in a text designed to support Masters-level research in geography education (Butt 2015), he takes issue with the popularity of what he refers to as 'Methodist' research. It is Morgan's contention that a considerable amount of discussion about research in geography education:

> is subject to an ideology which encourages an overly respectful and unnecessary attention to questions of method (p. 145)

He is forthright in his criticisms of 'methodism', pointing to a small industry of researchers and academics who produce research methods texts designed to feed the 'needs and insecurities of (mainly) practitioner researchers'. This criticism goes so far as to question the 'heady mix' of concepts, and terms, from philosophy and science—such as ontology and epistemology—that surround much educational research. These ideas are seductive in that they not only appear to offer the novice researcher ready conceptual and methodological solutions, but also render:

> our hunches, observations and interpretations into something more authoritative (p. 145)

Whilst some of Morgan's stated concerns are obviously somewhat 'tongue in cheek', perhaps even playful, there are important messages here that link back to the youth, insecurity and low status of much research in geography education (Kent 2000). As we have seen in Chap. 3, Furlong (2013) also rehearses the problems of education research in universities—with specific reference to the 'precarious positions' of many schools, departments and institutes of education, and of their academic staff. The insecurity of pre service teacher training, the fact that many education researchers only gain their doctorates relatively late in their academic careers—often having had careers in schools before entering academia, and having started their doctorates[1] years after most academics in other disciplines have completed theirs—are also concerns. Indeed, these 'conditions of service' may impact on the types of research, and research methods, that are favoured by geography education researchers.

The fact that much research in geography education is classroom-based and ostensibly geared to improving the quality of teaching and learning in the subject—with findings therefore predicated on evidence gathered in those classrooms—simply reinforces the apparent demand for research methods that are quick and easy to apply. Morgan's (2015) key concern is that this focus of 'Methodism' is corrosive:

> Methodism tends to encourage conformity and a narrow focus on the types of questions that can be asked and researched. This ensures that there are few philosophical, historical or sociological studies of geographical education. As such, much research is education research rather than educational research in that it seeks to 'problem-solve' rather than 'problem-pose'. (p. 146)

---

[1] Morgan also criticises the 'doctoral experience' of many education researchers, which he believes increase the problems of 'Methodism'.

The issues do not stop here. There are inevitable dangers that geography education researchers (indeed any subject-based researchers) unquestioningly 'carry across' methods of research from textbooks of generic research methods into their own context—without adequately thinking through whether there are uniquely *geographical* aspects of their research that need consideration. Consequently, there is a tendency for popular research methods to dominate research enquiries—with the inherent danger that geography education research becomes primarily focussed on the application of method and process, rather than on concepts within the disciplines of geography and education. Morgan (2015) refers to the 'invisible labour' that accompanies many research endeavours—where, ideally, the researcher reads not only the research already conducted by educationists in the relevant area of enquiry (not just geography educationists), considers related questions, thinks about geography's disciplinary contributions, and then honestly evaluates how their proposed research may add to the stock of knowledge generated through geography education research. In conclusion, Morgan, like Lambert (2015), questions whether the majority of research in geography education:

> is simply a specialist version of education research, or whether there are specific concerns and issues related to the nature of our field (p. 147)

4.3 Education Research Traditions

There are essentially two main traditions, indeed paradigms, of education research which dominate the approaches taken by most geography education researchers. These are the positivist/empiricist tradition and the interpretivist/relativist tradition—although an increasingly popular approach, which involves some combination of these, is commonly referred to as 'mixed methods' research (Onwuegbuzie and Teddlie 2003; Onwuegbuzie and Leech 2006; Teddlie and Tashakkori 2009). Sub categories exist within the two main paradigms, but these labels are certainly the most sensible ways to define the methodological framework for research in geography education, and indeed in any subject-based research context. Weeden (2015) correctly argues that choosing which research methods to adopt requires the researcher to have a knowledge of, and familiarity with, the place of theory and practice in research (see Chap. 8). The researcher also needs to have an appreciation of the intricate nature of social interactions in schools, and indeed elsewhere, to understand how difficult it is to make their 'truths' yield to research. When methods are discussed in geography education research what should perhaps be stressed more strongly are the ways in which the disciplinary roots of the subject may also influence research practice in education. It has been claimed that among researchers in geography education there is an attraction to undertake research:

> that has a close relationship to the issues concerning academic geographers and that employs research methods commonly used in mainstream geography (Williams 2003, p. 260)

Unfortunately, I can find relatively little evidence for this assertion and believe that for most researchers in geography education this link remains tentative, or non-existent. Furthermore, it is the nature of paradigms—in educational research or disciplinary research—to shift over time. Following appropriate academic scrutiny our previous ways of thinking, conceptualising and doing become inappropriate to our needs. Given that as geographers we are perhaps aware of the paradigm shifts that have affected our own discipline over the past half century, maybe we need to be more mindful of the related impact of these on our educational research methods. In this way there might be a case for arguing, as Williams (2003) does, that research methods in geography and geography education are closely aligned.

As Weeden (2015) reminds us, drawing on Robson's (2002) work, undertaking research in 'real world' contexts—such as in schools and classrooms, with human participants—can be daunting because of the inherent complexity of these situations. Here the multiplicity of types of social interactions inevitably make it difficult for the researcher to accurately describe and analyse social situations and experiences. Weeden identifies commonly used methods in geography education research as falling within the two main research traditions stated above (positivist/empiricist and interpretative/relativist), whilst also defending the 'mixed method' approaches adopted from a 'developed philosophical position' (p. 101). He concludes by arguing that researchers in geography education tend to adopt a relativist position: one which generates 'working hypotheses, rather than immutable empirical facts' (see Robson 2002, p. 25). The issue that mitigates against the employment of quantitative research methods—making them less popular with many geography education researchers, who as undergraduates were possibly also influenced by the impact of the quantitative revolution in geography—is that they have positivist roots. Quantitative methods imply that the researcher stands 'outside' the research they undertake—observing, measuring and investigating from a detached and impartial stance—objectively gathering quantifiable, numerical data with the intention of creating generalisations and possibly forming theories or laws. In the complex social contexts of educational research, particularly in classrooms, this is a position that is often untenable. Ultimately, whatever paradigm is chosen within which to select one's research methods, the researcher must be able to justify their position—by using knowledge of the philosophical roots and positioning of the research methods and methodology adopted (Clough and Nutbrown 2007). This also implies the adoption of a systematic, sceptical and ethical approach to research (Robson 2002, p. 18).

4.4   Being an Education Researcher

The role and approach of the practitioner researcher in school, compared with the academic researcher in university, is often very different. They are frequently motivated by different things. Furlong (2013) argues that being an educational researcher based in a university is a highly personal affair: driven by one's own ideas (at least initially) of what to research and the ways in which to undertake enquiries, often involving the investment of large amounts of professional and personal time. This, Furlong asserts, is reflected in the different approaches to training and development used to prepare academic researchers in education, compared with researchers in almost any other social science. In education, researchers are commonly encouraged to find their own research questions, to discover what sort of researcher they want to become, and indeed to clearly articulate their personal research context—customary questions in any doctoral examination. Education researchers also expect to detail the motivation for their research work and where they intend the results of this work to lead them next. In essence, they are asked *professional* questions linked to their research—as well as more standard questions about the research questions they ask, the methods they use, the theories they cherish and the outcomes they produce.

All well and good—but there is another side to 'being a researcher' which is strongly shaped in the modern 'enterprise university' by neo liberal forces and research assessment exercises such as (in the UK) the Research Excellence Framework (REF) and its forerunner, the Research Assessment Exercise (RAE). Furlong and Oancea's (2005) investigations of the types of research and researchers in UK universities recognises four different types of research identities in education faculties:

(i) research elites—staff who have either developed their academic careers within the faculty, or who have been attracted to the university as 'successful' academics in the run up to a research assessment exercise. These individuals might come from the national or international 'market' of researchers, being enticed by well-funded research opportunities, status and salaries often offered by prestigious universities. Interestingly, even these staff are under increasing pressure to perform 'for the institution' rather than 'for themselves'—being academically brilliant, and a talented and productive researcher, is in itself not enough if the institution does not see a financial return from one's research impact and visibility. Autonomy is therefore reduced, and strategic deliverability is expected. Some universities, with education faculties or departments that are predominantly 'teaching led', may struggle to appoint staff in this field.

(ii) career researchers—often on short term contracts (although some elite universities employ groups of researchers on permanent contracts, confident that they can continue to be employed on upcoming funded research projects) the opportunities for such staff have increased in universities that have done well in previous RAEs and REFs. These staff often work on projects where another, often senior, member of staff is the Principal Investigator (PI). This category might also include post-doctoral research fellows who are either partially, or fully, funded by the institution.

(iii) teaching staff—not usually entered into research assessment exercises, the status of these staff has often declined despite their utility to their institution as teachers, managers, leaders and administrators. In many institutions the work these staff do is responsible for the bulk of money that is earned by the education faculty. As a consequence of their more modest engagement with research, many staff in this category may have recently faced contractual changes, or redeployment. Staff who only have 'professional capital' in terms of their experience in schools, are often marginalised, untenured and part time.

(iv) teacher/researchers—numerically often the largest group, these staff have to attempt to juggle successfully both the primary demands of being a university teacher with those of being a published researcher. These traditional 'all-rounders', who (hopefully) thrive on a mix of teaching, research and even managerial work, are under increasing pressure in the university workplace. The expectation, in most universities, is that such staff steer their career paths towards taking on more research-oriented work, leading to the marginalisation of other previously legitimate and often prized forms of professional activity. This drive towards research performativity is now largely internalised, and indeed readily accepted, in most modern universities.

The research methods employed by each of these types of education researcher are often different. While it is not unfeasible that 'teacher/researchers' at some stage in their careers become engaged in large scale, externally funded research projects—usually as members of an extended research team, or possibly even as PI or Co-Investigators—this is less likely than for 'research elites' or 'career researchers'. This has implications for the types of research methods that teacher/researchers gain experience in using and for the size of the research projects they will typically become involved with: a situation which reinforces the rather more limited options for research that many who would describe themselves as 'geography education researchers', or indeed most subject-based researchers, will have in universities.

4.5   Establishing the Correct Steps

It is worthwhile considering the correct steps towards undertaking an education research project and the research methods that are usually associated with (small scale) projects in geography education. This enables us to subsequently consider issues concerning the selection of a qualitative, quantitative or mixed methods approach to enquiry. Planning research is not necessarily a neat linear process—research is messy and typically requires lots of 'forward and backward' steps—which Jones (2015) likens to a maze in which the proposed research plans will be 'revisited, reworked and refined' (p. 116). Citing the work of Crotty (1998), Jones states that there is:

> a hierarchical distinction between epistemology, theoretical perspective, methodology and methods where the researcher's philosophical stance is established by questioning their epistemological assumptions; these subsequently inform the selection of methodology and methods (p. 119)

Ideally, a proposal to undertake research forms a set sequence at whatever level and scale the research is undertaken. As Jones (2015) describes, this should start with considerations of epistemology and ontology, before deciding on methodology and methods (see Crotty 1998); however, new researchers often start from the other direction—driven by the research problem and the practicalities of 'doing research', rather than establishing a philosophical grounding from which to direct their enquiries. In a sense, one always has to start from the research problem and the research questions—but moving directly from these to favoured methods of data collection, without consideration of research paradigm or philosophy, is fraught with dangers. Similarly, the adoption of 'preferred methods', without contemplation of which methods and methodologies are best suited to the research theme, is also very dangerous: research questions tend to imply methodological stances, rather than vice versa. This is one of the key reasons why much practitioner research in geography education often fails to adhere to notions of best practice, therefore also failing to meet the quality benchmarks of rigour, originality and significance.

The correct steps towards ensuring high quality research are well established. These are, in effect, international—as the aims, issues, questions and often settings for each research enquiry make little difference to the procedure necessary to undertake research successfully.[2] The importance of starting with an identification of 'gaps' in knowledge, or the recognition of 'under researched' areas; the consideration of previous research endeavours (from the research questions posed, to the methods selected, to the resultant findings); the creation of clearly stated research aims and objectives, with associated research questions; a comprehensive literature review (covering both geography education and beyond)—followed by the research being undertaken; data gathering and analysis; and the agreement of findings, are all established steps. Within this deceptively simple, essentially linear, account of a research project there are inevitably hidden depths—the need to achieve conceptual, theoretical and methodological clarity when making decisions about the overall purpose of the research, leading to a 'fitness for purpose' in the overall research design, is essential (Jones 2015). There may also be a need to refer back to earlier steps taken if issues occur in the research process. The context in which the research is undertaken is also significant (Whitty 1997)—a significance that is often either ignored, or only given superficial consideration. This goes beyond a simple framing of the research project and the questions it pursues, for the intended audience for the research outcomes must also be kept in focus. The context may be political, philosophical, historical, locational or personal, and this should be acknowledged. Importantly the setting in which the research is undertaken can be used by others to question the choices made by the researcher. The bases for critique can be broad—as research can be censured if it does not include reference to recent political statements on education, to education policy debates, and to the outcomes of previous educational initiatives. There is also clearly a need for research to challenge existing orthodoxies, to question ideas that have become routinized, and to resist the temptation to succumb to fixed ideas about possible outcomes. It is acceptable for researchers to have a 'hunch' about what their research may find, but they

---

[2]Nonetheless, there may be very significant differences in the funding and staffing available to complete the research, the resources at the researcher's disposal, the existing support structures, and the means by which research findings can be disseminated.

must be open to, and excited about, the possibilities of finding things that do not simply confirm pre-existing views. Otherwise, why bother to do the research in the first place?

Choosing the most appropriate research questions to pursue encapsulates this whole process, and helps to determine the research methods used and the methodological frame in which these questions are set. Fundamentally, geography education researchers—like those in all other areas of research—need to spend time ensuring that they are asking the 'right' questions: 'what' and 'how' questions are open and consistent with exploratory and descriptive research; 'why' questions look for causation and explanation. Pring (2012) urges education researchers to achieve a much tighter appreciation of what they really mean when they use key words (such as 'learning' and 'skills') in the formulation of their research questions. He highlights the philosophical meaning of words, which must be understood if false assumptions are to be eradicated from the research process. The quality of research is therefore diminished if claims are made about answering research questions if these are not the ones that have actually been answered, but which occupy the same intellectual ground. This obviously affects whether legitimate claims can be made by the researcher to having created new knowledge. Reflexivity is important, with researcher(s) recognising their position in the research process and acknowledging their potential impact. It is generally expected that researchers should actively seek to reduce any possible bias in the research process, while also making consideration of any ethical issues,[3] as well as the values and beliefs of the researchers (see Wilson 2015). Importantly—and this may be a major consideration for much education research, not least in geography education—researchers need to be confident that their work is worthwhile (the 'So what?' question), that it can make legitimate knowledge claims, and that it stands up to external scrutiny from the wider research community. All of these considerations are captured in Jones (2015) diagram (see Fig. 4.1):

In conclusion:

> A good research proposal clearly conveys the author's philosophical, theoretical and methodological positions in justifying the conceptual connectivity of the research questions, methodology, methods and data (Jones 2015, p. 126).

4.6 Action Research

Action research, which is popular both among geography education and other subject-based researchers—particularly those with a practitioner background—is both pragmatic and practical. With its theoretical roots in Dewey's work, but achieving more practical realisation through the efforts of Lewin (1948), action research found its

---

[3]The British Education Research Association (BERA 2011) guidelines act, for British researchers in education, as a standard for the review of ethical matters before any research is undertaken.

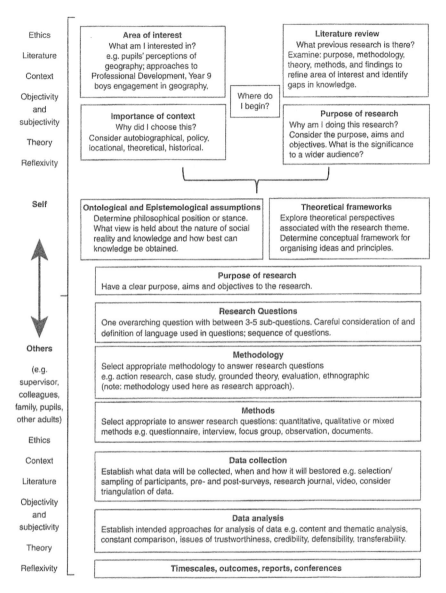

Fig. 4.1 Considerations when writing a research proposal. From Butt (2015), reprinted with permission of the publishers, Bloomsbury Academic, an in print of Bloomsbury Publishing Plc

first expression in an educational context through its association with Stenhouse's concept of the teacher researcher (Stenhouse 1975, 1981). This was the initial drive to establish practitioner enquiry within education. Elliott (1978, 1991), Carr and Kemmis (1986), Hillcoat (1996), Kwan and Lee (1994), Butt (2003), McKernan (1991), McNiff (1988) and Naish (1996) have further promoted the utility of action research in the contexts of schools and classrooms. The attractiveness of action research methodology, and its associated methods, is perhaps obvious—with its claims to solve real educational problems in situ it is obviously appealing to teacher-researchers, and others whose work brings them into the classroom. By its involvement with those closest to the issues researched—often striving to resolve problems by working collaboratively with others—the appeal to teacher researchers is perhaps obvious. Indeed, action research has made significant strides in removing the unhelpful divide between 'researcher' and 'practitioner', forging useful partnerships and encouraging 'practitioners who research' (Hammersley 1993). Action research usually involves a series of research cycles, which makes it open to a process of review and change with respect to research design, conceptualisation, questions pursued, implementation, intended outcomes and evaluation—this flexibility has an obvious appeal, although often the cycles are not completed or missed out as time becomes pressing to complete the research project. Action research has therefore been heavily criticised, even by some of the practitioner researchers who are keen to undertake it, with regard to its justification and validation of research methods. In conclusion, action research—because of its 'everyman' approach—has become popular among practitioner researchers in many subjects, but is often easily criticised: frequently with respect to its scale, limited expectations, shifting foci and poor standards of completion and execution. Its intention to find 'practical solutions to real problems' (Butt 2003, p. 274) has led to concerns about its lack of theoretical engagement; indeed, Gerber and Williams (2000) highlight its popularity within the international geography education research community because:

> there would be little interest among geography educators in undertaking research that is narrowly defined in terms of theory construction or the refinement of research methodologies (p. 212)

Bell (1993) provides an overview of different research paradigms, or 'frameworks', which helps to position action research within approaches to educational research that might be broadly described as scientific, interpretative or postmodern (Table 4.1).

When the range of research methods and techniques which fall within the boundaries of action research are explored, this produces a rather eclectic mix of possible methods to use. A further overview is provided in Table 4.2. This diversity is both a strength and a weakness of action research, where making a closer definition of more specific methodological approaches is almost impossible.

A major failing of action research methodology is its inability to regularly construct research outcomes from a solid base of what is already known. Much practitioner research has issues with 'over claiming' the originality of its findings and the significance of its results: precisely because the research process may focus

Table 4.1 An overview of different research paradigms

Framework	Aim	Methodology	Techniques
Scientific	To test relationships among variables, to understand interrelationships, to describe, explain and predict	Inductive or deductive, hypothesis formulation	Experimental, testing, pre-test/post-test formulae, hypothesis testing, observation and survey
Interpretative	To find meaning, to illuminate meaning in written and spoken accounts, past and present events and situations and interactions among people, to portray, to paint a picture	Anthropological, ethnographic, phenomenological, case study, context respecting	Observation, note taking, interviewing, (structured and unstructured), conversation, diary keeping, illuminative descriptions, thick descriptions
Action	To affect an improvement by action within a situation alone or with others	Critical stance, working within a situation	Acting, observing, refining and re-planning
Postmodern	To highlight the 'constructedness' or contingency of knowledge, to draw attention to the hidden agendas of knowledge claims	Working within a situation, case study, ethnographic, critical stance, self-reflexive, foregrounding or research subjectivity	Experiments with writing that blur the boundary between facts and fiction, textual analysis, collaborative research and collective authorship of research texts

After Bell (1993), with permission of the publisher, Open University Press

Table 4.2 An overview of action research methods

Observational/narrative methods	Non observational, survey and self-reporting methods	Discourse analysis and problem solving methods	Critical-reflective and evaluative methods
Participant observation, outsider observation, shadowing, case studies, diaries, journals, photos, video and audio recording, rating scales	Questionnaires, interviews, attitude scales, checklists, inventories	Content analysis, document analysis, episode analysis, brainstorming, group discussion	Triangulation, lesson profiles, student/teacher evaluation forms, critical trialling

After McKernan (1991), Elliott (1991), Naish (1996)

too closely on the interests of the researcher and participants, other important and equally valid perspectives may be ignored. In this way research can become blinkered, only recognising and valuing findings that coincide with the specific interests of the researcher. This has led some to argue that action research has more potential in terms of the professional development of teachers than in genuinely offering new, objective insights into research problems. These issues of partiality and lack of originality abound—often research enquiries are repeated, but in ways that do little to build new knowledge. With context specific frameworks, and problems concerning generalisability beyond the particular case, action research often has little impact and may face difficulties in stimulating future research. One wonders whether closer links to university education and the professional researchers engaged by research departments would help to iron out these issues. Hence practitioner enquiry, when detached from universities, will tend to remain modest and possibly promote findings of dubious wider relevance. As Butt (2003) concluded:

> Action research is not the solution to all our research problems in geography education. However, by enabling classroom-based practitioners to directly question and interpret their own educational situations, structures and ideologies it does offer hope beyond those methodologies which leave such aspects unquestioned or unresolved. Action research is primarily a vehicle for professional development, rather than a research process that can lead to generalisable results (p. 282)

4.7   The Quantitative-Qualitative Continuum

Even with the growing popularity of mixed methods approaches to research (see Sect. 4.9)—which essentially combine research methods from quantitative and qualitative approaches—implying that an unproblematic continuum exists between research methods may be risky. Some would argue that to combine methods places the researcher in danger of creating a methodological incoherence in their research. Quantitative methods tend to be less flexible, suggesting the possibility of larger scale studies and a set, pre-determined pathway for the research (often using questionnaires, survey and experiments in the hope of achieving greater objectivity). The promise is that the researcher may find generalisable results that show trends, patterns, correlations or connections. Qualitative research is often more flexible (within reasonable parameters), but lends itself to smaller scale, person centred research—which deals with the complexity, messiness and changeability of real-life situations more appropriately. It is best to have clear research plans, but some would question any overly strict pre-determination of research design and process (Maxwell 2013). Flexibility allows the researcher to justify possible changes to research design, method and approach, although liberality must be guarded against—for this can invalidate any claims to objectivity (Robson 2002). The attractiveness of qualitative approaches for geography education researchers is that they allow the researcher to (re) focus his or her efforts as new avenues for enquiry appear. A constructivist perspective may also allow theory building and the generation of novel research themes. This is often an iterative process. The *sine qua non* is that research methods must be determined by the research questions, purpose and context; it is wrong to chose the method and then tailor the questions to this (Grey 2004). There should, of course, always be a realisation that immutable, empirical 'facts' rarely, if ever, exist—particular in complex social situations such as those common in educational contexts.

4.8 Quantitative Research in Geography Education

It would be fair to claim that the tradition of quantitative research in geography education in the UK is relatively weak, compared to that witnessed across Europe and beyond. This owes something to the, often justifiable, scepticism with which researchers in the field have viewed the claims of quantitative researchers and research; to the traditions of preparation of researchers in geography education (from initial teacher preparation to doctoral research); and to the scale at which most research is undertaken. In the UK context the reluctance of researchers to engage with quantitative methods is not unusual, for subject-based research—indeed, the whole field of research into education has been dominated by qualitative research methods and methodology for many years. This has undoubtedly led to weaknesses in much research evidence—particularly research undertaken by practitioner researchers—as well as a tendency to 'over claim' on the basis of research data that might be highly selective, partial and subjective. Elsewhere the traditions of quantitative research in geography education are stronger: Schrettenbrunner (1993), for example, in response to the question of how to judge quality in research in geography education, high-lights the need for rigour and the correct application of theory, method and analysis predominantly in the context of empirical research. Here the focus is on hypothesis testing, correct choice of research instruments, standardised tests, statistical analysis, correlation and the possibilities of applying quantitative methods and then broadcasting results widely. Importantly, he notes the need for geography education research to be more outward looking—to strive to 'be of relevance to other subjects and not just to geography alone' (p. 74). Schrettenbrunner and van Westrhenen (1992) explore these issues more widely in their exploration of the links between empirical research and geography teaching.

4.9 Sources of Information

Lidstone and Williams (2006a, b) claim that researchers in geography education working in the school context have five principal sources of information: previous research studies; valid and reliable empirical data collected through systematic and rigorous research; anecdotal accounts of policy makers, students and teachers; official statements of governmental and quasi-governmental agencies; and curriculum publications such as textbooks, atlases and internet websites. They note that such information varies in quality from study to study, and from nation to nation—partly dependant on the strength of traditions of geography education research in the particular jurisdiction. The subjectivity of both the source and quality of information is acknowledged:

> Researchers may in some states be at liberty to choose their research topics and have relatively free access to the persons, institutions and activities they wish to study. Even where there is a high level of free choice, research will be strongly influenced by the research conventions pertaining in their local circumstances. It will also be strongly influenced by access to the media for publication everywhere there are editorial gatekeepers who are powerful in determining what is published in books or research journals. Where research is funded, funding bodies will set limits on the research that can be undertaken and may influence directly the findings that enter the public domain and the form of any presentation (Lidstone and Williams 2006b, p. 7)

Researchers are also under the 'control', or at least the influence, of the institutions that employ them—with many universities and schools having directive powers about what they value in terms of research activity and output. This links directly to what the individual researcher is encouraged, or indeed permitted, to work on. Restrictions may be ideological, or imposed by research structures (such as the institutions' established research groups or clusters), or both. In many countries such restrictions may filter down to the research questions and methods used, often influenced by research assessment exercises—which suggest certain approaches to research are more favourable than others in terms of research outputs and publications. For example, it is strongly suggested in England that 'action research' (indeed classroom-based, practitioner driven research) is less valuable than more objective, empirically rooted work.

Jones (2015) asserts that there are three main methodological approaches to research: analysis of documents; observing; and questioning. The first includes written materials (policies, prospectuses, plans), visual images (photographs, animations, videos), and electronic images (blogs, twitter, emails); the second, includes participant and non-participant observation (which may be structured or semi structured, and include a variety of mediums such as diaries, reflective journals, checklists, recordings, etc.); and the third includes surveys, focus groups, interviews and questionnaires. While not wishing to explore any of these too deeply—this chapter is not designed to do so—it is worth acknowledging that written materials, in whatever form, are produced for particular audiences and purposes and therefore may be 'situated', potentially revealing political, social and historical influences to the critical eye. Observations can provide 'thick descriptions' of social situations, but the researcher must guard against ethical and power issues. All researchers need to be mindful of the practical considerations of cost, access, scale, time and availability—all of which can unhelpfully steer the research methods and methodologies in directions that are not implied by the original research questions (Jones 2015). Honesty is necessary, with the researcher pointing out the benefits and dangers of the methods chosen.

4.10 Mixed and Multiple Methods Approach

It is undeniable that the use of mixed and multiple methods approaches in educational research has become increasingly popular (Jonson and Onwuegbuzie 2004; Onwuegbuzie and Teddlie 2003; Onwuegbuzie and Leech 2006; Teddlie and Tashakkori 2009). They have a strong appeal to practitioner researchers in that mixed methods can offer ready solutions to data collection problems, while tending to skirt over issues of applicability and relevance. Hence some major methodological difficulties can arise if a mixed methods approach is adopted without careful consideration of its limitations. Jones (2015), referring to the combination of qualitative and quantitative methods in much geography education research, states:

> they can appeal to teacher-researchers as a more practice-driven and pragmatic option than situating research within a quantitative or qualitative paradigm. However, adopting a mixed research approach does not guarantee fitness for purpose if the design itself lacks rigour and is fragmented and inconsistent ... justifying both why and when a quantitative or qualitative component has priority is an important requirement in mixed method research (p. 122)

However, too much flexibility in the research process can affect the quality of the data produced and therefore the applicability of the conclusions gathered. It is worth remembering that in educational research the use of mixed methods was not widely believed to be conceivable a few decades ago, and its application today still eschews some major methodological and epistemological issues. Essentially, mixed methods research is here to stay because of the utility it offers—we will see more of it, perhaps as inter and cross disciplinary work grows. Interestingly, Symonds and Gorard (2008) are less concerned with the mixing of methods, but have issues about whether methods are applied appropriately, and if the resultant data is of high quality and correctly analysed. The point about analysis is well made: data is not knowledge, it requires correct analysis by the researcher (Morrison 2007).

4.11 Conclusions

Yates (2004) notes that the quality of educational research suffers when researchers use a wide variety of methodological approaches and research methods (for example, from action research to Randomised Controlled Trials). She also questions the epistemological background and theoretical stance of much research, recognising the drift towards more atheoretical research approaches that were initially encouraged by postmodernism. The range of purposes of research is significant in that policy-driven research and applied research is very different in form, expectations and outcomes to that which is 'blue skies', or practitioner based. This is a concern to governments and funders, where the plethora of research approaches and outcomes become difficult to control. Education research is increasingly multi or inter disciplinary—creating methodological, epistemological, conceptual and theoretical complexities for both the creators and users of research evidence. Interdisciplinary research in education covers different sectors (for example, from early years to continuing education), theoretical frameworks (from applied linguistics to psychology), areas (comparative education to adult learners) and related methodologies (from quantitative to qualitative to mixed methods). Unfortunately, as Furlong (2013) points out, although this diversity and breadth might be seen as richness, it is increasingly viewed by those outside education circles as a muddle!

Lidstone and Williams (2006b) note that slipping between 'intellectualism, to partisanship, to action' (p. 10) is a feature of much educational research, but something that is rarely acknowledged. Essentially, this refers to the need for researchers to remain objective despite their positioning as geographers, environmentalists, activists, political representatives, and members of professional associations. Such allegiances are rarely stated, but may have a profound effect on what is researched, how it is researched, and to what end. This may be recognised when describing the *objects* of our research—for example, by recognising the age, sex, social background, grouping, ethnicity, ability and disability of the students we research—but is rarely stated about ourselves (for example, as say, white, male, middle class, middle aged, professional researchers). These characteristics are important, but are often taken for granted. Similarly, the concerns of geography education researchers working in the more economically developed, English speaking, world may not be the same as those researching elsewhere.

Who we are, where we come from, what experiences we have, and how we are perceived as researchers in geography education, are all significant. The institutional career route taken by the majority of 'subject-based' researchers in higher education institutions in the UK is a major influence on the research they produce. Most academics who would identify themselves as geography education researchers do not work in research institutes, government or non- government organisations, or even in professional associations: they work in higher education institutions and often have considerable experience of initial teacher education in their subject. These researchers, myself included, mostly started as geography teachers in schools (or primary geography specialists). For Lidstone and Williams (2006) this can be problematic:

> Recruitment on the basis of school teaching experience rather than advanced academic research training in the educational disciplines hinders authoritative research in geographical education (p. 10)

These factors define who geography education researchers are, what they do, and how they are perceived. They also tend to dictate the methods researchers will use and the questions they will pursue—determining the audiences they write for, and the attention taken of their work. Indeed, Williams (1998) argues correctly that the characteristics of geography education researchers define the impact of their research on the wider community of researchers, policy makers and other stakeholders in education. If researchers in geography education chose to position their work narrowly in the classrooms where geography teaching occurs, without concern for larger education questions and communities, then they will diminish the impact of their work. Research that is located immovably in classroom-based professional practice in geography—rather than research which speaks to broader theoretical and conceptual concerns in education (through geography)—will inevitably be less well regarded by the wider research community. Ironically, the very community that geography education researchers tend to write 'for' is the one that tends to have relatively little regard for the outcomes of such research:

> It is difficult to find evidence of research that impacts directly on classroom practice … there appears to be no great enthusiasm among schoolteachers to subscribe to and read journals and books concerned with research in geographical education (Williams 1996, pp. 12–13)

However, the inherent dangers of encouraging too many inward looking research projects are all too apparent. This brings us back again to the question of who geography education research is *for*—the answer, inevitably, is that there are potentially a large number of different audiences for such research. But here is where we need to be cautious: if research is carried out by a teacher practitioner to help solve a problem they are facing in the geography classroom the audience may be very small indeed— just the teacher, and his or her students. However, this audience may get bigger if other members of the geography department, the faculty, the senior leadership team or indeed other schools get to hear about, and value, the results. Further still, the findings may be of value to a wider professional community of geography teachers and geography education researchers who may seek to replicate and extend the research questions—with new data being disseminated through professional or academic journal articles, or through social media. Ultimately, if the research findings are believed to have wider purchase, they may be shared with other subject-based researchers, or within the broader field of education. We have moved from research audiences that are small scale, almost parochial, to those that may even be international in reach. And here is the dilemma. With much research in geography education conducted at the 'start' of this continuum—often of apparent relevance only to a small number of people—the audience tends to be encapsulated within an immediate community of practice. This is the very definition of inward looking, perhaps even self-serving, research: if the products of geography education researchers are to achieve real status this audience needs to be much wider.

References

Bell, J. (1993). *Doing your research project*. Maidenhead: Open University Press.

British Educational Research Association. (2011). *Ethical guidelines for educational research*. London: BERA.

Burke Johnson, R., & Onwuegbuzie, A. (2004). Mixed methods research: A research paradigm whose time has come. *Educational Researcher, 33*(7), 14–26.

Butt, G. (2003). Geography teachers as action researchers. In R. Gerber (Ed.), *International handbook on geographical education* (pp. 271–284). Dordrecht: Kluwer.

Butt, G. (Ed.). (2015). *Masterclass in geography education*. London: Bloomsbury.

Carr, W., & Kemmis, S. (1986). *Becoming critical: Education, knowledge and action research*. London: Falmer.

Clough, P., & Nutbrown, C. (2007). *A student's guide to methodology* (2nd ed.). London: Sage.

Crotty, M. (1998). *The foundations of social science research: Meaning and perspective in the research process*. St. Leonards, NSW: Allen and Unwin.

Elliott, J. (1978). What is action research in schools? *Journal of Curriculum Studies, 10*(4), 355–357.

Elliott, J. (1991). *Action research for educational change*. Milton Keynes: Open University Press.

Furlong, J. (2013). *Education: An anatomy of the discipline*. London: Routledge.

Furlong, J., & Oancea, A. (2005). *Assessing quality in applied and practice-based educational research*. ESRC TLRP seminar series. (December 1) Oxford University Department of Educational Studies.

Gerber, R., & Williams, M. (2000). Overview and international perspectives. In A. Kent (Ed.), *Reflective practice in geography teaching*. London: Paul Chapman.

Grey, D. (2004). *Doing research in the real world*. London: Sage.

Hammersley, M. (1993). On the teacher as researcher. *Educational Action Research, 1*(3), 425–445.

Hillcoat, J. (1996). Action research. In M. Williams (Ed.), *Understanding geographical and environmental education. The role of research* (pp. 150–161). London: Cassells.

Jones, M. (2015). Writing a research proposal. In G. Butt (Ed.), *Masterclass in geography education* (pp. 113–127). London: Bloomsbury.

Kent, A. (Ed.). (2000). *Reflective practice in geography teaching*. London: Paul Chapman Publishing.

Kwan, T., & Lee, J. (1994). A reflective report on an action research towards understandings of action research held by geography teachers. In H. Haubrich (Ed.), *Europe and the world in geography education* (pp. 387–406). Nurnberg: IGU.

Lambert, D. (2015). Research in geography education. In G. Butt (Ed.), *MasterClass in geography education* (pp. 15–30). London: Bloomsbury.

Lambert, D., & Jones, M. (Eds.). (2017). *Debates in geography education* (2nd edn). London: Routledge.

Lewin, K. (1948). *Resolving social conflicts: Selected papers on group dynamics*. New York, NY: Harper & Row.

Lidstone, J., & Williams, M. (Eds.), (2006). *Geographical education in a changing world: Past experience, current trends and future challenges*. Dordrecht: Springer.

Lidstone, J., & Williams, M. (2006a). Researching change and changing research in geographical education. In J. Lidstone & M. Williams (Eds.), *Geographical education in a changing world* (pp. 1–17). Dordrecht: Springer.

Lidstone, J., & Williams, M. (2006b). *Geographical education in a changing world*. Dordrecht: Springer.

Maxwell, J. (2013). *Qualitative research design: An interactive approach* (3rd ed.). London: Sage.

McKernan, J. (1991). *Curriculum action research: A handbook of methods and resources for the reflective practitioner*. London: Kogan Page.

McNiff, J. (1988). *Action research. Principles and practice*. London: Macmillan.

Morgan, J. (2015). Discussion to part III. In G. Butt (Ed.), *Masterclass in geography education* (pp. 145–147). London: Bloomsbury.

Morrison, T. (2007). Emotional intelligence, emotion and social work: Context, characteristics, complications and contribution. *British Journal of Social Work, 37*(2), 245–263.

Naish, M. (1996). Action research for a new professionalism in geography education. In A. Kent, D. Lambert, M. Naish, & F. Slater (Eds.), *Geography in education: Viewpoints on teaching and learning* (pp. 321–343). Cambridge, MA: Cambridge University Press.

Onwuegbuzie, A., & Leech, N. (2006). Linking research questions to mixed methods data analysis procedures. *The Qualitative Report, 11*, 3.

Onwuegbuzie, A., & Teddlie, C. (2003). A framework for analyzing data in mixed methods research. In A. Tashakkori & C. Teddlie (Eds.), *Handbook of mixed methods in social and behavioral research*. Thousand Oaks, CA: Sage.

Pring, R. (2012). Importance of philosophy in the conduct of educational research. *Journal of International and Comparative Education, 1*(1), 23–30.

Robson, C. (2002). *Real world research: A resource for social scientists and practitioner-researchers*. Oxford: Blackwell.

Schrettenbrunner, H. (1993). How to judge quality in research. *International Research in Geographical and Environmental Education, 2*(1), 73–75.

Schrettenbrunner, H., & van Westrhenen, J. (Eds.). (1992). Empirical research and geography teaching. *Nederlandse Geografische Studies, 142*.

Stenhouse, L. (1975). *Introduction to curriculum research and development*. London: Heinemann.

Stenhouse, L. (1981). What counts as research? *British Journal of Educational Studies, 29*(2), 103–114.

Symonds, J., & Gorard, S. (2008). The death of mixed methods: Research labels and their casualties. The British Educational Research Association Annual Conference, Heriot Watt University, Edinburgh, September 3–6.

Teddlie, C., & Tashakkori, A. (2009). *Foundations of mixed methods research*. London: Sage.

Weeden, P. (2015). Approaches to research in geography education. In G. Butt (Ed.), *Masterclass in geography education* (pp. 101–111). London: Bloomsbury.

Whitty, G. (1997). Creating Quasi-Markets in education: A review of recent research on parental choice and school autonomy in three countries. *Review of research in education., 22*(1), 3–47.

Williams, M. (Ed.). (1996). Positivism and the quantitative tradition. In M.Williams (Ed.), *Understanding geographical education: The role of research* (pp. 6–13). London: Cassell.

Williams, M. (1998). A review of research in geographical education. In A. Kent (Ed.) *Issues for research in geographical education. Research forum 1* (pp. 1–10). London: Institute of Education Press.

Williams, M. (2003). Research in geographical education: The search for impact. In R. Gerber (Ed.) *International handbook on geographical education* (pp. 259–272). XX Kluwer.

Wilson, M. (2015). Ethical considerations. In G. Butt (Ed.), *Masterclass in Geography Education* (pp. 129–144). London: Bloomsbury.

Yates, L. (2004). *What does good educational research look like? Situating a field and its practices*. Maidenhead: Open University Press.

Chapter 5
The Policy Context—Government Perspectives on Geography Education Research

5.1 Context

Criticism of the quality and relevance of educational research, both in the UK and internationally, has been marked since the late 1990s (see, for example, Hargreaves 1996; Hillage et al. 1998; Tooley and Darby 1998; Maclure 2003; Whitty 2005; Thomas 2007). Much research in the field of education has been disparaged, not only for its apparent lack of rigour and culmination, but also for its theoretical incoherence and ideological bias. It has been judged—and found wanting—with regard to its seeming irrelevance, poor user engagement, questionable quality and generally low value for money (Whitty 2005). The reaction of the education research community in the UK—including its flagship organisation, the British Education Research Association (BERA)—has been to deliberate on related issues, such as the expected priorities for educational research, how users of research might be involved in reviewing research aims, the extent to which research should lead to 'applied outcomes', and the overall relevance of educational research (see Rudduck and McIntyre 1998). In the field of geography education research, Roberts (2000) asserts that most researchers accept that there should be a link between educational research on the one hand and the 'policy and practice' of education on the other—although this conceptualisation can shift when the direct application of research findings to teaching is considered.

Dissatisfaction with educational research, and researchers, has not been without periods of relief. The New Labour government in the UK, which came to power in 1997, increased funding for educational research substantially up to the end of its office in 2010. Indeed, it doubled research funding in this period—setting up the National Educational Research Forum (1999–2006), presiding over the already established research assessment exercise (RAE) and new research excellence framework (REF), and establishing the Evidence for Policy and Practice Information and Co-ordinating Centre (EPPI-Centre)—the aim of which was to monitor and review educational research. Although it could not be claimed that the periodic assessment

© Springer Nature Switzerland AG 2020
G. Butt, *Geography Education Research in the UK: Retrospect and Prospect*, International Perspectives on Geographical Education, https://doi.org/10.1007/978-3-030-25954-9_5

of research outputs is universally loved by academics, the process has at least driven up the quality of much research in education in the UK and has been mirrored elsewhere globally (see Chap. 6). However, UK governments, even when supporting the creation of evidence-based policies in education (Sebba 2004), have regularly expressed their unease about the quality of research findings (Furlong and Oancea 2005). Government policy on education research has therefore tended to favour the generation of 'what works' solutions: research that has a practical, problem-solving focus targeted at directly improving school performance, raising standards, increasing effectiveness and enhancing performativity. As we have seen, this reflects a somewhat formulaic view of the process of recognizing, researching, analysing and 'solving' educational problems—often presupposing that 'off the shelf' solutions can easily be applied in a variety of different educational contexts and settings. The dangers of this type of thinking—for resolving complex, real world, problems by using reductionist approaches which oversimplify essential analysis—are palpable (Bassey 1981, 2001).

This chapter rehearses how education policy makers, in many countries, tend to prefer (and therefore fund) research which generates large scale survey data—which is reduced easily into 'action points' that fit neatly into policies and strategies. For obvious reasons governments will favour predictive research—which creates and utilizes generalisations—prizing it for its promise of potentially asserting control over the problems they are trying to resolve (Usher 1996). Nonetheless, by contrast, the most valuable educational research tends to be cautious, incremental and inherently slow-paced in its formulation and delivery—anathema to government expectations of researchers, who they seem to believe should be able to deliver urgent, innovative, 'what works' solutions which can simply be 'rolled out' to schools. The large-scale investment in educational research in the UK made by New Labour in the late 1990s, following criticisms about the purpose and outcomes of research, was therefore largely intended to fund 'what works' enquiries—to discover, and influence, factors that change teachers' practice, and to gather empirical research evidence that could underpin policy change. Linked to this research evidence was the prospect of (re)creating official teacher standards—which when established would be instrumental in influencing the formation of policy guidelines for teachers, to use as a source of information on which to audit, or inspect, educational practice.

We have seen that the impact of 'what works' research can be negative (Chap. 3). As Williams (2003) observes:

> Those who are drawn to immediate pedagogical problem solving may move from one research issue to another and from one research method to another over relatively short periods. What is important is the generally low level of interest in theory construction amongst researchers in geographical education (p. 261)

This is an important point: theory is seen as 'getting in the way' of research enquiry, rather than supporting it through helping to provide firm intellectual and philosophical foundations for action. The time and effort used in grappling with theory is seen as wasteful, both by governments and indeed some researchers, such that it is only paid lip service rather than regarded as integral to a thorough enquiry.

Williams also warns that practitioner-based research is 'commonly ignored by policy makers', often being too small scale, poorly focused and incapable of generating highly prized generalisations (Williams 2003, p. 261). Additionally, Furlong (2013) recognises the contextual importance of taking account of what teacher educators and researchers:

> actually do to advance their discipline, how they attempt to work within government frameworks, while at the same time realising their own vision of educational knowledge (p. 13)

I have acknowledged the rarity of large, well-funded, multi methods, international research in geography education—alongside the relative dearth of sophisticated research designs. Geography education researchers only experience such research frameworks if they work on large, often inter disciplinary, research projects with other researchers from *outside* the field of geography education. Much geography education research is therefore typified as personal, or small group, research, based at the classroom scale. This is not just an issue faced by researchers in geography education. Indeed, Furlong and Lawn (2010), reflecting on the state of subject-based education research in the UK, comment:

> only mathematics education and science education survive as strong curriculum-based areas of enquiry (p. 30)

In conclusion, it is important to remind ourselves that there is a significant area of research in geography education that is focussed on the policy context, rather than simply being *affected* by policy (Rawling 1991, 2001; Butt 1997; Tapsfield et al. 2015). Research into the ways in which government policy has impacted on geography education provides us with a means of understanding how we arrived at 'where we are now', in terms of the form and function of geography curricula, examination reforms, and revisions to syllabuses and specifications in national settings. Such research has been particularly popular during times of government-led reforms of curricula and assessment regimes—such as the formulation and revision of the national curriculum, the revision of Advanced level ('A' level) and General Certificate of Secondary Education (GCSE) standards, and the creation of the Educational Baccalaureate (EBacc) in England. Analysing how teaching and learning in geography education have changed over time is fascinating, but we must remind ourselves that these changes are often very heavily influenced by policy shifts. The past acts as an important landscape within which we can understand and question the present, and project into the future—it shows us whether we have achieved 'good practice' in education, and helps us to counter whatever is preventing this desired state of affairs.

5.2 Introduction

The close involvement of politicians and policy makers in educational matters has habitually appeared to educationists to be, at best, unhelpful. Lambert and Jones

(2013) comment on previous policy debates in education, in rather exasperated tones, as follows:

> politicians and ministers in particular appear compelled to 'meddle' and change things around, as if to keep us on our toes (p. 2)

But the challenges facing policy makers are real and profound in post-industrial nations, which all experience the pressures of globalisation: how, with restricted resource, does the state intervene in education to keep the nation competitive, internationally? how can education effectively respond to the challenges of a neo liberal, globalised world? how can the attainment and performance of young people be raised annually through education practice? how do we shift pedagogy to take account of our information dominated age? and what should we teach children about living in a world threatened by significant sustainability and environmental concerns? (Butt 2017a, b; Lambert and Jones 2013; Morgan 2011). One might argue that these questions reflect rather narrow, neoliberal and neoconservative agendas,[1] but they are nonetheless the sorts of questions that have driven debate in education, not least at policy level, in recent years. We must not forget that governments have a responsibility to resolve such issues.

By the end of the first decade of the 21st century it was apparent that the technical competence of teachers was being valued by government above all else—a point stated partly as a provocation in the Geographical Association's (GA's) manifesto *A Different View* (GA 2009). As such, the delivery of excellent examination performance by students was prized, undermining any notion of the importance of curriculum thinking and making—as a consequence the distinction between pedagogy and curriculum had become blurred, with discussion about skills and competences dominating any discourse about change and innovation in schools. This emphasis within policy thinking was certainly seen elsewhere, at the expense of any sustained focus on the importance of subject development and 'curriculum-making'. Deliberation about, and importantly research into, the curriculum was therefore largely absent. The policy agenda was clear—state resources were to be spent on improving student experiences and on the 'personalization' of teaching approaches, on prioritising 'learning to learn,' and on initiatives such as 'building learning power'—not least as an indicator of future performance against Programme for International Student Assessment (PISA) tests—rather than on debating the content, structure and construction of the curriculum. Such concerns were raised in one of the GA's periodic consultations with its members about revisions to the geography national curriculum. One question asked geography educators to consider the importance of dissociating curriculum issues (most notably 'what should be taught?') from those of pedagogy, while also seeking to define more clearly geography's goals and purposes. It was recognised in the survey that trying to 'nail down' the precise content, or even goals, of geography education was a largely Sisyphean task, and therefore not an entirely

---

[1]It is, of course, possible to configure such questions differently, or to shift entirely the focus for educational debate. For example, towards considerations of the extent to which education is still a public good; whether it is adequately serving the well-being of young people; whether education is delivering happiness, etc.

sensible objective to pursue. The previous experiences of official curriculum making, not least the creation of the first iteration of the geography national curriculum (DES 1991), served as a constant reminder of the difficulties of prescribing curriculum solutions that were based on limited research evidence (Butt 1997). Many believed that geography, maybe more than other subjects, still needed to justify its place in the curriculum—arguably rightly so: as it has no 'divine right' to be represented, despite what some geographers may think or believe. Those who fear that geography might easily be removed from the school curriculum, regularly press the geography education community to advance a case about the subject's continued relevance and the powerful nature of the knowledge that it brings. At different times, under different political regimes, these arguments have shifted in emphasis—sometimes rather uncomfortably—between justifications linked to vocational education, citizenship, life skills, and sustainable development (Butt 1997).

To an extent more recent education policy changes in England—not least those heralded by *The Importance of Teaching* (DfE 2010), which symbolised a radical shift in political thinking following the defeat of New Labour in 2010—ostensibly counter some of the concerns stated above. Legislation brought about a change of emphasis in educational practice heralding a switch from skills, competences and personalisation, towards traditional subject knowledge ('core' and 'essential') and promoting the neoconservative interest in motivation, well-being and the application of intellectual effort (McCulloch 2017).

5.3 Changes to the University Sector

The late 20th century saw the massification of university education, in terms of its growth in both student numbers and in the ways in which teaching was funded. Previous governments had largely supported the principle of state funding of universities in the UK—this changed following the Browne Report (2010) to a system whereby the 'consumer' would now pay directly for his, or her, education. Government intentions, which led to interventions, were to reduce the proportion of state funding for universities by over two thirds early in the 21st century. Government actions followed neo-liberal logic: which affected not only the ways in which universities would in future be governed, managed, administered and funded, but also the distribution of funding for teaching and research. For the latter, funding mechanisms would be underpinned by a 'quality research' (QR) allocation which would increasingly be driven by a competitive system of periodic assessment of excellence of research outputs and impact. The scale of this allocation of research money did not rise proportionately with increasing student numbers, in fact it did the reverse. This resulted in the most enterprising universities in the UK strongly pursuing new sources of research and teaching money, which were deemed to be within their capabilities to secure—often involving a prioritisation of scientific research, and research that aligned itself to the most stable funding sources (whether they be from research councils, charities, professional associations, industries, or business). The effects on educational research funding have not, in the main, been positive. In geography education, where research is often practitioner-led, research methods and interests have not dovetailed with those often witnessed in applied educational research—the type of research which is now favoured as a consequence of recent changes.

Universities have been described as institutions that have yielded to the expectations of 'new public management' (Hood 1995). Their organisational structures and administrative arrangements aping those of private sector companies and corporate business—particularly with respect to promoting internal competition, centralisation, profit maximization, the use of cost centres, and favouring hierarchical and bureaucratic management systems. A consequence of adopting this management model has been a reduction in trust and collegiality among staff, alongside an increase in competition and the need for staff to perform against measurable standards and key performance indicators (KPIs). This has changed the shape and governance of universities—which have never been fully part of the public sector, but which have now become more accountable to central government. Internal competition, resource allocation mechanisms (RAM) and reduced autonomy with regard to departmental finance are all now features of the modern UK university. The rise in the importance of the Research Excellence Framework (REF), and now the Teaching Excellence Framework (TEF), are key outcomes of such a philosophy—as universities comply with publicly stated, measurable, standards for research and teaching performance and forgo much of their previous professional independence and autonomy. Where performance gains are attributable to entrepreneurialism in the sector this is perhaps no bad thing: the important question is whether this spirit of competitiveness is necessarily always beneficial in driving forward what universities should be *for*. Furlong (2013) believes that the new public management of universities essentially avoids, or side steps, any question of what they are really intended to *do*, beyond rather simplistic utilitarian notions of the benefits they bring to government and industry through their applied research. Reflecting on the momentous changes to the UK university sector over the past three decades, Furlong concludes that it has:

> massively increased in size, lost its privileged and secure funding, been forced into developing an entrepreneurial approach to securing students, to securing research and to developing new third-stream forms of funding. It has had to diversify itself, moving into more and increasingly different markets and it has had to develop new and very different forms for managing itself internally; and all of this when, intellectually, it has lost its confidence about the value of objective knowledge (p. 118)

Arguably, as universities are now without a clear purpose, this has meant that governments have found it increasingly simple to control what they do. In education faculties change has led to diversification, or division, of staff—with those who have previously seen themselves as being driven by their teaching asked to increase their research output or, if they cannot deliver on this expectation, to change their contracts to 'teaching only', or part time, status. This has inevitably had a major impact on new staff and early career researchers; there is a sense of a ladder being 'pulled up' such that support for the development of academic staff becomes limited, with tenured staff who already have a research profile being expected to do more in terms of research productivity or teaching/supervision on higher degree programmes. Education, as a discipline, remains 'big business' for British universities. Higher Education Statistics Agency (HESA) data collected over the past decade reveals that around 10% of part and full-time students in higher education in the UK have been studying at least some element of 'education'. On this evidence, across the higher education sector, education is therefore currently the second largest discipline in the academy.

5.3.1 Reform to Initial Teacher Education (ITE) and Its Effects on Geography Education Research

There are a number of ways in which policy shifts can have both predictable and unpredictable effects on practice. Over the last 25 years initial teacher education in England has been subjected to significant reforms, aimed at shifting the locus of teacher preparation away from universities and into classrooms. Since the mid-2000s the designation of university-based courses as either professional courses, or post graduate courses bearing Masters credits, created a somewhat unexpected upturn in the numbers of beginning teachers who were required to undertake some form of research as part of their preparation for the classroom. The UK's New Labour governments (1997–2010) also established the groundwork for a Masters' degree in Teaching and Learning, with the intention that teachers would eventually form a 'Masters only' workforce (DCSF 2007)—similar to the professional expectations of teachers in some Scandinavian countries. Although these ambitious plans were ultimately not fulfilled, the numbers of teachers studying for awards and certificates that carried Masters credits increased in the UK. Whether their research was for small assessed projects, or for a Masters dissertation, the impact was similar, in that teachers were now expected to establish and pursue research enquiries. Although we know that such practitioner-led research has little impact on policy and practice (see Chap. 1), engagement with research is an important part of preparation for teaching and central to establishing the profession as being, in part, research-based or informed. Government rhetoric shifts regularly on this point—with eyes on the Finnish model of teacher preparation (but more importantly on their national levels of student attainment revealed in PISA assessments), UK governments have previously flirted with the idea of making new teachers part of a more highly qualified profession. This welcome expectation has now, sadly, been downgraded—indeed, almost reversed, as a 'craft' model of teacher preparation has taken centre stage in England.

The side-lining of 'traditional' partnership models of teacher preparation—which involved higher education institutions being closely linked with a range of partnership schools—has both reduced the amount of practitioner research in schools and, through loss of university courses such as the Post Graduate Certificate of Education (PGCE) with its attendant academic staff, decreased the amount of academic research into geography education. The 'anti-intellectualism' that underlies this decision has been decried (see Lambert and Jones 2013), as has the:

contrast with policies of other countries, including notably Finland and Singapore – often referred to in international comparisons – that put great emphasis on the professional training of teachers (Roberts 2015, p. 47)

There has always been a mismatch between the geographical themes and content that most beginning teachers have experienced in the academy and which they teach in schools. Graves (1972) noted this in a characteristically clear manner almost 50 years ago, when he reflected on his own progression from geography graduate to seasoned classroom practitioner, commenting 'we became no doubt more proficient practitioners in the classroom, but what we taught bore less and less resemblance to what current university geographers were doing' (Graves 1972, p. 10 cited in Rawding 2013). The degree to which this mismatch occurs tends to fluctuate— during the 1990s school subjects, including geography, experienced what Morgan and Lambert (2011) refer to as a policy-related 'pedagogic turn'; this took teachers' focus away from their subject and towards a greater concentration on the skills of teaching. The emphasis in education policy during this period was less on ground-breaking shifts in geographical content in the academy, and how these might be brought into the classroom, and more on how teaching skills could be improved— almost irrespective of any concerns about subject content. More recently, during the Conservative—Liberal Democrat Coalition government (2010–2015), and later Conservative administrations (2015 to date), we have seen a welcome shift towards the promotion of core knowledge and content.

Furlong (2013) asks why the study of teacher education, which is in many ways dominant in defining the discipline, has traditionally been so politically 'charged'. He concludes that this is because:

> behind debates about the form and content of educational knowledge in teacher education lie debates about teachers, what it is they know and what it is they do. In short, they are debates about the nature of teacher professionalism. (p. 11)

In the UK, governments have always sought to manage teacher supply through its direct involvement in ITE, and have also tried to secure control of teacher professionalism and professional development. This extends to the various concerns governments have had with the quality of teacher preparation and of the 'product' that has resulted. Previously, governments were happy to largely pass the 'quality control' of teacher preparation over to universities—but today their influence is more direct, being determined through funding regimes, inspections, validation arrangements and the creation of alternative school-based models of initial teacher education. This interference by government has even extended into changing the nature of what is accepted as legitimate knowledge—usually towards more practical, utilitarian, and skills-based forms of knowing.[2]

---

[2]It is worth noting that universities have also traditionally been nervous about the nature of educational knowledge, and the ways in which the pragmatics of teacher education take the focus away from research and scholarship. Here the influence of government, threatening the independence of the academy in matters of teacher education, has always been a concern for higher education institutions.

5.4   Policy Impacts on the Quality of Educational Research

There is a distinction between research that can be carried out by academics and that which is conducted by practitioners, each having their own expectations and audiences (Sachs 2003). Academic and practitioner research, and researchers, tend to have different foci: academic research being more theoretical or conceptual, practitioner research being geared more to practice. They also tend to use different methods and methodologies—although to claim any degree of exclusivity here would be foolish. Government policy in relation to education, particularly initial teacher education, obviously has a direct impact on research productivity. But there is also a question of readiness, and an issue about whether geography educationists are able to draw upon an existing research base to help inform the government about the likely impact on their subject of proposed educational changes. For example, when the Education Reform Act (DES 1988) indicated the government of the day's firm intention to develop a national curriculum for schools, subject working groups looked to draw on the extant research evidence in their fields to help influence the curriculum making process. In geography, the GA made reference to its *'Case for Geography'* (Bailey and Binns 1987), written in response to a previous Secretary of State for Education, Sir Keith Joseph's, 'six questions'—which were aimed at provoking geographers into making a justification for their subject's place in the curriculum. Indeed, the GA devised its own version of the geography national curriculum (Daugherty 1989) in an attempt to influence the future decision making of the Geography Working Group (GWG). All well and good, but it was apparent at the time that geography educators could not readily access enough research evidence in geography education to inform all the questions that arose—often the evidence base simply did not exist. It was also difficult to mobilise cohorts of geography education researchers to immediately respond to such questions—for obvious structural, organisational and financial reasons, but also because the cohort of available researchers was too small. Williams (2003), in consideration of Rawling's (2001) accounts of the workings of the GWG she was a member of, comments on the paucity of research input by geography educationists into the process of curriculum development:

> From a research perspective, what is striking is the absence of research findings to support any of the arguments promulgated in the group. No reference is made to any research input to the proceedings. Partly this is explained by the failure of the geographical education research community to articulate a coherent contribution to the discussions. It lacked the requisite organisation and, more importantly, a substantial body of research evidence. Further, this was reinforced by the failure of well-documented English curriculum development projects in geography to have a strong research dimension. They were *development* projects, rather than *research and development* project... It was this research evidence, validated and rendered reliable by various key sectors in the geographical education research community, national and international, that was seriously lacking in the intense debate about what should constitute school geography in England (Williams 2003, p. 261) [emphasis in the original].

These are serious claims. In part they are accurate—there is certainly some truth in the point that the geography education research community was incapable of making a more organised response—but Williams tends to ignore the tight timeframe, indeed immediacy, of the requirement to construct the first geography national curriculum. The sterling efforts made by the Geographical Association to steal a march on the official curriculum making process are also worthy of note.

Furlong (2013) identifies David Hargreaves' public lecture to the Teacher Training Agency (TTA) in 1996, where Hargreaves singled-out education research as being 'poor value for money' and 'frankly, second rate', as a stimulus to debate about the future of research in the field. This theme was enthusiastically taken up by the then Chief Inspector for Schools, Chris Woodhead, who stated that he no longer read any education research, indeed he had:

> … given up. Life is too short. There is too much to do in the real world with real teachers and real schools to worry about methodological quarrels or to waste time decoding unintelligible, jargon-ridden prose to reach (if one is lucky) a conclusion that is often so transparently partisan as to be worthless (Woodhead 1998, p. 51)

Pring (2004) acknowledges such concerns, particularly those raised by policy makers in government circles, questioning the usefulness of certain strands of educational research. The often stated gap between professional practice and academic research again comes to the fore, with Pring recognising that questions about the usefulness of research outcomes regularly focus on whether it successfully addresses the ways in which educational practice is best organised; whether it focuses closely enough on educational problem solving; whether it is coherent (it achieves a clear message or perspective, building successfully on previous work—often of a methodological and conceptual nature); and whether it is motivated more by politics and ideology than need.

A report by the Royal Society and British Academy in 2018—*Harnessing Educational Research* (Wilson 2018)—correctly asserted that given the huge government spending on education annually, better guidance on how to deploy these resources was needed: 'That is why educational policy and practice should be informed by the best available research evidence' (Wilson 2018, p. 5). The report assessed the state of educational research in schools and colleges in the UK, adopting the analogy of an ecosystem to monitor flows of people, funding and information. A key finding of the report was 'the need for the teaching profession to be research-informed' (p. 5), which was fashioned into guidance for the recently created UK Research and Innovation (UKRI) agency about ways in which excellent research could in future be identified and funded. Eight recommendations result: (i) to connect 'supply and demand', to achieve a better sharing of priorities between researchers, practitioners and policymakers and thus contribute to more sustained research effort (using a newly created Office for Education Research); (ii) to facilitate ways in which stakeholders in research could work together, particularly with regard to the distribution of educational research capacity across the UK, noting that collaborations may be regional and/or thematic. (It was recognised that many education research questions could yield to interdisciplinary approaches, but that barriers to intra-institutional collaboration needed to be removed to create a critical mass of researchers); (iii) to address issues affecting the supply of education researchers, whose age profile was predominantly over 50—which hindered the long-term health of educational research in the UK; (iv) to ensure that the training of educational researchers met the needs of mature learners (often teachers undertaking part time studentships), covered the full range of social science methods, and fostered better links between research students, policy and teaching communities; (v) to ensure that Quality Research (QR) funding remained a strong part of research funding, to enable the development of research infrastructures and prioritisation of research themes independent of the immediate priorities of funders and government; (vi) to support the use of research to inform teaching, with the DfE having a clear expectation that teachers should be informed by, and engaged in, research. (This would recognise the importance of research informed practice in the teaching standards, the requirements of ITE to be research informed, and the need to maintain engagement with research during newly qualified teachers' induction period, and during professional development); (vii) to facilitate the needs of policy makers through identifying the cultural and practical barriers to the use of research in education policymaking; and (viii) to support the production and synthesis of research evidence, such that teachers, researchers and policymakers benefited from sharing information about key research areas. The report helpfully identified ways in which governments and their agencies, UKRI and its constituent bodies, universities and other higher education institutions, and educational organisations could combine in the 'ecosystem' to harness research evidence and insights more effectively.

5.5 The Importance of the Subject

Arguably, over the last 50 years, the creation of the geography national curriculum for England and Wales in 1991 represented the most significant impact of education policy on the teaching and learning of geography in state schools. Machon and Ranger (1996) considered the impact of the Statutory Order for geography on the prospects for future curriculum development in geography, choosing to compare its impact to that caused by the 'conceptual revolution' in geography in the 1960s and early 1970s. However, we must recognise that there are significant differences between the two processes—the former driven by legislation linked to government policy, the latter by the discipline of geography itself, through 'an invigorating dialogue between geographers in higher education and in schools' (p. 40). Little dialogue between higher education and schools occurred either before, or after, the construction of the geography national curriculum—some observers referring to a 'decoupling' of geography in schools and higher education beforehand, and a 'fragmentation' of the geographic community after (Goudie 1993; Unwin 1996; Bradford 1996; Rawling and Daugherty 1996; Machon and Ranger 1996).

Consideration of the educational importance of school subjects found an urgency in UK government circles at the start of the millennium. By extension, this reinforced the need for teachers to be accomplished subject specialists, if their teaching was to be considered as 'excellent'. Policy directives, such as *The Importance of Teaching* (DfE 2010), indicated the government's concern about the appropriateness of subject content, although similar levels of anxiety were arguably not reflected among senior teachers in schools, or by the inspection service OfSTED. Both senior school leaders and inspectors appeared wedded to promoting the importance of generic teaching skills, rather than attempting to raise the quality of subject teaching (Lambert and Jones 2013). A number of geography educationists, and researchers, have recognised and pursued this agenda (Standish 2012, 2013; Marsden 1997; Lambert and Jones 2013).

Lambert and Jones (2013) recognise that a lack of policy support for geography education—despite the positive intervention through the Labour Party's Action Plan for Geography[3] (2006–11)—has been paralleled by the dismantling of other support agencies, assistance measures and funding. The persistent weakening of Local Education Authorities (now Local Authorities) in England, with the concomitant removal of advisory teachers for geography (and other subjects) has been a significant blow. The role and function of the advisory service was not without its critics, but the actual and symbolic effects of fileting out such support, advice and challenge for subject teachers in state schools is palpable. Alongside the re alignment of Her Majesty's Inspectorate (HMI) into a more punitive inspection force, re badged as OfSTED, and the downplaying of initial teacher education (ITE) in universities, support for geography teaching and teachers in schools was severely weakened. The criticism of the systematic unpacking of ITE, and its resultant impact on subject debate and support in education, is well rehearsed elsewhere (Tapsfield et al. 2015). Lambert and Jones (2013) note that, as a professional group, teachers *by themselves* do not necessarily have the wherewithal to make, and remake, the knowledge that defines their profession:

> In schools, which are intensely practically oriented places, often very inward-looking and with constant, urgent 'busy-ness' to attend to, it is very difficult indeed to do this, *especially* in the field of subject knowledge development (p. 6 emphasis in original).

---

[3]The Action Plan for Geography (APG) was initially a two-year project in England funded by the Department for Children, Schools and Families (DCSF) from September 2006. Led by the GA and Royal Geographical Society with Institute of British Geographers (RGS-IBG), its purpose was to raise standards in geography education in state schools. APG received further funding, for another three years, from 2008 following positive reports about its effectiveness (GA 2011). The effect of the Plan was to stimulate gradual, incremental improvement in geography education in primary and secondary state schools, particularly through improving teachers' subject knowledge and abilities as curriculum makers (OfSTED 2011). Geography teachers received online and 'face to face' mentor support, provided by the professional associations for geography and geography education—however, improvements in geography education were often sporadic and polarised, with inspections revealing that the gap between the best and the least satisfactory schools was growing steadily wider. The main areas of development, which largely remain, were targeted at primary level: relating to the development of pupils' sense of place, appropriate coverage of the geography national curriculum, clear identification of geographical elements in school curricula, fieldwork, greater use of technology, and the provision of subject specific support and professional development of teachers (OfSTED 2011).

5.6 Teacher Professionalism

The professionalism of teachers has been under sustained attack by governments in the UK for at least a quarter of a century, despite claims about improving both the professional status and standards of the teaching workforce. This has implications not only for the role of research in education, but also for the nature and focus of discourse about the professional standing of teachers, and for future government education policy. There are parallels between the experiences of the secondary and tertiary sectors: both teachers and academics have been faced with a rhetoric concerning their apparently greater freedoms and professional autonomy resulting from government actions, while being made to respond to more managerial, directive and prescriptive systems of accountability and control. Roberts (2015) puts this well when she states that recent government policies:

emphasize leadership, management, enterprise and choice at the same time as encouraging a return to a more traditional curriculum and pedagogy (p. 46)

The geography education research community must become more aware of the sustained pressures on teacher professionalism and make attempts to respond to it—not perhaps as daunting as it sounds, given that researchers are also caught up in similar systems of performance management, assessment and accountability. Teachers are currently 'positioned' by UK governments as 'craft workers', rather than professionals or experts, which also has implications for the status of practitioner researchers. There is a dangerous stereotype being advanced—which like most stereotypes contains a small kernel of truth, but is nonetheless not substantially accurate—that good teachers are 'born rather than made': part of the damaging conceptualisation of the 'charismatic teacher' (Moore 2004). Such simplistic notions of the characteristics of effective teachers are dangerous for the profession and reduce the perceived need for educational research. If beginning teachers are told that they simply need to watch and learn from other experienced teachers, and then 'have a go' themselves—rather than visualising themselves as specialists, experts, knowledge workers, and practitioner researchers capable of high-level questioning and making a significant contribution to their profession—their status is indeed reduced to that of craft workers. The rise of academies and free schools, which allow teachers to practice even if they do not possess a teaching qualification, again downgrades the kudos of the professional and the need for research to play a part in teacher preparation and continuing professional development (Hatcher 2011, 2012).

There is an acceptance that the modern teacher should be critically reflective, and able to assess their own performance based on established teaching standards and contributions from educational research. Lambert and Jones (2013) refer to this as a 'modern orthodoxy' (p. 5), which has inherent dangers in focussing too closely on teacher performance—rather than on subject specialisms, or the need to initiate learners into a 'world of ideas', intellectual debate and enquiry. This leads to a hollowed-out notion of educational practice, where 'effectiveness' is prioritised at the expense of the educational contribution made by subjects—which are seen simply as 'vehicles', or mediums, for learning activities. This is clearly an impoverished expectation of what subjects such as geography can offer young learners—geography teachers need to understand, and keep abreast of, conceptual and content shifts in the discipline of geography, or they will risk becoming outdated and redundant as subject educators. Policy that does not recognise and address this issue is not good education policy. Resolving this conundrum is not straightforward—but the consequence of not recognising disciplinary fluidity and complexity is to accept that geography, as a school subject, is now 'firm and fixed'. This has an implication that the parent discipline, and indeed our knowledge of the world, is essentially complete (Lambert and Morgan 2010).

5.7 Conclusions

Teachers need to work with a sense of moral purpose and responsibility; considerations that should override any policy debate (Lambert 2009; Lambert and Jones 2013). They also need to be able to understand, and endorse through their actions, appropriate concepts of both geography *and* education. As a professional body, teachers must question theory and practice in education, be confident of their knowledge-base as geographers, and not feel driven to be overly compliant in the 'delivery' of state sanctioned geography curricula. This implies that they must offer some resistance to externally imposed standards, inspection regimes and the use of performance 'league tables'. We have seen how a compliance culture in education has dramatically, and negatively, affected the professional skills of teachers in curriculum-making in recent years. Here is the prospect for education researchers in general, and geography education researchers in particular, to offer themselves in a supporting role for the professional teacher.

References

Bailey, P., & Binns, T. (Eds.). (1987). *A case for geography*. Sheffield: Geographical Association.

Bassey, M. (1981). Pedagogic research: On the relative merits of the search for generalisations and single study events. *Oxford Review of Education, 7*(1), 73–94.

Bassey, M. (2001). A solution to the problem of generalization in education: Fuzzy prediction. *Oxford Review of Education, 27*(1), 5–22.

Bradford, M. (1996). Geography at the secondary/higher education interface: Change through diversity. In E. Rawling & R. Daugherty (Eds.), *Geography into the twenty-first century*. Chichester: Wiley.

Browne Report. (2010). *Securing a sustainable future for higher education. Independent review of higher education*. London: BIS.

Butt, G. (1997). *An investigation into the dynamics of the national curriculum geography working group (1989–1990)*. Unpublished Ph.D., University of Birmingham, Birmingham.

Butt, G. (2017a). *Debating the place of knowledge within geography education: Reinstatement, reclamation or recovery?* In: Brooks, C., Butt, G., & Fargher, M. (eds.), *The power of geographical thinking* (pp.13–26). Dordrecht: Springer.

Butt, G. (2017b). Globalisation: A brief exploration of its challenging, contested and competing concepts. *Geography, 102*(1), 10–17.

Daugherty, R. (1989). *Geography in the national curriculum*. Sheffield: GA.

Department for Children, Schools and Families (DCSF). (2007). *The children's plan. Building brighter futures*. London: HMSO.

Department for Education (DfE). (2010). *The importance of teaching: The schools white paper*. London: The Stationery Office.

Department for Education and Science (DES). (1988). *Education reform Act*. London: HMSO.

Department for Education and Science (DES). (1991). *Geography in the national curriculum (England)*. London: HMSO.

Furlong, J. (2013). *Education: An anatomy of the discipline*. London: Routledge.

Furlong, J., & Lawn, M. (2010). *The disciplines of education: Their role in the future of education research*. London: Routledge.

Furlong, J., & Oancea, A. (2005). *Assessing quality in applied and practice-based educational research*. ESRC TLRP seminar series. (December 1) Oxford University Department of Educational Studies.

Geographical Association (GA). (2009). *A different View: A manifesto from the Geographical Association*. Sheffield: Geographical Association.

Goudie, A. (1993). Schools and universities: The great divide. *Geography, 78*(4), 338–339.

Graves, N. (1972). *New movements in the study and teaching of geography*. London: Maurice Temple Smith Limited.

Hargreaves, D. (1996). *Teaching as a research-based profession: Possibilities and prospects*. London: TTA.

Hatcher, R. (2011). The struggle for democracy in the local school system. *Forum, 53*(2), 213–224.

Hatcher, R. (2012). *The rise of academies, the end of the local authority?* Joint Oxford Brookes University/Oxford University seminar. Oxford Brookes University. 20 November 2012.

Hillage, J., Pearson, R., Anderson, A., & Tamkin, P. (1998). *Excellence in research in schools*. London: TTA.

Hood, C. (1995). Contemporary public management: A new global paradigm? *Public Policy and Administration, 10*(2), 104–117.

Lambert, D. (2009). *Geography in education: Lost in the post?* Inaugural lecture. Institute of Education, University of London.

Lambert, D., & Jones, M. (2013). Introduction: Geography education, questions and choices. In D. Lambert & M. Jones (Eds.), *Debates in geography education* (pp. 1–14). Abingdon: Routledge.

Lambert, D., & Morgan, J. (2010). *Teaching Geography 11-18: A conceptual approach*. Maidenhead: McGraw Hill/Open UP.

Machon, P., & Ranger, G. (1996). Change in school geography. In P. Bailey & P. Fox (Eds.), *Geography teacher's handbook*. Sheffield: Geographical Association.

Maclure, M. (2003). *Discourse in educational and social research*. London: McGraw Hill.

Marsden, W. (1997). On taking the geography out of geographical education—Some historical pointers on geography. *Geography, 82*(3), 241–252.

McCulloch, G. (2017). *Education and the new Conservatism: Social wellbeing, national character and British values*. Oxford University Department of Educational Studies. Public lecture, 27 February 2017.

Moore, R. (2004). *Education and society*. Cambridge: Polity Press.

Morgan, J. (2011). *Teaching secondary geography as if the planet mattered*. Oxford: David Fulton.

Morgan, J., & Lambert, D. (2011). *Geography and development. Development education in schools and the part played by geography teachers*. Research paper 3. DERC. London, Institute of Education.

OfSTED (Office for Standards in Education). (2011). *Geography: Learning to make a World of Difference*. London: HMSO.

Pring, R. (2004). *Setting the scene: Criticisms of educational research in philosophy of educational research*. London: Continuum.

Rawding, C. (2013). How does geography adapt to changing times? In D. Lambert & M. Jones (Eds.), *Debates in geography education* (2nd ed., pp. 282–290). London: Routledge.

Rawling, E. (1991). Making the most of the national curriculum: The implications for secondary school geography. *Teaching Geography, 16*(3), 130–131.

Rawling, E. (2001). *Changing the Subject. The impact of national policy on school geography 1980–2000.* Sheffield: Geographical Association.

Rawling, E., & Daugherty, R. (Eds.). (1996). *Geography into the twenty-first century.* Chichester: Wiley.

Roberts, M. (2000). The role of research in supporting teaching and learning. In A. Kent (Ed.), *Reflective practice in geography teaching* (pp. 287–295). London: Paul Chapman.

Roberts, M. (2015). Discussion to part 1. In G. Butt (Ed.), *Masterclass in geography education* (pp. 45–50). London: Bloomsbury.

Rudduck, J., & McIntyre, D. (Eds.). (1998). *Challenges for educational research.* London: Paul Chapman.

Sachs, J. (2003). *The activist teaching profession.* Milton Keynes: Open University Press.

Sebba, J. (2004). Developing evidence-informed policy and practice in education. In G. Thomas & R. Pring (Eds.), *Evidence-based practice in education* (pp. 34–43). Maidenhead: Open University Press.

Standish, A. (2012). *The false promise of global learning. Why education needs boundaries.* London: Continuum.

Standish, A. (2013). What does geography contribute to global learning? In D. Lambert & M. Jones (Eds.), *Debates in geography education* (pp. 244–256). London: Routledge.

Tapsfield, A., Roberts, M., & Kinder, A. (2015). *Geography initial teacher education and teacher supply in England.* Sheffield: Geographical Association.

Thomas, G. (2007). *Education and theory: Strangers in paradigms.* Maidenhead: Open University Press.

Tooley, J., & Darby, D. (1998). *Educational research: A critique.* London: OfSTED.

Unwin, T. (1996). Academic geography: The key questions for discussion. In E. Rawling & R. Daugherty (Eds.), *Geography into the twenty-first century* (pp. 19–36). London: Wiley.

Usher, R. (1996). A critique of the neglected epistemological assumptions of educational research. In D. Scott & R. Usher (Eds.), *Understanding educational research.* London: Routledge.

Whitty, G. (2005). *Education(al) research and education policy making: Is conflict inevitable?* Presidential address to BERA conference, University of Glamorgan, Pontypridd.

Williams, M. (2003). Research in geographical education: The search for impact. In R. Gerber (Ed.), *International handbook on geographical education* (pp. 259–272). XX Kluwer.

Wilson, A. (2018). *Harnessing educational research.* London: Royal Society and British Academy.

Woodhead, C. (26 March, 1998). *Academia gone to seed* (pp. 51–52). New Statesman.

Chapter 6
The Consequences of Assessment of the Quality of Research Outputs

6.1 Context

The previous chapter revealed how government concerns about the quality and relevance of educational research have deep roots (Butt 2006, 2010a, 2015). Indeed, recent changes in education policy has dramatically affected the functioning of universities, initial teacher education (ITE), professional development and subject-based education: with concomitant effects on education research. Demands for increased research accountability—with particular regard to issues of both quality and excellence—have resulted in periodic Research Excellence Framework (REF) assessments in the UK against criteria for 'originality', 'significance' and 'rigour'. These criteria are applied to each published research output to help determine its quality. The assessment of research quality also involves recording contextual information about the institution in which the return of information was produced, including its staffing levels; research income; numbers of research students; nature of the research environment; and various indicators of esteem. More recently, assessments of research productivity have included evaluations of the impact of the research produced. On the basis of these REF findings—benchmarked against international standards of excellence—public money is distributed to support research activity in the UK, with the most significant allocations inevitably being directed to high ranking, elite institutions. Money used to finance research comes predominantly from two sources—public money awarded through the REF, called QR ('Quality Research') money, which broadly supports research infrastructure; and funding which is competitively gained from various sources by bidding for grants to finance individual research projects. We must note that creating fair systems to assess, rank and reward research at the national and international level—which are inexpensive, effective, easy to administer and 'work' across all university departments—is an extremely tall order (Oancea 2010a). Nonetheless, most higher education institutions in the UK believe

The later sections of this chapter draw significantly upon previously published chapters in 'Master-Class in Geography Education' (see Butt 2015).

© Springer Nature Switzerland AG 2020
G. Butt, *Geography Education Research in the UK: Retrospect and Prospect*, International Perspectives on Geographical Education, https://doi.org/10.1007/978-3-030-25954-9_6

that REF panel judgements are fair and that the overall exercise of assessing the quality of research, and researchers, is legitimate. For these reasons, perhaps, many governments across the world have implemented their own versions of a 'research excellence framework'.

This chapter explores how, unfortunately, much research in geography education still sits rather uneasily in this research assessment arena. Having traditionally failed to attract significant amounts of government, research funding council or project money, we have seen that the awarding of large-scale research projects in geography education in the UK is infrequent—with most research being produced as more minor, unfunded work, often reflecting the interests of only a few individuals. This type of research activity can perhaps be viewed, at best, as offering opportunities for a limited number of researchers—rather than opening up more strategic, cost effective and high impact lines of research activity (Butt 2006). The possibilities that in the next few years educational research might be concentrated into only a handful of higher education institutions in the UK is tangible, with obvious implications for the prospects of geography education research and researchers.

The recent drivers for similar research assessment exercises to be conducted in other national jurisdictions, and the potential effects on geography education research elsewhere, must be considered—particularly in the light of the tendency for such frameworks to restrict academic freedoms, reduce autonomy in terms of research focus, and supress emergent research cultures. We must also be aware that in the UK around two thirds of education staff are *not* entered for the periodic REF reviews, and that for geography education researchers this figure is even higher—with particularly low scores being returned for research quality in teaching-intensive institutions (BERA UCET 2010). This trend appears to be repeated internationally.

6.2   The Demands for Research Accountability—Quality and Excellence

In the UK the first Research Assessment Exercise (RAE), organised by the Higher Education Funding Council (HEFCE), was conducted in 1986[1] primarily as a way of increasing the accountability of higher education institutions with respect to their research outputs. Subject specific panels and sub panels agreed on assessment criteria in the light of principles, standards and generic measures that were applied across all the panels—using a numeric rating[2] system to record the quality of individual publications. The fledgling RAE was largely supported by English universities who recognised that their recent expansion, coupled with low accountability measures for attaining block grants of research money, made the existing national system of resource allocation unsustainable. It was felt that the RAE would also support the

---

[1]RAE was conducted in 1986, 1989, 1992, 1996, 2001, 2008 and, as the REF, in 2014 in the UK.

[2]Ratings, which have shifted slightly within the grading system over the years, now range from 'unclassified' to 'internationally excellent' on a five-point scale.

development of research in the post 1992 universities, encourage better research management, aid the growth of graduate schools, and stimulate the completion of higher quality research and publication outputs—although resultant outcomes were by no means universally positive across institutions. By periodically repeating this exercise it was believed that the international standing of UK research would be improved, also making researchers more publicly accountable and increasing the overall quality of their publications. Unfortunately, there is a strong sense—backed by empirical evidence—that periodic research assessments have also added uncertainty, perhaps even crisis, into research activity across parts of the HE sector (Oancea 2009, 2014, 2016).

Furlong (2013) identifies that the main driver for increasing research productivity in education from the 1980s onwards was that schools and departments of education increased their demands on staff to become more 'research active'. Undoubtedly, the national research assessments have provided a significant and forceful stimulus to produce better research—whether academics wanted to engage in research work, or not—and have meant that academics in education departments have had to perform in similar ways to those employed in other university departments:

> But with the diversity of intellectual frameworks available, with the strong emphasis on qualitative research, and with the lack of a strong research tradition in many departments and faculties, this rapid growth in educational research laid it open to criticism, criticism that it was a 'cottage industry', small scale, unsystematic and largely noncumulative. (Furlong 2013, p. 39)

While there are good reasons why the research activities of universities should be more publicly accountable—after all, they are using taxpayers' money—the resultant effect of suppressing academic freedoms and creativity have been widely criticised. The impacts of the RAE, and of its subsequent iteration as the Research Excellence Framework (REF), have been dramatic and widespread. Alongside other quality control and auditing measures—which affect both research and teaching, each now employing heightened systems of bureaucracy, quantification and accountability— these exercises have tightened the previously enjoyed freedoms of academics. This led Ranson (2003) to refer, unsympathetically, to the measurement of academics' new responsibilities as having gradually shifted from using systems of traditional, professional accountability, to employing newer, neo-liberal systems for 'holding to account'. Here the emphasis is on target 'setting and getting', establishing and maintaining competitive advantage, and increasing profits and 'deliverables'—rather than ensuring intrinsic support for individual staff towards achieving research or teaching excellence. Essentially these are systems designed to facilitate the competitive allocation of resource, with research assessment being seen as a method of benchmarking and positioning research outputs (from institutions, disciplines and individual members of staff) in the wider higher education landscape (Oancea 2010b). The increasing competition for research funding between institutions has led to calls for greater central funding—indeed, as Furlong (2013) reports, centralised research funding *doubled* for education between the 2001 and 2008 RAEs,[3] with around

---

[3] Only around 15% of this money goes to the 'new', post 1992, universities in the UK.

60% of this money coming directly from government. Significantly, this enabled UK governments to establish a new 'social contract' for research through which they could increasingly determine what was researched, the methods and methodologies applied to undertake such research, and the ways in which research findings might be used. With specific reference to geography education, Andrew Goudie—in a seminal editorial titled *Schools and universities: the great divide* (Goudie 1993)—also highlighted that the (then) RAE was one of the main reasons why geographers in universities now had less contact with schools. He reported that academics now spent more of their time conducting research, which took them away from forging stronger links with schools.

6.3 Research Selectivity

The purpose of any research selectivity exercise is to apportion and distribute public resource to the best performing institutions. In the UK this resource is referred to as Quality Research ('QR') money. It is hard to deny that universities should be held to account for the ways in which they spend public money, and for the outcomes and cost-effectiveness of such spending. Whether research assessments are the best way to carry out such accounting—for geography education research, or indeed any other area of research—is open to debate. The benefits of research assessment exercises are reasonably clear: they provide a means by which UK research can be benchmarked against itself, but also increasingly internationally, using agreed standards of excellence. Striving to create common assessment criteria and mechanisms to measure research excellence is laudable—the problem is whether this system is fair. There are unseen impacts of assessment on certain types of educational research, including research undertaken by subject-based education researchers—such as those employed in geography education research. Achieving measures of research performance that can achieve parity—across disciplines and subjects, fields and subfields, different types of research endeavour, and national and international contexts—is clearly not straightforward. Like any form of assessment, the REF has now spawned its own industry of consultants, theorists and methodological experts, accompanied by a diverse and expanding literature base which seeks to advise on how to get the best out of the assessment process. The REF has also attracted the attention of an army of critics: not least because the exercise takes considerable time and money to undertake, skewing activities within institutions for years; membership of a 'nonreturned' school or department can be detrimental to academics' career progress; and the allocation of QR money is not primarily designed to 'bring on' emergent research areas, or researchers. Such is the common experience of geography education researchers, for it appears that the old maxim of 'publish or perish' has now become manifest, forcing individuals into making difficult career decisions (or, more likely, having institutions make these decisions for them). Schools, departments and institutes of education—often already on the periphery of their establishments in terms of their status—may be forced into dire circumstances by a poor performance in the REF, which in turn affects future staffing, levels of research activity and rep-utation. Coupled with the shift of initial teacher education into schools in the UK, the effects of a poor research assessment can be catastrophic. Indeed, some Russell Group universities have changed academics' contracts of employment to improve the REF returns of their departments—often making emergent research staff accept 'teaching only' contracts if they do not appear to be on track to be 'REF returnable' in time for the next assessment. This has certainly impacted directly on the careers of many young academics in geography education, whose research activities have been drastically curtailed, or even removed from them completely. Efforts to avoid these outcomes were evident in the review of the research assessment processes made by Lord Stern in July 2016 (Stern 2016). His twelve recommendations urged that all research active staff should now be 'returned' to the REF and suggested measures to lower the assessment burdens on HEIs and review panels; to reduce 'game playing; to reduce the personalisation of assessment to make judgements of research qual-ity more institutionally focused; to support HEIs in making investment in research; to provide a more rounded view of research activity; to increase the emphasis on interdisciplinary research; to broaden impact: and to develop public engagement fur-ther to include impact on curricula and pedagogy. These recommendations have had some positive effects, but essentially there remains a league table of research active departments and academics, a transfer market for elite researchers, and a legacy of negative impacts on many subject-based researchers in education.

6.4 Does Research Assessment Improve Research Quality?

The question of whether research assessment actually *improves* the quality of research produced by institutions is an easy one to pose—but deceptively difficult to answer. The rather unhelpful response is that it does for some institutions, and individuals, but not for others; importantly, predicting the nature of these effects is hard. Some individuals find ways to perform well in terms of research productivity, despite being in relatively hostile environments, and vice versa. What is increasingly clear is that wherever institutions have attempted to 'gear up' their research performance this has had a concomitant impact on teaching and other activities. In geography education, there has often been an expectation among managers and senior leaders that staff will not be able to produce research outputs that will be highly rated—which has resulted in geography educators being asked, or told, to increase their commitment to teaching. This obviously represents a downgrading of the research status[4] of academics. So, the REF has certainly served to sharpen up the quantifiable measures used in the allocation of research money, the assessment of research outputs

---

[4]This is readily seen in the various workload management systems that are now used in univer-sities, where staff have specific allocations of time for researching, teaching and for administra-tion/management activities. Most geography education researchers in UK institutions have seen a reduction in the amount of time afforded for their research in recent years.

(by number and type), the descriptors of quality used, and has had an impact on where academics choose to publish their work. As a consequence, it has also adversely affected the overall balance of academics' work in some institutions. Interestingly, the introduction of the Teaching Excellence Framework (TEF) in England has partially helped to redress this perceived imbalance. Seeking to hit the externally created research targets of *significance, rigour* and *originality* is also difficult for all staff—particularly so, given that no official feedback is provided which scores assessed publications *individually*. It is therefore not clear whether specific returns from staff have performed well—only whether the overall assessment of *all* publications from a department has been successful in their given unit of assessment.

6.5   Significance, Originality and Rigour

The assessment of individual pieces of published research in the REF is driven by panel judgements as to their significance, originality and rigour. One of the main problems with much small-scale, practitioner research in geography education is that the researcher can often become too 'close' to what is being researched—particularly if the research is ethnographic in form, adopts an action research methodology, or has been conducted with a very limited sample of respondents over a short period of time. It is not uncommon in these circumstances for bias to creep in, with claims being made from resultant data that overstep the acceptable parameters for such research. The whole point of most qualitative, ethnographic and action research is, of course, that it enables the researcher to get closer to the objects being studied; this does *not* allow for unrestricted subjectivity and it is generally acknowledged that the findings from research of this kind will not be generalizable. This also has obvious implications for the significance, originality and rigour of research. The issue concerning the assessment of much of the research produced in geography education is that it is 'strongly qualitative' in nature—existing on the borders of research work that is too 'personal' and which is based on particular views and opinions that may be highly individualistic, or which may struggle to stand up to even the most rudimentary scrutiny. Here issues of validity, reliability and fitness for purpose come to the fore, particularly when the research outcomes were either predetermined, strongly suspected from the outset, or even strived for. Examples of research data being made to fit the expected (hoped for?) outcomes are rather too common in small scale research in geography education.

When we consider the official measures of significance, originality and rigour of research outputs, which are currently central to considerations of the quality of research publications in the UK, there is unfortunately limited official guidance as to their meaning. These measures can be described in various ways—let us consider some advice on each of these criteria given to Masters students in geography education in a recent publication:

Significance can be best understood with respect to whether an important and meaningful problem, or question, is being researched. You might consider whether the theme chosen for your research is relevant at the small scale (just you, and your classroom practice), or at a much larger scale (a 'universal' problem considered by geography education researchers, or indeed across other disciplines). Is the significance of the research increased because it investigates a problem that has relevance to *other* subject areas (and is this made explicit?), or to educationists *in general*? Put crudely, someone weighing up the significance of your research might ask 'Why should I care about this?', or 'Is this a problem that is really noteworthy and important?', or indeed 'So what?'. *Rigour* is usually considered to be a measure of the thoroughness, precision and accuracy of your research method, methodologies and presentation—one might ask whether there is an intellectual meticulousness, robustness and appropriateness to your work? Does the research display integrity, coherence and consistency? Lastly, we might introduce a consideration of the *originality* of your research—does it provides new (empirical) material or insights? Does it seek to develop innovative research methods, or methodologies, or analytical techniques? Or has it successfully employed already established research techniques? Is there a possibility of generating new theory? These are quite challenging criteria … nonetheless, they do present valid questions which can be considered whenever the quality of any research work is assessed (Butt 2015).

Much of the research conducted in geography education can be criticised with respect to methodological and research methods issues. Research may make claims about following strict guidelines for research practice, but on closer inspection often only pays lip service to these requirements; similarly literature reviews may be partial and biased towards findings that support one 'side' of an argument. This will, of course, affect the formulation of research questions and sub questions: if an appropriate literature search has not been carried out, and the conceptual and theoretical foundations for the research are weak, then the subsequent findings will also be undermined.

6.6 Reviewing the Impact of Research Assessments

It is obvious that finding a 'silver bullet' solution to identifying the most accurate and reliable indicator(s) of research quality and excellence—applicable to all subjects and disciplines, and across all forms of research—is unlikely. A satisficer solution is the best we can hope for, as any notion of identifying a perfect indicator of research quality is surely a fiction—all methods of measuring excellence will consist of a combination of techniques and approaches, each with their own 'positives and negatives', and each with different degrees of validity, reliability and fitness for purpose.

The established measures of research assessment, trusted by most academics, favour the use of some form of peer review. Peer review is well understood by academics, already being 'part and parcel' of their daily work—particularly in relation to publishing in academic journals and having research bids and reports assessed by fellow academics. The belief in experts who judge the quality of one's work—if this process is free from political control, or special interests—creates a sense of trustworthiness, fairness and accountability. As such, systems where academics tend to 'police' themselves are, not surprisingly, popular in higher education, but less so with politicians—academics will favour systems that preserve their autonomy, but governments are not so keen! We must acknowledge the dangers prevalent in most assessment systems, some of which are particularly acute for small communities of researchers such as those in the field of geography education research—to be effective, peer reviewers must have a shared sense and understanding of what constitutes *quality*. Here is a source of contention, as measurements of quality reflect a number of facets which, put crudely, may prioritise selectivity over sensitivity (with the favouring of different types of research activity, or different fields of research); innovation over tradition; and merit over promise (Hackett and Chubin 2003, cited in Oancea 2008). Peer review is understandably costly in terms of resource, time and effort—it ties up senior academic staff and creates a considerable administrative burden. The opportunity cost of the REF should therefore be considered through cost effectiveness exercises—which attempt to find the most appropriate review methods to reduce expense, but maximise reliability of outcomes. Without these tensions being adequately resolved there may be errors, bias and a lack of trust in the system—especially if the reviewers are placed in competition with individuals, institutions and projects they are employed to review objectively. Inherent conservatism may also work against innovation—with smaller, newer, less established research groupings, such as those in geography education research, often performing poorly. Additionally, if there is pressure for a particular type of research to be favoured— say, large scale, policy-driven, 'what works' research using Randomised Controlled Trials (RCTs); compared with small scale, longitudinal, ethnographic research—this will have an impact on more modest, non-funded, research.

When considering the most effective means of judging quality and impact of research outputs, the possible use of citations often arises. However, the 'best' peer-reviewed research may not perform particularly well in terms of its number of citations. Using citations also has issues in that research that is poor, or controversial, may receive large numbers of citations—precisely because it attracts legitimate academic criticism—compared to work which is secure, but unspectacular. Establishing correlations between research quality and numbers of citations is therefore not straightforward. There is also a tendency that recently published work gets cited more than pieces that are a few years older, that some authors choose to cite more than others (and may be very selective in their citations), that self-citation and erroneous citation are common, and that idiosyncratic or very specialist work (that may still be of high quality) does not always receive the number of citations it deserves. Bibliometrics, based on publication outputs, are quantitative measures of research outputs whose perceived objectivity and cost effectiveness in assessing impact is an obvious attraction. They also have the advantage of being easy to compare against other institutions, researchers and research communities, nationally and internationally. But they too fall into the 'citation trap'—which some research journals acknowledge—because heavily cited work is not necessarily of high quality, importance and impact; citing other academic's work is prone to game playing; and using citations may not provide robust measures of quality and team working. As with any quantitative measure, the accuracy of the output is driven by the appropriateness of the databases—if these are flawed, then the analysis based on them is also flawed. The experience among smaller sub-units of assessment, such as geography education research, is that using citations can result in considerable inaccuracy—it is also claimed that this may impact negatively on early career and part time researchers (van Raan et al. 2007).

Because of such problems, Oancea (2008), calls for a better appreciation of the socio-cultural, historical and philosophical nature of research assessment. Referring to the instrumentalism of research assessment, she notes the:

> narrowing of the 'official' concept of *quality* to 'measurable performance', of *research* to 'production and delivery', and of *research assessment* to 'quantification' (p. 126, emphasis in the original).

She concludes:

> If assessment is defined exclusively by its use as a means towards pre-determined ends (distribution of funding and holding to account), the consequences of this are likely to be the subordination of appraisal to measurement, and of process to output (p. 127).

This requires a shift from the hunt for a *perfect* indicator of research assessment, to a more nuanced understanding of what assessment can and cannot do, and what constitutes quality. It implies moving from a narrow technical framework of measurement, to a more historical, discursive, democratic, ethical and philosophical understandings of the assessment process (Oancea 2008, 2009) (see Table 6.1).

Such a change would recognise that the accurate measurement of research excellence for geography education, or indeed for any other subject-based research, does not yield easily to the crude application of standards of performance.

In conclusion, the nature of the impacts of research assessment exercises on higher education institutions and academics in the UK has been known for some time. Indeed, the UK government acknowledged in 2006 that the RAE imposed some unacceptable burdens on universities—as evidenced by the Chancellor's budget statement

Table 6.1 Current and alternative research assessment discourses, from Besley, A (ed) *Assessing the Quality of Educational research in Higher Education*, reprinted with permission of the publishers, Sense

Current discourse	Alternative discourses
Hierarchical relationship between modes of research	Complex entanglement of research and practice and different modes of knowledge
Quality assurance and quality assessment	Nurturing excellence/virtue (epistemic, technical and phronetic)
Quantification, measurement and ranking of performance	Deliberation and judgement
Assessment techniques unquestionably produce externally-specified outcomes	Assessment techniques help research communities to gain increased control over the contingencies of their practice without stifling diversity

After Oancea (2008)

that year, which announced that research funding allocation would be radically simplified after RAE 2008, with the assessment of research quality and allocation of QR monies becoming more metrics-based (Oancea 2008). Subsequent consultation by the Department for Education and Skills (DfES) suggested the consideration of five possible research funding models based on the use of metrics[5]—which eventually led to the nomination of a double assessment system, varying between Science, Technology, Engineering and Maths (STEM) and humanities subjects, and partly based on a combination of metrics and light-touch expert reviews. Further consultation[6] by the Higher Education Funding Council for England (HEFCE) (2007) led to proposals for the new 'Research Excellence Framework', again split between a predominantly quantitative assessment of the sciences, and a light-touch, metrics-informed peer review elsewhere. The suggestions for change were clearly moving towards a 'two track' approach, but with increased use of quantitative measures and the inclusion of citation indices to determine quality of research outputs. As Oancea (2008) pointed out at the time, the aim was to create a system that allowed cross-disciplinary measures of 'relative' research excellence that would have currency internationally. The response from universities, the British Academy, the Research Councils of the United Kingdom (RCUK) and the Royal Society revealed concern about the divisiveness, and potential unfairness, of a dual assessment system. The importance of maintaining peer review and of using esteem indicators selectively was therefore reinforced. HEFCE's (2008) subsequent analysis of their consultation exercise strengthened the differential allocation of QR monies and the identification of research 'excellence', with a tacit approval for the increased use of quantitative measures and metrics in the future. However, there was clear concern that the differentiation of assessing excellence between disciplines and subject groupings was unpopular, as well as an awareness that inter-disciplinary research and emergent research groupings might be threatened. Importantly for geography education research, it was recognised that small groups of researchers who might potentially exhibit research excellence could be over looked.

The Independent Review of the Role of Metrics in Research Assessment and Management, launched on 9 July 2015, stated that existing metrics were not yet robust enough to replace peer review—particularly in terms of assessing the impact, originality and significance of research publications. This review strongly suggested that the widespread application of research metrics to make assessments would be flawed: the use of more qualitative 'indicators' to measure research processes, products and environment, tailored to institutional and (inter) disciplinary needs, would therefore still be favoured (Oancea 2015). The complexity of the factors that account for good research output were seen as highly consequential. Indeed:

---

[5]For Science Technology Engineering and Maths (STEM) subjects these were based on external research income; for the arts, humanities and social sciences these were to be a 'basket of metrics' including: research grant and contract income; research volume; Ph.D. completions; quality and output of research; bibliometrics; user impact; peer esteem; research council evaluation; and institutional assessment.

[6]Earlier reviews of RAE had been conducted by Dearing in 1997, and Roberts in (2003).

> With or without metrics, higher education needs a healthy ecology of funding sources, institutional structures and career paths; structural conditions for responsible governance; and independence in the exercise of scholarship (Oancea 2015, p. 21)

Internationally, in an arena where some would claim that research assessment in the UK has led the way, there is a common insistence among policy makers about making the research performance of higher education institutions more public. Governments and other research users have increased their demands on academics to be more accountable and to abide by given criteria for the assessment of the quality of their research—a not unreasonable request with respect to public funded research. However, the focus of such assessments has shifted:

> from traditional concerns with scientific worth and contribution to the advancement of the discipline, towards considerations of quality, productivity, performance, economic efficiency, impact and use, feasibility and capacity (Oancea 2008, p. 110)

There has been movement away from modes of assessment which are 'professional'—which seek to encourage dialogue about findings between assessors and assessed—to more hierarchical, technical and managerial forms of assessment designed to increase accountability (Oancea 2009). It is also evident that assessment of research excellence has been increasingly stratified at the macro (international, national, multidisciplinary, and disciplinary); meso (organisations, research units, programmes); and micro levels (teams, individuals projects, products and outcomes). Using data from six countries (Australia, Denmark, France, Netherlands, New Zealand, United Kingdom) Oancea (2008) indicates the ways in which national authorities and governments have increasingly attempted to stratify and specialise their research assessment measures. The aspiration is for assessment to increase accountability and quality assurance in research, towards performance-based fund allocation; unsurprisingly this has had rather negative impacts on the nurturing of research cultures. Oancea's (2008) conclusion, citing Slowey (1995), is that although academics and research communities have an input into such assessment criteria, ultimately decision making tends to be centralised—reflecting shifts in policy and the rise of externalised measures of quality.

6.7 How Does Education Fare?

When the grading of research outputs in the REF unit of assessment 'Education' is weighed against grades awarded elsewhere, it is apparent that fewer optimum grades are given compared with cognate fields. This has a harmful influence on wider perceptions of research quality in education, both publicly and within universities—the effects are witnessed through subsequent revisions and re-grouping of research structures within many departments; the increase of internal and external scrutiny of research activity, outputs, and performance; and changes to the substantive and methodological focus of educational research. Overall, in many post-1992 universities, modest or poor performance in research assessment exercises has increased strategic thinking about research—often putting educational research 'on the agenda' for the first time. Entrepreneurial institutions have seen how research activity can be lucrative and self-sustaining, rather than a drain on resources. However, the nurturing of research activity, and the preparation for the assessment of research, tends to be very 'top down' in most institutions—which often creates tensions between those deemed 'research active' and those classified as 'teaching' staff. There is also evidence that research assessments can downplay the importance of pedagogic research. The benefits of high performance in the REF, specifically with respect to increasing staff involvement in research, are tangible—these include, the widening of career prospects of research active academics, the improvement of their professional identity and confidence, and the enhancement of their performance as a consequence of better support and encouragement (Oancea 2010b). But these impacts are complex and unpredictable—research assessment can have both positive and negative effects in the same institution, even within the same sub groups of research-active staff.

The main effects of research assessments are therefore not always those intended—an inevitable consequence of how assessment data is used. REF results are manipulated for managerial purposes by most higher education institutions—being used to decide on, and change, the contracts of academic staff; to identify academics who might be enticed to join the institution in the expectation that they will significantly contribute to, and possibly raise, its research performance; and to decide on future research funding priorities. Inevitably, research assessment has also led to considerable 'game playing'—head hunting of staff to boost research ratings, the creation of a 'transfer market' of academic staff, use of short or fixed term contracts to employ staff only during the immediate assessment period, and practices such as naming staff on publications for which they have made little academic input. With respect to the publication of research findings there has been a favouring of peer reviewed journals, over chapters, monographs, edited books and professional publications. Finally, the greater emphasis on theoretical work endangers pedagogical research, while smaller pockets of research expertise may suffer (Dadds and Kynch 2003; Giddens 1984). Many of the effects of research assessment may not be tangible and quantifiable, such as the creation of climates of divisiveness, demoralization and a sense of unfairness (AUT 2002).

It is known that most schools and departments of education in the UK have returned fewer members of staff to the research assessment exercises, year on year—this reflects fears that standards of work submitted are not high enough, with concomitant effects on reducing QR funding. The result is that the QR budgets for many faculties, institutes, departments and schools of education have steadily reduced. The experience of geography education research mirrors this trend—indeed, virtually all subject-based research has been included in this decline. The picture is interesting in that over the last 20 years QR money allocated across units of assessment has dropped by almost one half, even though education (because of the size of its staffing and returns) has always received one of the largest QR pay outs nationally. Nonetheless, Furlong (2013) describes how education has seen the biggest drop in QR funding of all the social sciences, many of which have experienced significant increases in recent years.

6.8 The Effects of REF on the Quality of Geography Education Research

There have been perennial concerns about the quality of research in geography education and about the nature of the publications and products that have resulted from research activity (see Lidstone and Gerber 1995, 1997). Such concerns have been heightened by the outcomes of the periodic assessment of research environments, outputs and impacts conducted in the UK. Geography education researchers are open about their anxieties—almost thirty years ago, a special edition of the journal *International Research in Geographical and Environmental Education*, posed the question 'How shall we judge the quality of research in geographical and environmental education?' (Naish 1993). Respondents tended to outline similar measures, characteristics and features of high-quality research, although their emphasis on the importance of using qualitative or quantitative methods varied (see Benejam 1993; Boardman 1993; Clary 1993; Purnell 1993; Schrettenbrunner 1993). The focus on trying to define the characteristics of outstanding research in the field has endured, being sharpened by the magnitude of the consequences of the REF assessments (see Butt 2006, 2010a, b, 2015, 2018).

It is possible that research assessments serve to encourage academics to 'raise their game' in terms of the amount and quality of the research they produce. It is equally possible, of course, that they have the opposite effect. Harley (2002) reports that early and mid-career academics have felt under the most pressure to adapt their working practices to respond to the demands of research assessments—making changes with which they often do not agree, and which do not necessarily enhance the nature and status of the research they undertake. There are inevitable questions in geography education about what constitutes research standards and about what forms of research are prioritised in research assessment exercises—with McNay (2003) suggesting that empirical research is favoured over non-empirical; academic research more than professional or applied; funded rather than non-funded; and disciplinary more than inter-disciplinary. The issue of maintaining *quality* in research in geography education, as in other subject-based research, is obviously of great importance. The demands of achieving high quality research in geography education are always with us, particularly given recent changes in teacher education in the UK and the resultant effects this has had on research productivity.

Oancea (2009), in her analysis of the publication patterns of three major British education research journals,[7] has provided tentative indicators of some of the effects on research quality of the periodic research assessment exercises in England. In essence, she believes that:

> while effective at screening out poor quality research ... the subsequent funding decisions based on the results, may have endangered pockets of expertise and emerging research cultures (p. 3)

---

[7]The total percentages of articles analysed from the three sampled journals were: 63% from high ranking institutions, 15% from low ranking institutions, and 22% from 'other'.

This is most apparent in the reduction of funds for institutions that are lowly rated, with consequences for the kinds of research conducted, for who is considered appropriate to promote as research active staff, for where research funds will be concentrated, and on the behaviour of researchers and research groups. The importance of being 'research active' and 'returnable', in career terms, is very significant in many institutions, leading to an implicit downgrading of those academics who would regard themselves as university *teachers*. This may have been an intended consequence of the actions of governments who wished not only to establish a system of national research assessment—but also to reduce wasteful funding, and to reward research excellence (Gillies 2007). Less conventional research and the work of subject-based research groups, such as geography education research, may have been most strongly affected by this. The relationship between teaching and research—with the suppression of the former and promotion of the latter—has also changed funding and resource provision, staffing and student recruitment, mobility of staff and the continuation of research plans. As Oancea (2009) states:

> It seems quite unacceptable these days for researchers to spend their time in reflection, critique, meaningful interaction with others, and long-term pondering of evidence, if while doing this they fail to keep up with the required cadence of publication and proposal writing (p. 4)

Her research into the expected audiences[8] for research makes for interesting reading. Oancea analysed these audiences against the research assessment ratings of institutions—the highest rated institutions (predominantly Russell Group) produced research which was targeted at other researchers and policymakers, while the lowest rated institutions had marginally more research directed towards audiences such as teachers, students, school administrators and non-British interest groups. The research cultures of low achieving institutions seemed to favour producing research for practicing teachers, aimed at supporting their development—a feature of much geography education research. This finding was, not surprisingly, reflected in the types of research groupings observed in high and low ranked institutions—high ranking institutions could invest in longitudinal research and tended to explore methodological issues. The focus of academic research in these institutions was on areas such as: the economics of education, the politics of education, educational psychology, learning out of school, assessment, philosophy of education, Special Educational Needs (SEN), comparative and international education, and child development; by contrast, low ranking institutions tended to favour research into teacher supply and retention; problem-based learning; secondary education; school-based learning, very able pupils, Physical Education (PE), business education, Information and Communications Technology (ICT), Further Education (FE) and Higher Education (HE), and continuing education/lifelong learning. These institutions also favoured the use of action research. Oancea (2009) is at pains to point out that her analysis does not necessarily accurately map the education research landscape of the time, but

---

[8]These were: researchers, teachers, policymakers, administrators, students, employers, general public and international audiences.

seems indicative of the impact of managerial definitions of research and of practices of accountability in research reporting throughout the home countries of the UK (p. 12)

Significant increases in QR funding, following successful REF performance, have been transformative for some higher education institutions: investment in professorial appointments, capacity building, professional development, funded doctoral studentships and post-doctoral fellowships—as well as supporting visiting and emeritus fellows—have all been consequences of REF success. But there is huge diversity: the variation in quality of education research, not least in its 'sub fields' such as geography education, is highly significant. Where education research is of the best quality, it stands shoulder to shoulder with that produced within the other social sciences, however there is often an acute contrast between 'the innovative and the pedestrian' (Furlong 2013, p. 98).

6.9   The Research Context with Reference to Initial Teacher Education (ITE)

ITE in the UK has previously established that the principles of teaching, and teacher education, are at least partly driven by research. Consequently, ITE in HEIs contains strong elements of research-informed and research-generating practice. In 2013, the British Education Research Association (BERA) with the Royal Society of Arts (RSA) commissioned a paper on policy and practice in ITE (Beauchamp et al. 2013), to report on the relationships between the revised teacher standards (DfE 2013) and research-informed teacher education provision across the four jurisdictions of the UK. This discovered a distinct 'turn or (re)turn to the practical' (Furlong 2013) in teacher education, highlighting the increasingly school-based and craft (rather than research) driven nature of ITE—especially within England, which was recognised as an 'outlier' in terms of beginning teachers' experiences of ITE. The UK Coalition Government's education policy from 2010 onwards indicated a decisive shift away from the idea of teaching being a research-based profession, and the steady removal of research from teacher preparation—in favour of promoting practical performance and experiential knowledge over 'theoretical, pedagogical and subject knowledge' (Beauchamp et al. 2013, p. 2). The knowledge base essential for teaching therefore became dominated by 'good subject and curriculum knowledge', with little explicit reference being made to teachers' engagement with research. This was seen by some as a reflection of the teaching standards, which were described by Mahony and Hextall (2000) as regulatory, rather than developmental, in intent.

6.10   Advancing Research Quality

When considering how geography education research, and researchers, can sensibly respond to the research environment they find themselves in, it is tempting to be somewhat down hearted. The policy and research assessment mechanisms, combined with the shift of initial teacher education into schools, mitigates against individuals who wish to forge their careers within geography education research. Nonetheless, in a spirit of striving to advance research quality, it is worthwhile to briefly consider how large-scale research organisations view their research profiles. This has implications for how the geography education community might evaluate, and change, its own research practices and outputs.

It is obvious that the levels of funding, organisational structures, and staffing profiles associated with prestigious research projects often lie outside the day-to-day experiences of geography education researchers. For Pollard (2005), large scale research projects embrace considerations of research quality which extend beyond the usual methodological and theoretical issues most researchers face, embracing: (i) user engagement for relevance and quality; (ii) knowledge generation by project teams; (iii) knowledge synthesis through thematic activities; (iv) knowledge transformation for impact; (v) capacity building for professional development; and (vi) partnerships for sustainability. Such quality indicators are difficult for individual researchers to address, but provide helpful yardsticks against which to measure current performance. Furlong and Oancea (2005) also note that 'research quality' is no longer solely defined within theoretical, ideological or methodological parameters, but by multidimensional criteria which embrace how others might engage with research findings. They outline four major dimensions of research quality:

(i)   Epistemic—traditional theoretical and methodological robustness. These qualities are usually most visible in research reports and are the key dimensions of excellence used in assessing social research. Sub dimensions: trustworthiness, contribution to knowledge, explicitness in designing and reporting, propriety, paradigm dependent criteria.

(ii) Technological—provision of facts, evidence, new ideas. Sub dimensions: timeliness, purposivity, specificity and accessibility, impact, flexibility and operationalisability.

(iii) Capacity building and value for people—including, potential for collective and personal growth, changing people through collaboration and partnership, increasing receptiveness, moral and ethical growth. Sub dimensions: partnership, collaboration and engagement; plausibility; reflection and criticism; receptiveness; personal growth.

(iv) Economic—whether the money invested in a research project is well spent, or adds value. Sub dimensions: cost-effectiveness; marketability and competitiveness, auditability; feasibility; originality; value efficiency.

These quality indicators are illustrative, rather than prescriptive, and do not suggest any procedures that might link to their provision or assessment. Evaluation of research quality is obviously dependent on the purposes, intended audiences and aims of the research—factors which vary according to the type and scale of the research undertaken, but which should be considered for any research endeavours in geography education.

6.11 Conclusions

The UK has played a major role, nationally and internationally, in developing advanced research evaluation and accountability measures—although whether the effects of such assessment have been wholly beneficial remains debateable (Oancea 2008). In education, not least in geography education, RAE and REF have both focused attention on the overall performativity and value for money of research. Undoubtedly, REF has become 'big business'—in the last research assessment (2014, at the time of writing) the unit of assessment for Education 'returned' around 1800 active researchers in the UK, with research-driven institutions benefiting from the distribution of an annual budget in the region of £85 million. However, these figures mask a complex and multi-faceted range of outcomes—institutions are now increasingly selective about which staff they enter into the REF, resulting in a reduction in the numbers of staff 'returned'. In the period between RAE 2001 and 2008, there was a 17% reduction in staff 'returned' in the UK, and a 23% reduction in England; the total funding awarded to education research was reduced by £3.4 million (BERA/UCET Review 2010).

The ways in which research quality is evaluated are obviously important, for these have a direct impact on the scale and nature of future research activity. Oancea (2008) identifies a mismatch between economic, strategic, professional and academic indicators, noting that:

> As we move from internal evaluations, professional contexts, and small-scale assessments towards external evaluations, administrative and policy contexts, and large-scale reporting, productivity, competitiveness and medium-term impact become core values of research assessment (p. 116)

This ties with raised expectations of achieving greater objectivity in assessment, possibly through the use of metrics. The assessment panels involved in making judgements of research quality help to set a blueprint for the shape of future research—especially for small research communities such as those involved in geography education research. This may be problematic if these panels—either by design or by chance—make pejorative decisions which shape future research questions, methods and methodologies for research communities. Assessment panels clearly set the tone for the evaluation of 'high quality' research. As Oancea (2009) states, this has a differential impact on accountability:

> Hence 'corporate' answerability may be advocated in public management circles, but deemed dangerous and unacceptably narrow in academic ones; while 'collegial' accountability may be perceived from outside the inner sanctum of academia as too weak and unstructured (p. 16)

As a community which has traditionally valued its collegiality, geography education researchers may have been forced onto the back foot. More importantly, there are
dangers that the research assessment mechanisms, by encouraging competition and
selectivity, create quasi markets within state funded research. This places an onus on
research leaders in education, and in other disciplines, to be able to accurately 'read'
the impact of high stakes assessments to assess the quality of research across different sub fields, institutions, cultures and communities (Oancea 2010a). Assessment
encourages us to consider the ways in which research decisions impact (i) internally in
HEIs, across different staff, schools, departments and research teams, and (ii) externally, across different competitors, partners, users, beneficiaries, and national and
international research structures and organisations (Oancea 2010b). This complex
landscape of 'winners and losers'—and the questions raised about where geography
education research, and researchers, sit within it—is more complicated than REF
data shows, or QR distributions suggest.

Participants in workshops organised by BERA and UCET to review the prospects
for educational research in the UK HEIs were invited to reflect on the challenges and
opportunities posed by national research assessment exercises. Table 6.2 summarises

Table 6.2 SWOT analysis, based on BERA/UCET review workshops, from *Prospects for Education Research in Education Departments in Higher education Institutions in the UK*, reprinted with
permission of the BERA/UCET

Strengths	Weaknesses
Size and diversity of education sector and its market	Fragmentation
Some departments recognised within HEIs for their income generation	Dependence on limited government funding
Education still priority (economic rationale), partly protected from cuts	Conflicted strategic vision across the sector and pressures towards increased inter-institutional competition
Tradition of cross-disciplinary cooperation	Difficult internal positioning of departments within HEIs—pressures to downsize and diversify
Established links between research, policy and practice—recognition of public service role	
Opportunities	**Threats**
Current good position/rewards for international and interdisciplinary research across the board	Scarce funding sources in current economic and political climate
Impact agenda—education well placed > retying research and practice	Increased selectivity and changes in student funding
Teacher educators to undertake practice-based research	Abolition of public bodies
Growth in publication outlets and outputs	Completion for limited sources from increasing number of alternative providers (commercial, think tanks, social entrepreneurs)
Institutional support to develop capacity and improve quality in research	HE research's position of authority challenged
Philanthropists' interest and ESRC priority areas—opportunities to pitch	Pressure to separate research and teaching activity
	Institutional morale/culture
	Staff profile (late entry/older profile)

Oancea et al. (2012)

participants' suggestions in a Strengths (S), Weaknesses (W), Opportunities (O), and Threats (T) format, providing a useful culmination to this chapter.

The *impact* of educational research—the monitoring of which is part of the assessment process—is often downplayed or simply not realised. This issue is not uncommon among the social sciences, for evaluating effective impact is challenging—most research has an intangible, or indirect, impact on policy and practice, only building incrementally on what we already know and do. Here is a dilemma which faces much research in education: inside universities, educational research is generally seen as being of low value and status, linked too closely to practice as an applied field; outside universities, in schools and colleges, it is seen as 'overly academic' and 'out of touch', being largely irrelevant to the day-to-day needs and challenges faced by students and teachers alike.

In conclusion, we must acknowledge that research in most schools, departments, faculties and institutes of education is actually funded, in the main, through *teaching*. The amount of research money gained through competitive bidding for research contracts, and from QR, is much less than the income generated from teaching—in the past, much of this teaching was carried out within initial teacher education programmes. This raises a serious issue: as initial teacher education in England and elsewhere shifts into schools, and staffing and teaching income from ITE shrinks, there is an inevitable impact on subject-based research. In essence: as teaching income falls there is a direct, 'knock on' effect on levels of research activity. In the UK, lectureships, Readerships and Professorial posts in geography education have always been small in number—the academic staff fortunate enough to occupy these positions have traditionally been the 'foundations' for research in the field, but these secure footings are rapidly being removed. Like other sub disciplines in education (such as sociology, psychology, philosophy) it is increasingly likely that geography educationists and researchers will have to look away from their institutions—to professional associations, learned societies, specialist journals and conferences, Council of British Geographers (CoBRIG), and Geography Education Research Collective (GEReCo)—for future lifelines of support.

References

AUT (Association of University Teachers). (2002). *The future of education.* London: AUT.

Beauchamp, G., Clarke, L., Hulme, M., & Murray, J. (2013). *Policy and practice within the United Kingdom: Research and teacher education: The BERA RSA enquiry.* London: BERA RSA.

Benejam, P. (1993). Quality in research in geography education. *International Research in Geographical and Environmental Education, 2*(1), 81–84.

Boardman, D. (1993). Evaluating quality in research: Asking why? As well as how? *International Research in Geographical and Environmental Education, 2*(1), 85–88.

Butt, G. (2006). How should we determine research quality in geography education? In K. Purnell, J. Lidstone, & S. Hodgson (Eds.), *Changes in geographical education: Past, present and future* (pp. 91–95). *Proceedings of the IGU-Commission on Geographical Education Symposium.* Brisbane: IGU CGE.

Butt, G. (2010a). Perspectives on research in geography education. *International Research in Geographical and Environmental Education., 19*(2), 79–82.

Butt, G. (2010b). Which methods are best suited to the production of high-quality research in geography education? *International Research in Geographical and Environmental Education, 19*(2), 101–105.

Butt, G. (2015). What is the role of theory? In G. Butt (Ed.), *MasterClass in geography education* (pp. 81–93). London: Bloomsbury.

Butt,G. (2018). *What is the future for subject-based education research?* Public seminar delivered at the Oxford University Department of Education. 22 October www.podcasts.ox.ac.uk.

Clary, M. (1993). Quality criteria in action research in environmental education. *International Research in Geographical and Environmental Education, 2*(1), 76–80.

Dadds, M., & Kynch, C. (2003). The impact of the RAE 3B rating on educational research in teacher education departments. *Research Intelligence.*

Department for Education (DfE). (2013). *Teachers' standards.* http://media.education.gov.uk/assets/files/pdf/t/teachers%20standards%20information.pdf.

Furlong, J. (2013). *Education: An anatomy of the discipline.* London: Routledge.

Furlong, J., & Oancea, A. (2005). *Assessing quality in applied and practice-based educational research.* ESRC TLRP seminar series. (December 1) Oxford University Department of Educational Studies.

Giddens, A. (1984). *The constitution of society.* Oxford: Polity Press.

Gillies, D. (2007). Lessons from the history and philosophy of science regarding the research assessment Exercise. *Philosophy of Science. Supplement to Philosophy, 61,* 37–73.

Goudie, A. (1993). Schools and universities: The great divide. *Geography, 78*(4), 338–339.

Hackett, E., & Chubin, D. (2003) *Peer review for the 21st century: Applications to education research.* Paper to National Research Council (SUA) workshop, Washington DC, 25 February 2003.

Harley, S. (2002). The impact of research selectivity on academic work and identity in UK universities. *Studies in Higher Education, 27*(2), 187–205.

HEFCE (Higher Education Funding Council for England). (2007). *Research excellence framework. Consultation on the assessment and funding of higher education research post 2008.* London: HEFCE.

HEFCE (Higher Education Funding Council for England). (2008). *Analysis of responses to HEFCE 2007/34, the research excellence framework.* London: HEFCE.

Lidstone, J., & Gerber, R. (1995). Editorial: Quality as a feature of research reporting. *International Research in Geographical and Environmental Education, 4*(1), 1–2.

Lidstone, J., & Gerber, R. (1997). Editorial: Academic paper or personal essay: The qualities of quality research. *International Research in Geographical and Environmental Education, 6*(1), 1–3.

Mahony, P., & Hextall, I. (2000). *Reconstructing teaching: Standards, performance and accountability.* London: Routledge Falmer.

McNay, I. (2003). Assessing the assessment: An analysis of the UK Research Assessment Exercise 2001, and its outcomes, with special reference to research in education. *Science and Public Policy, 30*(1), 47–54.

Naish, M. (1993). 'Never mind the quality—feel the width'—How shall we judge the quality of research in geographical and environmental education? *International Research in Geographical and Environmental Education., 2*(1), 64–65.

Oancea, A. (2008). Standardisation and versatility in the assessment of education research in the United Kingdom. In A. Besley (Ed.), *Assessing the quality of educational research in higher education* (pp. 105–135). Sense.

Oancea, A. (2009). Performative accountability and the UK research assessment exercise. *Access: Critical perspectives on communication, cultural and policy studies, 27*(1/2), 153–173.

Oancea, A. (2010a). *Research assessment in the United Kingdom. World social science report: Knowledge divides* (pp. 259–261). Paris: UNESCO Publishing/ International Social Science Council.

Oancea, A. (2010b). *The BERA/UCET review of the impacts of RAE 2008 on education research in UK higher education institutions*. London: BERA/UCET.

Oancea, A. (2014). Research assessment as governance technology in the United Kingdom: Findings from a survey of RAE 2008 impacts. *Zeitschrift fur Erziehungswissenschaft, 17*(6), 83–110.

Oancea, A. (2015). Metrics debate must be ethical as well as technical. *Research fortnight* (22 July) (p. 21).

Oancea, A. (2016). *Challenging the grudging consensus behind the REF*. https://www.timeshighereducation.com/blog/challenging-grudging-consensus-behind-ref. Accessed March 25, 2016.

Oancea, A., McNamara, O., & Christie, D. (2012). Evolution and direction of the field. In D. Christie, M. Donoghue, G. Kirk, O. McNamara, M. Menter, G. Moss, J. Noble-Rogers, A. Oancea, C. Rogers, P. Thomson & G. Whitty (Eds.), *Prospects for education research in education departments in higher education institute in the UK*. London: BERA/UCET Working Group on Education Research.

Pollard, A. (2005) *Taking the initiative? TLRP and educational research*. Educational Review Guest lecture. 12 October 2005. University of Birmingham.

Purnell, K. (1993). Evaluating quality geographical and environmental education research. *International Research in Geographical and Environmental Education, 2*(1), 70–72.

Ranson, S. (2003). Public accountability in the age of neo-liberal governance. *Journal of Education Policy, 18*(5), 459–480.

Roberts, M. (2003). *Learning through enquiry: Making sense of geography in the Key Stage 3 classroom*. Sheffield: Geographical Association.

Schrettenbrunner, H. (1993). How to judge quality in research. *International Research in Geographical and Environmental Education, 2*(1), 73–75.

Stern, N. (2016). *Building on success and learning from experience an independent review of the research excellence framework (Stern Review)*. London: Department for Business, Energy and Industrial Strategy.

Slowey, M. (Ed.). (1995). *Reflections on change: Academics in leadership roles. Implementing change from within universities and colleges*. London: Kogan Page.

van Raan, A., Moed, H. & van Leeuwen, T. (2007). *Scoping study of the sue of bibliometric analysis to measure the quality of research in UK higher education institutions*. Report to HEFCE by Centre for Science and Technology Studies, Leiden University. London: HEFCE.

Chapter 7
International Perspectives on Geography Education Research

7.1 Context

This chapter offers an analysis, at the international scale, of 'why things are the way they are' in geography education research—although it comes with a caveat that achieving an overview of research across different national jurisdictions is problematic, and possibly even unhelpful. In a previous publication, David Lambert and I acknowledged that attempting to create a passable account of almost any aspects of the global 'state of play' in geography education was fraught with difficulties (Butt and Lambert 2014). Nevertheless, this chapter aims to identify the key drivers and inhibitors of geography education research and how these characterise research in national settings—noting the ways in which they are influenced by each country's educational trends, societal concerns, government priorities and historical research traditions.

Sahlberg (2012) coined the term Global Education Reform Movement (or GERM) to describe how education policies and practices are now expanding globally 'like an epidemic that spreads and infects education systems through a virus' (Sahlberg 2012). These education reforms, to which a number of countries previously seemed immune, have seen national education policies become articulated around three key principles: accountability, standards and decentralisation. Globalisation had undoubtedly accelerated policy dissemination and reform, particularly where nations appear to have similar problems and priorities, allowing for the adoption of common policy rationales (so called, 'policy borrowing'). Indeed, Salhlberg:

> has identified the principal features of GERM as increased standardisation, a narrowing of the curriculum to focus on core subjects/knowledge, the growth of high stakes accountability and the use of corporate management practices as the key features of the new orthodoxy (Fuller and Stevenson 2019, p. 1)

The process has led to the adoption of a melange of education policies and practices, creating education systems that feel 'cracked' for the teachers, students, researchers and other stakeholders who experience them (Ball and Olmedo 2013,

© Springer Nature Switzerland AG 2020

G. Butt, *Geography Education Research in the UK: Retrospect and Prospect*, International Perspectives on Geographical Education, https://doi.org/10.1007/978-3-030-25954-9_7

p. 85). These policies first became visible in education reforms introduced 30 years ago in the United States, UK and Chile—countries whose political leadership followed the ideological direction of the New Right—spawning, in the UK, the 1988 Education Reform Act and the first national curriculum for state schools. Since the early 1990s, the adoption of similar education policies increased around the world, although their influence may vary in form, content, pace and feature according to the national context in which they are introduced. The impact of neoliberal policy on geography education, and geography education research, worldwide needs to be recognised and responded to.

Many of the major challenges that geography education and geography education research face, internationally, are obvious and persistent: reductions in the numbers of specialist geography teachers and researchers; the dominance of assessment and performance-led systems in schools; the growth of research into technological applications (internet, multimedia and GIS), possibly at the expense of other areas of research; and the demands of public/policy/parental expectations of geography as a school subject (after Rawling 2004). It is evident that the condition of geography education in schools, and of geography education research wherever it occurs, is a complex one, both nationally and internationally. While it is possible to highlight some of the widespread, generic international concerns that attract research interest and funding in education—such as, inter alia, assessed student performance; the impacts of Programme for International Student Assessment (PISA) scores; the competences/skills agenda; teacher workload, recruitment and retention—the manner in which these affect the geography curriculum in schools, and the preferred themes for geography education research, will vary from country to country. Such themes often reflect societal and political concerns about how schools prepare young people for life in particular national contexts—for example, how education should be employed to promote intercultural understanding, moral and spiritual education, and citizenship—which can have implications for the focus of geography education research.

This chapter also discusses the significance of the IGU CGE International Declaration on Research in Geography Education (IGU CGE 2015)—which was written almost a quarter of a century *after* the original International Charter for Geography Education (IGU CGE 1992). In this context, the historical dominance of the Euro American traditions in geography education research, theory and thought—and how this influence is now changing, with the growth of research in Korea, Singapore, Hong Kong, Japan, China and beyond—is considered. The research priorities for geography education identified by the USA Roadmap are highlighted and their applicability at the international scale discussed.

7.2 Introduction

It is perhaps sensible to start with a broad overview of the ways in which education research, as a field, has developed across the world. Wagner et al. (1993) note how, in the nineteenth century, different professional groups sought to define themselves partly through persuading universities to offer aspects of their specialist bodies of knowledge to a wider population. In diverse national contexts this played out in different ways, influencing the components and qualities of discrete professions. In the domain of education, Furlong (2013) offers the thought that university-based research into the field—which helped to underpin its professional status—has had a variety of drivers in different national contexts:

> in the French and German traditions (educational enquiry) is understood primarily as a science, focusing on theory and basic research, in the Anglophone world it is built on an enduring but unstable pragmatic compromise, a compromise between theory and practice, between knowing that and knowing how, in the commitment of the academy to make both an intellectual and a practical contribution to the advancement of the field (p. 7)

Education, as a discipline, is currently well represented in universities in the UK and increasingly so around the world, exhibiting strong 'institutional reality':

> In the UK today there are some 5,000 academics who are employed in university education departments; it is the UK's second largest social science and is equally large in other English speaking countries: the US, Australia, Canada, Malaysia, Hong Kong, and New Zealand (Furlong 2013, p. 11)

Furlong also notes that there are large numbers of academics who research into educational themes in other university departments and faculties, as well as a plethora of journals, learned societies, professional associations and conferences that position themselves under the label of 'education'.

Unfortunately, for a number of valid reasons, analysis of the international picture of educational research, and therefore of research into geography education, is convoluted and complicated. The very nature of 'research in education' is described differently across different jurisdictions—in the UK such research tends to be described as either 'education research' or 'educational research' (with the distinction being broadly linked to research endeavours that are either beyond, or within, the classroom), but in Europe the favoured terms may be 'pedagogical research', 'didaktics research', or even 'educational science' (Furlong 2013; Hofstetter and Schneuwly 2012). Each of these terms is somewhat problematic, as none are precise enough to capture the gamut of research enterprises in education—which vary from the practical to the epistemological. Gerber and Williams (2000), in their brief overview of international perspectives on geography education research, noted at the beginning of the new millennium that the field had experienced three decades worth of 'dramatic increase in the numbers of researchers' (p. 209)—although no hard, empirical evidence for its rate or scale of growth, or for its geographical spread, is offered. Their assertion is based on the record of publications collected in a British bibliography of geography education research (Foskett and Marsden 1998), and an American bibliography on the internet, collated by the National Council for Geographic Education (NCGE 2000). These are valid, but rather piecemeal and selective, measures of research outputs that understandably could never completely capture the global picture of geography education research. In passing, Gerber and Williams (2000) also refer to an increasing breadth of research methods and ways of reporting research in geography education, in a variety of different languages—although again these points are not pursued further.

Williams (1998) had previously explored a number of features of the global research culture in geography education, suggesting that there are three periods in the development of research from an incipient, to an intermediate, and then a mature stage (see Table 7.1):

Despite Gerber and Williams' (2000) claim that their positioning of research in geography education worldwide firmly in the incipient stage did *not* necessarily adversely affect the 'quality and quantity of the work being undertaken' (p. 209), the factors within this stage almost certainly colour the views of *other* researchers, curriculum policy makers and classroom practitioners. From the perspective of a further 20 years of the development of geography education research since Williams (1998) wrote, it is now difficult to agree with his claim that a 'fundamental shift … from the relatively isolated individual functioning in one institution to an international group of researchers collaborating on a single project' (p. 2) has occurred since the late 20th century. There *is* evidence of greater international collaboration in research in geography education—within some impressive international research projects, noted later in this chapter—which builds on the strong traditions of conferences, symposiums and seminars where geography education researchers meet, exchange ideas and arrange to produce research-based publications. But the reported *fundamental* shift is hard to recognise. It is, however, difficult to argue with the rather self-evident point that to progress, research in geography education still requires a strong research infrastructure and to contribute to 'frequent research-focused symposia and conferences with publications appearing in refereed specialist and general journals and on websites' (Gerber and Williams 2000, p. 211); unfortunately Gerber and Williams do not attempt to offer a strategy for how to achieve sustained growth in geography educational research—they simply make a plea for governments and non-government funding agencies to include geography education research higher up in their funding priorities. On the one hand, we should not be too harsh in judging Gerber and Williams efforts—it is extremely difficult to fashion a strategic plan across jurisdictions which might exemplify the case for further support for geography education research. On the other, there is neither a justification, nor even an argued case, presented as to why geography education research *matters*—simply a caution that research into revisions of geography curricula globally often occurs without recourse to previous research findings, and that geography education research tends not to be funded nationally. Despite what might be regarded as an unfortunate naivety associated with making such relativist statements—essentially, expressing the expectation that geography education research and researchers should be recognised and funded, regardless of limited evidence of their capacity to deliver anything of wider value and appeal—there is an honesty and directness in the way in which Gerber and Williams (2000) comment on the nature of much geography education research, globally:

> Generally, there would appear to be little interest among geographical educators in undertaking research that is narrowly defined in terms of theory construction or the refinement of research methodologies. There appears to be a strong interest in practitioner-based research (p. 212)

Table 7.1 Stages of growth in the culture of research in geographical education (after Williams 1998), from Kent, A (ed) Issues for Research in Geographical Education

Incipient stage	Intermediate stage	Mature stage
Individuals researching in isolation	Intra-institutional groups	International groups
Idiosyncratic and changing substantive focuses	Stable substantive focuses though subject to personnel changes	Enduring substantial focuses that are unaffected by personnel changes
Unfunded	Funded by local and national organizations	Funded by international organizations
Unsupported by a professional body	Supported on the margins of a national professional body	Central to the work of an international body
Dominated by immediate practical issues	Linked practical and theoretical issues	Dominated by universal theoretical issues
Focused largely on a single sector of a national education system	Focused on more than one sector of a national education system	Focused on lifelog learning in an international context
Undeveloped specialist geographical education research language	Emergence of a specialist geographical education research language	Use of a sophisticated geographical education research language
Absence of textbooks on geographical education research	Introductory textbooks on geographical education research	An array of established textbooks on geographical education research
Lacking close ties with conventional educational disciplines	Developing ties with a number of educational research disciplines	Closely integrated into the educational research community.
Lacking any subdiscipline strengths within geographical education	Emergent subdiscipline strengths within geographical education	Established subdiscipline communities within research in geographical education
No infrastructure of research-focused symposia, conferences, web pages, journals and other publications	Developing nationally based infrastructure of research-focused symposia, conferences, web pages, journals and other publications	Well developed international infrastructure comprising symposia, conferences, web pages, journals and other publications
Few opportunities for training in research in geographical education	Limited national opportunities for training in research in geographical education	Many international opportunities for training in research in geographical education

Research Forum 1, reprinted with permission of the publishers, Institute of Education Press, University of London and IGU-CGE.

They indicate that much research in geography education is classroom and practitioner-based, essentially of a 'what works' variety and not closely linked to any social science discipline. This is research intended to improve professional practice, but not to step beyond it. Williams (1998) indicated the dangers of this approach, and of geography education research that was too policy driven, when he stated:

> Research in geographical education ought not, in my opinion, become a political football used in a game where the goal posts are ever changing. Far better would it be if the research agenda was defined according to concerns for improved research methodologies and substantive issues of an enduring and universal kind (p. 6)

7.3   The Role of the International Geographical Union (IGU)

Internationally, the IGU, and its commission for geography education (IGU CGE), has been influential in progressing the cause of geography education research. Worldwide meetings, symposia and conferences of geographers, geography educationists and researchers date back over a century—the first geographical congress being held in Antwerp in 1871, with the IGU established in Brussels in 1922 (Chalmers 2006). The IGU has since consisted of a tripartite structure—a General Assembly of delegates sent from member countries; an Executive Committee; and numerous Commissions (such as the IGU CGE). The IGU-CGE, established in 1952 as the IGU's Commission on the Teaching of Geography in Schools, is a dual language (English and French) organisation, being the longest standing commission of the Union—as well as one of its largest and most active. It is also consistently one of its least well-funded (Wise 1992). The original intention of the first Commission was to establish the case for the teaching of geography in different countries, particularly as an aid to developing international understanding after the Second World War. It was also believed that the promotion of geography education in (so called) 'developing countries' could help to improve environmental knowledge and raise standards of living; in the 'developed world' the methods of geography teaching—and indeed the aims of the subject itself—were always better understood across national borders, however greater information sharing might be afforded. In many countries, geography as a school subject in the post war period was under pressure to integrate with social studies, as had occurred across the US, such that Wise (1992) refers to the launch of the IGU CGE as having occurred at a time when it was important for:

> an international drive to present the value of the subject as an educational discipline, to illustrate the contribution that it could make to national educational aims and to urge again its potential value to international understanding (p. 234)

The IGU CGE was assisted in its aims by collaborating with the United Nations Educational, Scientific and Cultural Organization (UNESCO), whose first—arguably rather prescriptive and didactic—*Handbook of Suggestions on the Teaching of Geography* appeared in 1951, followed in 1965 by a *Sourcebook for Geography Teaching* (originally published in French, English and Spanish, and eventually translated into nine languages including Hindi, Polish and Slovene). In 1982 these publications were followed by the *New UNESCO Sourcebook for Geography Teaching* (which was translated into Russian, as well as other languages) (Wise 1992). The first two books had a distinctly practical edge and were instrumental in raising debate about appropriate 'ideas and approaches' within geography education that might be adopted in the African, Latin American and Asian contexts. The final book was more obviously research-driven. Wise (1992) concludes that the IGU CGE, like all organisations, changed its aims and objectives over time, but that the importance of the research dimension in geography education only increased incrementally:

> The early emphasis on the lowly place of the subject in education and on rectifying the obsolete nature of concepts and teaching methods has given place to an emphasis on the educational strength of the subject and to positive demonstrations, through the Source Books, of the vitality of the subject, of its value as an educational discipline and of its possible applications (p. 244)

This rather upbeat appraisal, made in the last decade of the 20th century, may have been justified at the time—but it would be hard to concur wholeheartedly with it today, particularly when compared to the international status achieved by other subjects and subject-based research groups. Although links with UNESCO certainly supplemented the IGU CGE's funds significantly, providing additional means of communication and distribution of research findings, they also meant that the work the Commission undertook was more formally contracted, and of a different character, to that completed previously. It is important to remind ourselves that the Commissions' work has required the dedication and selflessness of many international scholars, who have given their time voluntarily and without remuneration, in an effort to build international collaborations, and to coordinate research efforts, worthy of 'inspiring a commonality of purpose in colleagues from different lands and from varied traditions' (p.245).

In New Delhi in 1968 a review of all the IGU Commissions had led to the re launch of the 'Commission on the Teaching of Geography in Schools' as a 'Commission on Geography in Education', with a wider remit that included a role for research. At this time the British Sub Committee of the IGU-CGE was launched, becoming:

> the most active of the national sub-committees, meeting twice a year with Norman Graves as convener (Wise 1992, p. 240).

The British Sub Committee[1] remains unique—no other country has established a standing committee of the IGU CGE that meets regularly as a national, or indeed cross national, body. By initially creating links between geographers, geography

---

[1]At the time of writing, the (renamed) UK Committee of the IGU CGE is in close discussion with another UK group—the Geography Education Research Collective (GEReCo)—with a view to a probable merger. This is both practical and sensible, given that the membership of the groups overlaps considerably. Nonetheless, each group has different strengths and intentions—particularly with regard to international reach, research ambitions and publication interests. Significantly, neither group receives any form of sustained funding.

educationists and researchers the Committee was instrumental not only in producing advisory and teaching materials for schools, but also in promoting research into geography education—although this was arguably not its primary aim. The presence of its members at geography education symposia, held before the main conferences of the IGU World Congress every four years, has always been strong—with the British often eager to organise national, continental and global symposia and conferences under the auspices of the IGU CGE. Among these, their regular London conference—hosted for many years by the British Sub Committee at the Institute of Education, University of London—has a longevity and influence that stands out. This gathering has often resulted in post conference publications—which usually present re-worked research papers, in the form of chapters, based on the conference theme. Alongside the conference proceedings, 'spin off' publications—either dedicated books (see, for example, Brooks, Butt and Fargher 2017), separate book chapters, and scholarly or academic journal articles—has given the work of geography education researchers greater global impact.

In 1992 the IGU CGE launched its own peer reviewed academic journal, *International Research into Geographical and Environmental Education*[2] (commonly shortened to 'IRGEE'). The original journal editors, Rod Gerber and John Lidstone, were based in Australia, with the American academic, Joe Stoltman, taking over Rod's editorial duties following his death in 2007. This important journal for disseminating geography education research is currently edited by Chew-Hung Chang (from Singapore) and Gillian Kidman (from Australia)—significantly, this has again positioned its editorial leadership away from direct Euro-American influences.

7.4   Three International Texts—30 Years Apart

The IGU CGE has presided over the production of at least three[3] influential edited books, each with international authorship, which were published approximately every decade for the past 30 years (in 1996, 2006, and 2017).

The Foreword and Introduction to Williams' (1996b) *Understanding Geographical and Environmental Education: the role of research* is interesting for two reasons. Firstly, Hartwig Haubrich—then Chairman (sic) of the Education Commission of the International Geographical Union (IGU), perhaps rather optimistically, states in the book's Foreword that:

> Over the last twenty years research in geographical and environmental education has increased in quality and quantity. Substantial progress has been made in many directions. Not

---

[2]There are other specialist journals dedicated to research in geography education, including *Research in Geographic Education*, founded in 1999, edited by Richard Boehm and David Stea in the US and based in the Gilbert Grosvenor Center for Geographic Education at SW Texas State University, at San Marcos.

[3]A number of other publications of influence—not least the proceedings from the IGU-CGE's international conferences and symposia, and related spin off books such as 'Teaching Geography for a Better World' (Fien and Gerber 1988)—have also been produced.

only have researchers developed small-scale and larger-scale projects which have addressed a wider range of topics, they have also employed increasingly sophisticated research methods (Haubrich 1996a, p. xi)

He proceeds to exalt how greater numbers of geography education researchers from different countries attend the regional and global symposia and conferences of the IGU, and observes that they 'engage in international collaborative projects' as well as bilateral and multilateral research. Haubrich (1996a) reports that he has witnessed:

over a short period of time the blossoming of the research careers of many colleagues drawn not only from universities and other institutions of higher education but also from schools and colleges (p. xi)

From today's perspective on the state of geography education research, globally, this statement appears to be a rather over optimistic description of a 'golden age' of research in the field. It is doubtful whether the extent and longevity of the 'blossoming' period was that remarkable—indeed, the period described by Haubrich was, in terms of research productivity, more modest than he suggests.

Secondly, in Williams' (1996a) Introduction to a text that boasted a reasonably wide international authorship—including 25 authors from the predominantly English-speaking world (14 based in England, 4 Australia, 3 Wales, 2 USA, 1 Hong Kong, 1 Germany)—he comments on the breadth of chapter themes, which are taken to exemplify contemporary research in geography and environmental education. The intention of the publication was to draw on a range of social theory and methodologies to illustrate:

how research is conducted at a variety of scales (classroom, school, regional, national, international), and focused on documentation (historical records, legislation, policies, plans, tests, curriculum materials), persons (pupils, students, teachers, student teachers), ideologies (classical, vocational, reconstructionist, postmodernist) and curriculum (policies, plans, implementation, evaluation (p. 1)

A decade later Michael Williams, with John Lidstone, edited a further publication *Geographical Education in a Changing World* (Lidstone and Williams 2006a). What is interesting about this book is that 10 of the contributing authors are the same as those who wrote for William's publication in 1996—which, if viewed positively, attests to the stability and longevity of many researchers publishing on the international stage in geography education, but can also be read as reflecting a lack of new talent emerging to offer their research findings. Although the authors are from a wider spread of countries, with better representation globally[4] (5 from USA, 4 England, 4

---

[4]The breakdown of authors is taken from the book's stated 'List of Contributors' (pp. ix–xii); however, many chapters include sections contributed by authors not stated in this list—including section contributions from Taiwan, China, Brazil and Singapore—such that the actual total of authors is 33, not 26. We must also remember that for texts such as this some prospective authors will have had their contributions rejected by editors, and that the authorship of publications that emerge from conferences is sometimes rather 'weighted' by numbers of delegates from particular countries having being attracted by the conference theme. Ease of access to funding is a significant factor in attendance, of course.

Australia, 2 Germany, 2 Canada, 1 each from Holland, Chile, Argentina, Wales, Hong Kong, New Zealand, Venezuela, Spain, Finland), there is again a dearth of contributions from researchers and academics in Less Economically Developed Countries (LEDCs). What I present here is a deeply un scientific and superficial analysis—authors who contribute book chapters are chosen in various ways: personal networks are influential, author and editor preferences occur, and often established authors are given opportunities that others are not afforded—but the finding is interesting, nonetheless.

Finally, *The Power of Geographical Thinking* (Brooks et al. 2017), published in 2017, is an offering arising from one of the regional conferences of the Commission held in London. The authorship is again international (7 from Chile, 4 England, 3 Germany, 2 USA, 2 Singapore, and 1 each from Australia, Sweden, Portugal and Korea) with this publication being the first in a series of three for the IGU CGE under the joint editorial supervision of Clare Brooks and Joop van der Schee. Interestingly, the frontispiece to this book makes a clear statement about it being led by 'the priorities and criteria set out in the Commission's Declaration on Geography Education Research' (see Sect. 7.10), with an intention to make geography education research 'accessible to the global community'. The aim to publish research that explores contemporary developments in the geography education community, and also to support the advancement of early career researchers towards publishing high quality and high impact research, is laudable. The main theme of powerful thinking in geography education was topical at the time, with contributions that also related to popular 'key topics', here stated as: the geography curriculum, spatial thinking, GIS, geocapabilities and climate change.

Although I have specifically chosen just three publications from the past 30 years to highlight the research intentions and productivity of members of the IGU CGE, others have at different times made their own choices. Williams (2003), for example, considers two IGU CGE publications 'to illustrate the range of work undertaken by researchers in geographical education' (p. 263)—specifically those by van der Schee et al. (1996) and Houtsonen and Tammilehto (2001). Each of these publications claims to convey innovative research and practice in geographical education, with one of the books (van der Schee et al. 1996), reviewing current international educational journals to identify key themes. Williams contends that of the chapters contained within these publications (n = 74), 45 were 'conceptual and descriptive', 16 'quantitative' and 13 'qualitative'. These definitions may not be entirely helpful, but they do indicate a wide and eclectic range of research activity—although Williams asserts that 'It was difficult to find evidence of novel methodological approaches to research in geographical education' (Williams 2003, p. 265). Perhaps more concerning are his observations that few articles appear to attend to issues of reliability and transferability, with research often being undertaken using very small 'samples of convenience'. There is a commonality of themes in geography education research across each of the publications, including: curriculum change and associated policies; developing learners' skills and abilities; new technologies; developing and using resources for teaching and learning; evaluation and assessment; sustainability; and 'the future'. More recent international publications, such as Brooks et al. (2017), have moved towards the analysis of *one* key theme—but through the examination of a breadth of relevant sub themes.

Williams' (2003) conclusions to his chapter *Research in geographical education: the search for impact* do not conceal his concerns about the general quality of the research produced in the field of geography education. There is an uncertainty surrounding the 'great variety of topics being studied and research methods being employed' (p. 270), and although the sustained interest in particular aspects of geography education over a number of decades is positive, Williams observes that geography education research is often conducted by individuals rather than groups, or teams, of researchers. Evidence of theory building is applauded—particularly where this is then applied to curriculum design and pedagogy—as are attempts to develop and deploy innovative research methods to explore the links between teaching and learning geography. The key issue for Williams is that examples of each of these are limited. Gaps in research are identified in: assessing the impact of geography education research on national curriculum policy making, decision making in 'government curriculum development centres' (p. 270), and how the work of publishers, examination boards and classroom teachers affect the process of curriculum making. Similarly, Slater (2003), at the same time, observed that only rarely does research in geography education 'flow back into, modify or become part of the body of general educational research. We may recognise that a one-way relationship prevails' (Slater 2003, p. 289). Both commentators are concerned about the lack of impact, nationally and internationally, of research in geography education. Recognising the difficulties in establishing any causal relationships between educational research and the behaviour of others (including policy makers, pedagogical workers and researchers), Williams (2003) laments that there are few ongoing, large scale research studies in geography education that have been sustained by 'internationally respected research groups.' Again, a related observation is that researchers in geography education tend to work alone, or in small groups—often on short term, unfunded or personal research projects. Such work, if published at all, is usually found in conference proceedings, or in edited books, or less frequently, in specialist academic journals (Williams 1998). In conclusion, Williams' overall assessment is that the impact of geography education research is modest, although he acknowledges the existence of some limited evidence of research findings extending beyond the intended geography education audiences to 'unanticipated recipients'. What counts as research in geography education is recognisable by its breadth, *but* lack of focus—it ranges from descriptive, non-peer reviewed accounts of what happens in classrooms (say), to publications from a small number of more considered, larger scale international projects. The quality of the products of geography education research—whether they result from individualised, personal accounts, or international, collaborative research studies—are also of some concern (Williams 1998).

7.5   Conferences

There are currently a variety of national and international research conferences for geography educators. These occur most frequently in more economically developed countries, which have established a tradition of research in geography education, but nonetheless are spreading to other less economically developed countries and continents. Such conferences are paralleled by larger, generic, education conferences such as (among others) those of the British Education Research Association (BERA), the American Educational Research Association (AERA), the Australian Association for Research in Education (AARE) and the European Educational Research Association (EERA). These are important gatherings—not least because they offer academics in education opportunities to publicise and promote their research to wider networks beyond the narrow confines of their subject groups. They may also open up the prospects of publishing more widely in associated proceedings, publications and journals, and through professional associations and their networks. It is essential that geography education researchers present their work to their subject peers, but also have their ideas tested, debated and challenged on the 'broader stage' of educational research. In this way geography education research becomes less introspective, and hopefully achieves greater status and impact.

But in education in general, and in geography education in particular, we have a problem:

> Although the quality of our best work is as good as any other comparable discipline, there is still a long tail of average and poor-quality work, as any large conference programme will demonstrate. And although things have improved substantially, there is still far too little quantitative research capacity within the discipline. It is not right that the discipline representing such a major field should still be so dominated by small scale (often untheorized) qualitative research studies. And finally, there is a need for more considered epistemological work in relation to mixed methods research (Furlong 2013, p. 197)

Promoting our research work is beneficial only if it is of high quality and relevance,[5] not only to our own community of geography education researchers but also beyond to policy makers, teachers, students and other stakeholders. We must have something important and relevant to say about the significance of good geography education to the educational 'offer' that young people receive in schools—and to be credible, accessible, forceful and well-rehearsed in the ways in which this information is presented.

---

[5] Brooks (2017) makes the interesting point that we may be guilty of striving to be 'overly inclusive' in our small community of geography education researchers worldwide, given that conference papers that are lacking in rigour and culmination may be accepted for presentation at international research conferences. The reasons for this vary—from noble attempts to support academics from jurisdictions that do not normally contribute papers to international conferences (for reasons of finance, organisation of research, lack of support, etc.), to less altruistic reasons of wanting to maximise conference attendance and revenue!

7.6 Educational Research Systems

Internationally, questions about what types of research universities and governments favour, and therefore seek to prioritise, rapidly increased at the start of the new millennium. Broadly, the preference is for the production of high-quality research (however determined), that is enterable for research assessment exercises and awards (such as the Research Excellence Framework (REF) in the UK), and which will attract government and research council funding. University leaders would argue otherwise, perhaps, in the interest of maintaining their rhetoric about academic independence, sovereignty and choice—but most are acutely mindful of where their research funding comes from, and how this funding stream can be maintained. The 'formal' types of research—as opposed to scholarship—are certainly recognised by the parameters set by governments and research funding agencies. Given that universities are also increasingly awake to issues of competitiveness, both nationally and internationally, the production of this type of research is welcomed because it helps to create a global 'shop window' for their research endeavours and outcomes. This 'high end' research is increasingly positioned to make international comparisons easier—rendering critical engagement simpler and favouring research that builds upon an established body of work, or which is defensible with respect to its innovative approaches and findings. There is an understandable conservativism about educational research—research arguably gains recognition more easily if it builds on established and agreed theoretical and methodological positions—but an increasing awareness of the importance of the international research 'market'.

In many countries research funding gained by academics and institutions comes from various organisations: a significant player is usually the national government, but much research money is sourced by academics applying to funding bodies for competitive tenders and awards. Most countries have some form of national research council, often government funded, which may have a variety of 'arms'—as well as sources of research money residing in charities, businesses, professional associations and unions. Where national research assessment exercises exist—and these are growing internationally, as we have seen in Chap. 6—these are primarily used to divide 'the sheep from the goats' in terms of research quality. Here the proportions of government money distributed to universities is dependent on their performance in periodic research assessment exercises—the highest quality of research, usually deemed to be of 'international standing' (in some fashion), with measures for research environment and impact also coming into play. The scale of government funding, and its targeted allocation, is therefore immensely important in most countries: this is one of the major sources of research funds for universities. We must also recognise the global influences on higher education institutions with regard to research activity. Globalisation has, arguably, convinced governments, businesses and university leaders of the imperatives adopting neo liberal attitudes and approaches—where the only way forward appears to be by the adoption of the principles of the market. In schools, departments and faculties of education the effect can be accentuated for, as Furlong (2013) argues, the neo liberal 'framing' of activities subsequently carries across to the work of educationists in the academy. Such 'enterprise universities' often consider their education departments to occupy relatively weak positions: driven financially by the imperative to sell their courses to government [see initial teacher education (ITE) in the UK], these departments are in the unenviable position of being strongly influenced by external, rather than internal, forces. If the government of the day does not look to the university sector to provide courses for ITE, or teachers' professional development, then alternative sources of funding (through research, consultancy, etc.) must quickly fill the gap. This has had 'a major impact on the courses that are taught, the student and staff that are recruited and the research that gets undertaken' (Furlong 2013, p. 110).

Brooks (2017), drawing on her experience as a doctoral supervisor of geography education students from a number of different jurisdictions—as well as her previous role as Secretary of the IGU CGE—notes the difference in aims and objectives of geography education research in jurisdictions around the world. In some countries, her students visualise geography education research as 'solution focussed' (for example, geography curriculum research in the USA), which contrasts with research which is focused on 'understanding the problem'—but which does not necessarily consider solutions (arguably, research in the UK). A third type of geography education research is 'practice focussed'—seeking to understand practice better and then to improve it (arguably in China). This alters the approach to, and perspectives on, the *point* of research in geography education. In different national contexts, Brooks (2017) also observes that education and geography education research have different status and specified roles. This has a related effect on individual researchers' academic career development and prospects of tenure. In some jurisdictions the importance of grant capture, number and quality of publications, and one's 'positioning' in the field of geography education research may be important. In others, patronage and mentoring are more significant—as esteemed professors may closely direct the work that their more junior 'academic acolytes' undertake. In some countries, China for example, there are the additional political drivers which may dictate the collective focus for research and blur distinctions between professional and academic research work. This contrasts with the relative academic freedom experienced by researchers in UK, USA and across Europe—even if some research efforts in institutions are directed strategically, often towards eventual research assessment.

The research themes pursued by senior academic staff—such as Senior Lecturers, Readers and Professors in Geography Education—tend to achieve prominence, both within their host institutions, and possibly nationally or even internationally. Similarly, the selection of conference themes also has an impact on subsequent research activity and production: 'it's not just about networks of people, but also networks of ideas—and which are the dominant ideas coming up now' (Brooks 2017). Common themes are difficult to establish, indeed in some jurisdictions issues that appear problematic to some geography education researchers are viewed as essentially non-problematic to others. These processes combine to help define the field of research themes in geography education more narrowly, arguably leading to the generation of a relatively modest number of ideas that are of actual, or symbolic, importance amongst researchers, internationally. The small scale of the geography education community worldwide adds a further complication: there is 'less room for rigorous critique and debate, and dangers of a confirmatory bias in the field' (Brooks 2017). Brooks considers that this may be a significant problem among those who might be thought of as 'research elites'—whose work sometimes appears to be accepted too readily by others, or has a disproportionately large impact, due to the small size of the field in which it has been published.

7.7   Global Research Issues in Geography Education

Tasking themselves with finding the 'enduring and universal' issues pursued by geography education researchers, Gerber and Williams (2000) indicate the main areas for research, globally, as follows:

- development of policy in geographical education
- learning and teaching in geographical education
- geography curriculum development
- assessment and evaluation in geography
- teaching resources
- technology in geographical education
- geographical education in differing social contexts
- geographical and environmental education.

Their list was achieved by looking back at journals and databases of geography education research over two decades from 1980 to 2000. Unfortunately, it does not provide us with much support in the quest to determine which issues geography education researchers have previously, or should now, research. It is an extremely broad list—and one which, almost twenty years later, could simply be re-stated, as it contains the main themes that researchers in geography education currently choose to pursue. This might suggest that little or no progress has been made in researching many of these areas, or that these are 'generic' areas that will always require research effort in geography education (and which might equally apply to almost any other curriculum subject).

7.8   Research in ITE in Geography

It is often the case that in many countries the 'home' for research in geography education is co located with that for initial teacher education, frequently with the same people being responsible for both. The providers of the best quality ITE are normally also those who undertake serious research, with at least some of this work being based in a partnership between schools and universities. The Organisation for Economic Co-operation and Development (OECD), reflecting on international PISA results, espouses this relationship—arguing that to raise the standards of teachers such that they mirror those visible in other professional groups their preparation should ideally be school based, research-informed and with significant input from the academy. Using oft quoted examples of high performing jurisdictions (China— Shanghai, Finland, etc.), the OECD suggest that ITE should be forged through school practice, but also indicates that academic excellence among teachers is important. In this way, the university—and its educational research—still has an important role to play in ITE, using research into excellent school-based practice to inform the training process. In many countries the use of what, in England, would be called University Training Schools is paramount in the preparation of teachers, fostering and forging innovative research into the day-to-day practice of teaching and learning. The place of disciplinary knowledge in ITE is also seen as crucial—particularly in ITE systems that combine teacher preparation between faculties of education and disciplinary departments—ensuring a focus on subject knowledge. In some countries, notably Finland, teachers are expected to be public intellectuals, enhancing their subject and disciplinary knowledge as they gain professional qualifications beyond those provided by pre-service training, up to Masters level and potentially beyond. The expectation that teachers should be research-engaged professionals, linked to university departments of education, is arguably at the pinnacle of models of teacher preparation—looked on enviously by educators in many other jurisdictions who are aware of the enhanced status and financial rewards that are afforded. The expectations that teachers engage with, and produce, research which is based on their own practice, in both Finland and China, is noteworthy. In turn teachers are expected to support state education by trialling and developing innovative educational ideas and approaches in the classroom. This is interesting in that university-based education departments may deny themselves a future if they falsely claim that they must take sole responsibility for, and control of, the research preparation of teachers—rather than embracing a more open partnership model that engages the strengths and traditions of both schools and HEIs (see Chap. 10).

7.9 Research into Geography Curricula in Schools[6]

A popular and enduring theme for international research in geography education, perhaps unsurprisingly, is the analysis of the ways in which geography is expressed in national curricula, or standards, for primary and secondary schools. This theme has the advantage of being relatively easy to consider, nationally, by researchers in their home context and is 'part and parcel' of the work that many geography education academics undertake. However, achieving comparability of both methods and findings across jurisdictions is not straightforward. There is little commonality about the ways in which school geography is defined in different countries, and divergence in how nations choose to represent geography in their schools add further

---

[6]This section draws significantly on Butt and Lambert (2014).

complications (Butt and Lambert 2014). Such choices reflect political, cultural, social and philosophical customs and behaviours.

In August 2004, as a contribution to the Symposium of the CGE in Glasgow, the British sub-committee of the IGU-CGE (in association with the Scottish Association of Geography Teachers) published *Geographical Education—Expanding Horizons in a Shrinking World* (Kent et al. 2004). This 'progress report on current trends in geographical education world-wide' (p. 9) attempted to highlight common characteristics in the provision of school geography from a range of countries, featuring fourteen short chapters which presented the current 'state of play' for school geography in particular countries, or regions. Rawling (2004) analysed these for 'key issues and challenges' in concert with considering their statements for 'current and planned developments for the future' (p. 167). Her overview of their findings on 'challenges' is listed below:

- geography's uncertain place in the school curriculum, particularly at primary level.
- decline in the opportunities for high quality teacher education in geography, with a concomitant decline in the number of specialist teachers of geography.
- problems arising from the growth of assessment and performance-led systems, often at the expense of curriculum development.
- the need to ensure geography is involved in technological developments (internet, multimedia, GIS).
- the need to address the public image of the subject, such that the public, students and policy makers recognised its potential as a school subject.

With hindsight some of the 'opportunities' identified may have subsequently proved problematic—with geography being used as a vehicle to deliver other aspects of the curriculum, rather than being respected as a subject in its own right, with a distinct knowledge base. This is the case when geography is valued, for example, *only* because it can assist in educating young people about (say) the environment and sustainable development; about national identity, cultural heritage and the strengths of diversity; about numeracy or literacy skills; and about ICT knowledge, skills, and competencies. While Gerber's (2001) previous survey of geography in 31 countries led him to a 'cautious optimism' about geography's place in the school curriculum, Rawling's (2004) subsequent assessment was less confident—stressing the need to 'recognise the real threats to the subject' and the need to 'take immediate action within our own communities' (p. 169). Recurrent among these perceived threats were the growth of integrated studies, the rise of vocational education and the tendency towards more 'skills-based' (rather than 'subject-based') education in schools. Interestingly, this narrative of introspection was acknowledged within the context of a vibrant and dynamic discipline of geography in universities and research institutions in several countries, leading to a separate discourse on the 'gap' between academic communities in schools and higher education institutions—and some soul searching over the types of geography taught (see, for example, Goudie 1993; Castree et al. 2007; Stannard 2003; Bonnett 2008; Butt and Collins 2013, 2018).

In a special edition of *International Research in Geographical and Environmental Education* on geography curricula and standards, Butt and Lambert (2014), assembled analytical accounts from seven national settings that reflected a Euro-American 'sphere of influence'—both in their approaches to education and their traditions of geographical study (see Johnston 1979). In choosing the countries to include in the edition, they acknowledged the equally strong case for undertaking an international analysis of school geography from 'under-represented' countries in the Islamic world; or more systematically from Asian countries; or from the continents of Africa and South America. There was an expectation that each contributor to the special edition would explore aspects of core/essential knowledge, powerful knowledge, the rationale applied for the selection of curriculum content, and the conceptualisation of the curriculum (see Lambert 2011a, b), as appropriate to their national contexts. Embedded within these accounts were considerations of three of the main forces which have shaped geography, and other subject-based curricula, internationally:

- educational processes, including international concerns to express competences and 'transferable skills' in the curriculum designed to promote the generic idea of 'learning to learn';
- social and cultural issues, including the positioning of environmental concerns (such as climate change), society matters (such as intercultural understanding) and more generally the role of geography in supporting moral and spiritual education; and
- the discipline of geography, including an exploration of the relationship that exists between geography expressed as a school subject on the one hand, and as a research discipline in Higher Education on the other.

Attempts had previously been made either to provide an international overview of the state of geography education in schools (see Gerber 2001; Rawling 2004); or to offer commentaries on school geography provision in different continents (see Lidstone and Williams 2006); or (more frequently) to comment on the status of teaching geography in various national jurisdictions (see Kent et al. 2004).[7] The ambition of such accounts and collections is laudable—but in practice the results are inevitably flawed, or incomplete. Firstly, their scope of coverage tends to be problematic—at the 'macro' scale it is difficult to provide a concise, scholarly account of the global 'state of play' with respect to geography education in schools, while at the 'micro' scale national subject updates occur frequently, but are so context specific that they can only be understood readily in an inconsistent or partial manner. Secondly, although national updates tend to be completed by individuals who have considerable experience and knowledge of their own educational setting, these only afford limited possibilities for direct comparisons between *other* national jurisdictions. Lastly, issues of both focus and research rigour abound. The countries that were selected by Butt and Lambert (2014)—Australia, England, Finland, New Zealand, Singapore, South Korea and USA—exhibit a certain level of common understanding of traditions and influences with respect to provision of geography education

---

[7]Descriptions of geography education and geography education research within different national contexts are relatively common; see, for example, Reinfried (2001a, b, 2004) on Switzerland, Graves (2001) on France, Birkenhauer (2002a, b) and Schrettenbrunner (1990) on Germany, Butt (2008) on the UK, etc.

in their schools. A decade earlier, Marsden's (2005) provocation that geography was now 'the worst taught subject' in English schools—which echoed a famous statement at the time by Her Majesty's Chief Inspector of Schools, with respect to geography teaching in primary schools (see Bell 2004)—prefaced the publication of a number of national perspectives on school geography (including comments from Argentina, Canada, England, Germany, Hong Kong, Hungary, the Netherlands and USA). Despite its eclectic coverage of themes, the picture of geography education that emerged was *not* entirely negative, with Marsden himself concluding that he was left 'seriously inclined to doubt …. (the) general validity' (p. 1) of Bell's statement at the international scale. Lidstone (2005) similarly noted correlations in the responses from different countries, identifying geography education's positive contribution to an 'enhanced concentration on transferable skills and the creation of ever more technologically based knowledge societies' (p. 61). Others note the enduring popularity of geography in schools, examples of upturns in student understanding and attainment, the growth of professional subject associations and the modernisation of curriculum development in geography.

John Lidstone's introduction to a substantial section devoted to 'Contemporary School Geography' in *Geographical Education in a Changing World* (Lidstone and Williams 2006a) provides a similar commentary to Rawling's, completed two years earlier. Contributions to this text were made 'regionally' (Europe, North America, Australasia, East Asia and South America), with each section including a number of national perspectives. Lidstone noted that trends in geography education across jurisdictions might owe much to 'policy borrowing' from particular countries—for example, former colonies of the UK were seen to teach separate disciplinary subjects for longer, whilst those under the influence of the USA tended to adopt 'social studies' in their school curricula. Furlong (2013) has commented:

> the UK provides a very important 'case'; some of the messages from the UK experience have been widely influential, particularly in the English-speaking world, though, as with all 'policy borrowing', those ideas and practices have been changed as they entered other specific policy contexts' (p. 6).

It was felt that certain countries wished (perhaps subconsciously) to emulate the social, educational, economic and political status of others, leading to the shape and content of their school subjects being strongly influenced by the education systems, curriculum structures and subject content imported from elsewhere. Indeed, Lidstone and Gerber (1998) asserts that this often meant that 'the names of a relatively small group of geographical education theorists are cited in the work of curriculum designers and researchers in (so) many countries' (p. 87). English-speaking jurisdictions have therefore been highly influential in spreading curriculum and pedagogical ideas in geography education.

7.10　School Geography in the Wider Policy Landscape

It is apparent that looking only 'inwardly' on the structure and content of the school geography curriculum—or indeed the curriculum of any school subject—is useful, but only to a limited degree. As Lambert (2015) states:

> The principles guiding the selection of what to teach in schools vary significantly between national jurisdictions, depending partly on how geographical knowledge is conceptualised and valued (p. 21)

We may obsess about the specific details of each geography curriculum, but this is a wasted effort if we ignore the wider policy landscape in which the subject curriculum 'sits'. Political and ideological influences on the curriculum, as well as the school system and accountability environment in which it operates, must be understood. This does not condone constant subservience to an external agenda—endlessly trying to justify the place of geographical knowledge in the curriculum, in terms of its contribution to literacy, numeracy, transferable skills or indeed 'good causes' (such as citizenship or sustainability), invariably leads to a diminution of subject knowledge. However, recognising that school geography is not a 'given' entity—particularly when compared to core subjects such as mathematics, or the sciences—may be important. So, whilst geography as an idea will never disappear—unless we believe that the job of making sense of the world is complete, or unnecessary—it is possible to imagine geographical knowledge becoming undermined, or reduced, in an era of rapid policy change. There is therefore a necessity to encourage robust and irresistible arguments for keeping geography in schools, based on empirical research evidence.

Nonetheless, the power of national policy objectives to disregard even the most carefully researched arguments, resulting in the removal of geography from national and regional curricula, should not be ignored. Across much of Australia geography was absent from schools for a generation and although it has now returned there is still a challenge to implement it to a high standard, given the existence of a work-force of teachers who have only had a very limited exposure to geography in their professional training. In the USA, despite the *Geography for Life* standards being in their second, fully revised edition, the subject struggles to secure Federal funding support—leaving it in a diminished condition in all but a minority of schools. One of the most significant influences on the future of state schools in England, where secondary school teachers receive specialist training, is the maintenance of a high-stakes accountability and inspection system. Here published league tables of examination results, and regular school inspections, have led to a 'de-professionalisation' of teaching and teachers, alongside pressures to 'teach to the test'. The marketisation of schooling in England, and elsewhere, is assisted by the rise of academy chains and trusts which directly control school curricula, and have troubling powers over teachers' salaries and conditions of employment. In conclusion, the future of all school subjects is always closely tied to national policy contexts. What might initially appear to be relatively benign shifts in education policy—perhaps related to changing school funding and governance, rather than curriculum matters—also has the potential to radically affect the future teaching of subjects in schools. In England, with question marks against the ability of academy schools to raise educational standards—and no legal requirement for pupils to study geography, within what is now (as in federal jurisdictions such as USA and Australia) a 'national' curriculum in name only—there are few *guarantees* that the future of geography education in schools is secure. Indeed, evidence suggests that poorer performing academies tend to offer narrower, vocationally driven, curricula rather than prioritising subject-based education (Wrigley and Kalambouka 2012).[8]

7.11   IGU CGE International Declaration on Research in Geography Education[9]

Past attempts to create a sense of international solidarity in geography education have met with only limited success: in some respects, geographers are seemingly divided by a common language. Considerable effort was expended almost thirty years ago to create an international charter for geography education which is, in many ways, still an impressive document (IGU CGE 1992). The updated version (IGU CGE 2016) improves on the legacy of its first iteration, being both more approachable and pertinent to a wider range of stakeholders. But any charter has to stand the test of time—in attempting to impose a singular view of geography, charters are arguably rather insensitive to the contexts and traditions of geography in different national settings. They may also not be able to take account of the local educational context in their pronouncements—to accommodate whether geography has previously been represented as a discrete subject in the curriculum, or has been taught as part of social studies, the humanities, or the sciences. Inadvertently, the international charter for geography education was arguably too 'internalist': by getting an agreement about how to express 'geography' in the school curriculum, it was oblivious to broader trends and influences that needed to be recognised and understood. Attempting to achieve international agreement about curriculum matters, including the form and function of subject-based education, is notoriously hard given the special circum-

---

[8]Wrigley and Kalambouka's (2012) research identifies that in a case study of four academies, only 12% of pupils studied History or Geography at GCSE, compared to 57% in other schools locally. In one academy their research found that only a single pupil took history, with no one taking geography, at GCSE.

[9]Other notable declarations and agreements have, of course, been signed internationally that have had a bearing on research in geography education. For example, (i) the Association of European Geographical Society (EUGEO) signed the Rome declaration on Geographical Education in Europe, in 2013, alongside the International Geographical Union—Commission on Geographical Education (IGU-CGE), EUGEO, and the Association of Italian Teachers of Geography (AIIG). (ii) The Lucerne Declaration, signed by IGU-CGE in 2007—which supports the aims of the UN Decade of Education for Sustainable Development (UNDESD)—has similarly had an impact on research in geography and environmental education.

stances that abound in different national contexts. As Lidstone and Williams (2006b) pithily observe:

> Perhaps the whole notion of an international community of scholars with common aims for their field is unsustainable in the face of local agendas and priorities (p. 2)

The Declaration on Research in Geography Education (IGU CGE 2015) inevitably suffers from some of the same drawbacks experienced by most charters and declarations. It states that those who teach geography in primary and secondary schools, and in further and higher education, need to be supported by research intelligence in order to:

- clarify the purposes and goals of geography education, no matter how the geography curriculum is expressed locally;
- refine curriculum, pedagogic and assessment practices used in the teaching and learning of geography;
- deepen collective understanding of learning progressions in geography;
- improve ways in which high quality materials and resources for geography teaching and learning can be developed and provided;
- develop understanding of learners' geographical knowledge and experience, including their misconceptions, to enhance geography's teaching and learning;
- improve the teacher education of geography educators, linking innovative teaching practices to empirical research in geography education.

These are laudable and helpful statements, which outline clearly a reasonable expectation of research in geography education for the future. What is less obvious—although it is perhaps churlish to expect a Declaration to offer concrete suggestions for how these points might be delivered—is exactly how these goals might be achieved. The document is admirably short, being one page long (although the Appendix and Annex run to a further 16 pages[10]), with the advantage that it can be presented to politicians and policymakers in a form that is easily digestible. In the Declaration the IGU CGE states that it seeks to both support and promote research in geography education in all nations and cross-nationally. It aspires to develop an international culture of research in geography education that will enable the development of policy and practices that enhance the quality of geography teaching and learning for all. By deepening our understanding of the current state of research, and seeking to identify future research intentions, the IGU CGE attempts to support geography educators in their efforts to recognise educational priorities and the means by which these may be brought to fruition. Essentially, the Declaration is an honest attempt to provide support to all those involved in geography education research worldwide, whatever the circumstances and contexts in which they work, and should be applauded as such.

---

[10]The Appendix includes the following sections: Executive summary, Preamble; Nature, Value and focus of geography education research; Contexts of and challenges in geography education; The development of researchers; Strategic development; Methodology; Dissemination; and Impact. The Annex includes some examples of topics for research in geography education.

7.12 Roadmap for 21st Century Geography Education Project

As has been recognised, the setting for geography education research—its priorities, antecedents and prospects—are all shaped by national contexts (Gerber 2000). In the US the teaching of geography in schools is largely subsumed within the social sciences, reflecting something of the historical development of the discipline at university level (see Johnston 1979). This is very significant, given that at secondary school level the majority of geography teachers in the UK are geography (or geography-related) graduates, while in the US this level of geographical qualification for teachers is unusual. This is particularly apparent when one is concerned with teachers' disciplinary knowledge, and therefore their preparedness for teaching a subject—in the UK geography teachers may have very different backgrounds regarding their disciplinary knowledge (there may be geography teachers who still describe themselves as broadly 'physical' or 'human' geographers), but their general understanding of the philosophical nature of the subject will perhaps be similar. In the US, making assumptions about teachers' levels of subject knowledge are more difficult. Lambert (2015) asserts that these differences may not be so significant as they first appear—given that many geography teachers in the UK will have little interest in the advancement of their subject after they have gone through pre-service training and into teaching, or about how their discipline interconnects with the geography curriculum in schools; while in the US some social science teachers may be very excited by the prospect of advancing geography education in their schools.

Stoltman (1992) reflects on a decade of 'renaissance' of geography education in the US during the 1980s, pointing to ten developments that prompted political leaders, professional geographers, societies of geographers, teachers and students to 'become engaged in a major revitalization of the discipline within the curriculum' (p. 262). Under the optimistic heading of 'Outlook Bright', Stoltman details how the 1980s saw a more promising future for geography education in the US than any other time in the 20th century: highlighting the 'surprisingly insignificant' position that geography in US schools had previously occupied (see Natoli 1986); the traditional lack of proactive influence of professional and academic societies of geographers; and the almost negligible impact on schools of the High School Geography Project (in contrast to its more substantial influence on curriculum development in the UK). Geography was brought to greater prominence in schools by the publication of the Survey of Public Perception of Education (Gallup 1980)—which indicated that half of its respondents believed that more time needed to be devoted to children learning about other nations; and 'A Nation at Risk: the imperative for educational reform' (National Commission on Excellence in Education 1983)—which claimed that the educational attainment of a large proportion of US citizens had regressed to levels that represented a threat to the fabric of society and its governance.[11] In addition,

---

[11] Morgan (2017) interestingly considers the concept of 'geographical ignorance', through reflecting on the need for a 'knowledge turn' in school geography. By applying the work of US geographer Martin Lewis, he discusses that a certain level of public ignorance of geography is inevitable, but

various national and international test results revealed strikingly similar intelligence about the lack of geographical knowledge, skills and understanding of students across the spectrum of education (Hill 1981); this increased the publication of geography textbooks in response to a demand for materials to combat the geographical illiteracy of school students, albeit within a regional paradigm (Stoltman 1992). The publication of official reports that specifically focussed on geography education was also instrumental in spurring interest and action: such as the 'Guidelines for Geographic Education: Elementary and Secondary Schools' (Joint Committee on Geographic Education 1984).

The formation of geographic 'alliances'—between teachers, primary and secondary schools, academic geographers in HE, private industry and public agencies—gained pace in the 1980s, resulting in a significant spur for the development of geography education in US schools. This process started in California with the launch of the California Geographic Alliance, which had strong links to the University of California at Los Angeles, and gained momentum nationally after 1986—creating a series of alliances supported and funded by the National Geographic Society. The Geographic Education National Implementation Project (GENIP), established in 1985 to implement the recommendations of the 1984 'Guidelines', represented all the geographical societies of the US, with the intention of promoting and implementing geography education in US schools. GENIP, alongside other organisations, was instrumental in establishing a National Geography Awareness Week, starting in 1987, while the 'Guidelines' instigated discussion about the need for pre-service teachers to have some training in geography methods.

These 'significant developments' are recorded in some detail because although many specifically relate to the practicalities of geography teaching and teacher preparation, rather than geography education research, they indicate a direction of travel that has successfully raised the profile of geography education in the States. These developments have also had a direct impact on research in geography education elsewhere in the 21st century. From a foundation of extremely modest amounts of geography being taught in US schools, the Americans have succeeded in motivating political interest and action in geography education—attracting private and governmental funding to their educational mission, stimulating official reports from commissions and study groups, and motivating the learned societies of geographers to support developments in geography education. Stoltman (1992) is bold, but probably accurate, in referring to this as a 'renaissance' for geography education—for it increased the amount of geography teaching in schools, raised the public profile of geography education, and 'directed the attention of policy makers to geography in the school curriculum' (p. 274).

The US *Roadmap for 21st Century Geography Education Project,* launched in 2010 and hosted by the Geography Education Research Committee (GERC) (Bednarz et al. 2013), was in some respects a logical consequence of the 'renaissance' Stoltman observed in geography teaching in schools from the 1980s. This project

---

highlights that modern versions of the geography national curriculum in England and Wales may ironically serve to promote this state of affairs.

addressed the issues faced by the contra-positioning of, theoretically speaking, 'what works' research and 'higher' research into geography education that took place in the academy. Addressing the issues directly, the project leaders found that geography education research in the US was unfocused and parochial, largely disconnected from what other education researchers were striving to achieve. The dangers of research into geography education becoming generic educational research—what Lambert (2015) refers to as research with a 'geographical hue'—is a problem that GERC was keen to identify, isolate and eradicate. As a consequence, the five chosen areas of focus for geography education research were defined as follows: learning progressions; effective teaching; exemplary curricula; impact of fieldwork; and teacher preparation. Embedded within these research priorities, GERC suggested that researchers should also concentrate on the contribution of their geography education research to at least one of four stated priorities (or subfields):

- how do geographic knowledge, skills, and practices develop across individuals, settings and time?
- how do geographic knowledge, skills, and practices develop across different elements of geography?
- what supports or promotes the development of geographic knowledge, skills, and practices?
- what is necessary to support the effective and broad implementation of the development of geographic knowledge, skills, and practices?

The ambitions of the US Roadmap project, and of the enquiry questions that the project coordinators set themselves, can be set as a yardstick against which the priorities for geography education research could be established in other jurisdictions. With sufficient funding and organisational flair higher education-based research into geography education—in a context where geography education in schools did not have an impressive history—has become significant and impactful in the States. It is also important to recognise the 'ground work' that preceded the project and helped to make it successful, not least the efforts of people like Michael Solem (Director of Geography Education at the AAG), Dick Boehm (Director of the Gilbert Grosvenor Centre) see Boehm (2002), Sarah Bednarz (Professor of Geography Education at Texas A+M University) and others in geography education such as Golledge (2002)—who pointed out the unique qualities of geographical knowledge; Hanson (2004)—who promoted the value of the geographical perspective; and Gersmehl (2006)—who has suggested various ways to teach geography successfully. While recognising the insights offered by the project to geography educators in other countries, Rutherford (2015) also highlights enduring issues affecting the direction taken by geography education in the United States: indeed, the project's research report 'documents how little is really known about key aspects of geography education .... There is much more to learn' (Rutherford 2015, p. 33).

It also important to remind ourselves that despite the steps forward for geography education research forged by the International Declaration on Research in Geography Education and US Roadmap project these are not enough, on their own, to assure the future for research in the field:

While the Road Map report and IGU Research Declaration provide sound rationales for geography education research, in and of themselves the publications guarantee nothing in terms of how researchers will do research and how research might inform practitioners in the future. Action is needed to transform the considerable amount of latent energy contained in both reports into tangible and kinetic lines of research addressing the big educational challenges of our times (Solem and Boehm 2018a, b, p. 191)

7.13   Conclusions

The need for geography to justify the curriculum space it occupies has, for some, always been apparent. There are no guarantees that geography will always be taught in schools across the globe (Beneker and van der Schee 2015). In many jurisdictions, at particular times, a strong case has had to be made for its continued inclusion in the school curriculum—which has often necessitated the support of a cohesive and vibrant subject specialist community (such as, inter alia, the Association of American Geographers (AAG) and the National Council of Geographic Education (NCGE) in the USA, Australian Geography Teachers' Association (AGTA) in Australia, EURO-GEO in Europe, and the Geographical Association (GA) in England). Recently the geography education community, both in national contexts and occasionally across the globe, has begun to find ways to promote and popularise its work in areas of curriculum development and teacher support. Geography education, and geography education research, take various forms—but there are generalities in research and teaching which can be commonly understood, despite the fact that these 'play out' slightly differently in local contexts. As geographers know better than most, the uniqueness of locality does not imply that patterns and processes can only be understood as singularities. There are global processes at work, both in schools and in research communities, and enduring 'big questions' for geography educators to consider, almost regardless of their national context (see, for example, those posed by the US Roadmap Project—which we need to acknowledge and understand, even if their resolution has a distinctive, national flavour). Arguably, geography education researchers now need to identify some common themes to pursue, based on a meta-analysis of their current contributions to the field and the enduring research questions that still need to be answered.

Finally, let us remind ourselves of the review of articles published in *International Research in Geographical and Environmental Education* in the decade following the journal's inception (Lidstone and Williams 2006b). This discovered that these articles had been written by researchers based in 43 countries (n = 324 major articles and *Forum* articles, in 28 issues), but that just three countries dominated. Of the authors of major papers, 29% were from the UK (mainly from England); 17% from Australia; and 12% from USA (a combined total of 58% of all articles). Eighty five percent of *Forum* papers came from researchers based in these same three countries. This is, of course, shocking for the global representation of geography education research—implying that research into geography education is either not being carried out in any substantive form in large areas of the globe (but more particularly in LEDCs), and/or that the research produced in many countries is not of a high enough quality to be represented in an international, peer reviewed academic journal. As mentioned earlier in this chapter, there is a danger that a Euro-American tradition in geography and geography education research is imposing a potentially unhealthy hold on global efforts to produce research. Bagoly-Simó's (2014) longitudinal analysis of research publications in geography education (between 1900 and 2014) additionally reveals a persistent focus on similar topics, methods and challenges. Interestingly, he concludes that the majority of geography education research publications are limited to specific journals (such as *International Research in Geographical and Environmental Education*), communities (such as in Germany, and USA), and 'display a rather prescriptive and textbook-like approach' (Bagoly-Simó 2014). Applying a systematic, comparative, meta-analysis of 500 publications in geography education, from both regional and international contexts, Bagoly-Simó shows that while non-English language publications made links to a range of international literature, articles developed in English-speaking countries tended to ignore completely scholarly work published in other languages. Another finding was that geography education research was more content-oriented in regional publications, whereas international literature often addressed subject-specific and cross-curricular educational objectives, including themes in cartography and map skills.[12] As Lidstone and Williams (2006a) observe in their introductory chapter of *Geographical Education in a Changing World*:

> There is a danger for researchers from English language communities world-wide assuming that researchers working in English language contexts have conducted all the best work. In reading through research reports in the English language, the absence of references to non-English sources is remarkable. It would appear that most researchers working in an English language context are quite unfamiliar with work published in, say, Chinese, French, Spanish or Japanese (p. 9)

The 'last words' in this chapter, on international perspectives on geography education research, go to Gerber and Williams (2000). In their outline of 'Research Futures' for geography education, they refer to Delors (1998) influential '*Learning: the treasure within*' report from the International Commission on Education for the Twenty-First Century. For Gerber and Williams (2000) the way forward for geography education and geography education research, across the globe, is to strive to achieve the following:

1. Improving teacher education through searching for new perspectives in geographical education.
2. Using the resources of the information society to broaden geographical learning and teaching.

---

[12] The international publications in English analysed by Bagoly-Simó (2014) encompass, as one sample, the journals *International Research in Geographical and Environmental Education, Journal of Geography* and *Geography*. In addition, publications in German from *Geographie und ihre Didaktik, GW-Unterricht, Praxis Geographie, Geographie und Schule,* and *Geographie Heute,* along with journal articles, monographs and book chapters from Hungary, Romania, Spain and Latin America (including Argentina, Chile, Mexico, Colombia, Cuba) were represented in other samples.

3. Emphasising geographical education that is aimed at improving the situation of people living together.
4. Refocusing curriculum development at all levels of formal geographical education to promote greater autonomy, judgement, and personal responsibility.
5. 'Rediscovering geography' as a basis for effective lifelong learning in both formal and non-formal education (after Gerber and Williams 2000).

Perhaps, rather grandly, these points are offered as a 'future agenda' for researchers in geography education—with a note that we should recognise that the worldwide geography education community is made up of groups and individuals that function in very different contexts, nationally and regionally. As such, it is claimed, achieving a better understanding of how groups, networks and research communities in geography education operate 'locally to globally' would prove beneficial.

References

Bagoly-Simó, P. (2014, August 22). *Coordinates of research in geography education. A longitudinal analysis of research publications between 1900–2014 in international comparison.* Paper delivered at IGU Regional Conference, Krakow.

Ball, S., & Olmedo, A. (2013). Care of the self, resistance and subjectivity under neoliberal governmentalities. *Critical Studies in Education., 54*(1), 85–96.

Bednarz, S., Heffron, S., & Huynh, N. (Eds.). (2013). *A roadmap for 21st century geography education research. A report from the Geography Education Research Committee.* Washington DC: AAG.

Bell, D. (2004). David Bell urges schools and the geography community to help reverse a decline in the subject. OfSTED press release, November 24.

Béneker, T., & van der Schee, J. (2015). Future geographies and geography education. *International Research in Geographical and Environmental Education, 24*(4), 287–293.

Birkenhauer, J. (2002a). Secondary level geography: A comparative analysis of study programmes in Germany. *International Research in Geographical and Environmental Education., 11*(3), 262–270.

Birkenhauer, J. (2002b). Proposals for a geography curriculum '2000+' for Germany. *International Research in Geographical and Environmental Education., 11*(3), 271–277.

Boehm, R. (2002). Fostering research in geographic education: The Gilbert M. Grosvenor center for geographic education. *International Research in Geographical and Environmental Education, 11*(3), 278–282.

Bonnett, A. (2008). *What is geography?.* London: Sage.

Brooks, C. (2017, February 27). Interview with Clare Brooks at UCL Institute of Education, London.

Brooks, C., Butt, G., & Fargher, M. (Eds.). (2017). *The power of geographical thinking.* Dordrecht: Springer.

Butt, G. (2008). Is the future secure for geography education? *Geography, 93*(3), 158–165.

Butt, G., & Collins, G. (2013). Can geography cross 'the divide'? In D. Lambert & M. Jones (Eds.), *Debates in geography education* (pp. 291–301). Abingdon: RoutledgeFalmer.

Butt, G., & Lambert, D. (2014). International perspectives on the future of geography education and the role of national standards. *International Research in Geographical and Environmental Education., 23*(1), 1–12.

Butt, G., & Collins, G. (2018). Understanding the gap between schools and universities. In M. Jones & D. Lambert (Eds.), *Debates in Geography Education* (2nd ed., pp. 263–274). London: Routledge.

Castree, N., Fuller, D., & Lambert, D. (2007). Geography without borders'. *Transactions of the Institute of British Geographers, 32,* 129–132.

Chalmers, R. (2006). International geographical education past, present and future. In J. Lidstone & M. Williams (Eds.), *Geographical Education in a changing world* (pp. xiii–xv). Dordrecht: Springer.

Delors, J. (1998). *Learning: the treasure within. Report to UNESCO of the International commission on Education for the Twenty-first Century.* Paris: UNESCO Publishing.

Fien, J., & Gerber, R. (Eds.). (1988). *Teaching geography for a better world.* Edinburgh: Oliver and Boyd

Foskett, N., & Marsden, B. (Eds.). (1998). *A bibliography of geographical education 1970–1997.* Sheffield: Geographical Association.

Fuller, K., & Stevenson, H. (2019). Global education reform: Understanding the movement. In *Educational review. Special Issue: The Global education reform Movement: Contemporary developments and future trajectories* (Vol. 71, Issue 1, pp. 1–4).

Furlong, J. (2013). *Education: An anatomy of the discipline.* London: Routledge.

Gallup Report. (1980). The twelfth annual Gallup poll of the public's attitudes towards the public schools. *Phi Delta Kappa, 61*(1), 45.

Gerber, R. (2000). International research in geographical education: Reflections from the retiring chair of the commission on geographical education. *International Research in Geographical and Environmental Education, 9*(3), 195–196.

Gerber, R. (2001). The state of geographical education around the world. *International Research in Geographical and Environmental Education., 10*(4), 349–362.

Gerber, R., & Williams, M. (2000). Overview and international perspectives. In A. Kent (Ed.), *Reflective practice in geography teaching.* London: Paul Chapman.

Gersmehl, P. (2006). *Teaching geography.* New York: Guildford Press.

Golledge, R. (2002). The nature of geographical knowledge. *Annals of the Association of American Geographers, 92*(1), 1–14.

Goudie, A. (1993). Schools and universities: The great divide. *Geography, 78*(4), 338–339.

Graves, N. (2001). The evolution of research in geographical education in France. *International Research in Geographical and Environmental Education, 10*(1), 4–19.

Hanson, S. (2004). Who are 'we'? An important question for geography's future. *Annals of the Association of American Geographers, 94*(4), 715–722.

Haubrich, H. (1996a). Foreword. In M. Williams (Ed.), *Understanding geographical and environmental education. The role of research* (pp. xi–xii). London: Cassells.

Hill, D. (1981). A survey of global understanding of American college students. *Professional Geographer., 33*(2), 235–237.

Hofstetter, R., & Schneuwly, B. (2012). Institutionalisation of educational sciences and the dynamics of their development. *European Educational Research Journal, 1*(1), 1–24.

Houtsonen, L., & Tammilehto, M. (Eds.). (2001). *Innovative Practices in Geographical Education. Proceedings of the Helsinki Symposium of the IGU-CGE.* Helsinki: IGU-CGE.

IGU CGE (International Geographical Union Commission on Geographical Education). (1992). *International charter for geography education.* Washington: IGU CGE.

IGU CGE (International Geographical Union Commission on Geographical Education). (2015). *International declaration on research in geography education.* Moscow: IGU CGE.

IGU CGE (International Geographical Union Commission on Geographical Education). (2016). *International charter for geography education.* Beijing: IGU CGE.

Joint Committee on Geographic Education (1984). Guidelines for geographic education: Elementary and secondary schools. Washington, DC: Association of American Geographers and the National Council for Geographic Education.

Johnston, R. (1979). *Geography and geographers: Anglo-American human geography since 1945.* London: Edward Arnold.

Kent, A., Rawling, E., & Robinson, A. (Eds.). (2004). *Geographical education: expanding horizons in a shrinking world.* Glasgow: SAGT with CGE.

Lambert, D. (2011a). Reviewing the case for geography, and the 'knowledge turn' in the English National Curriculum. *The Curriculum Journal, 22*(2), 243–264.

Lambert, D. (2011b). Reframing school geography: A capability approach. In G. Butt (Ed.), *Geography, education and the future* (pp. 127–140). London. Continuum.

Lambert, D. (2015). Research in geography education. In G. Butt (Ed.), *MasterClass in geography education* (pp. 15–30). London: Bloomsbury.

Lidstone, J., & Gerber, R. (1998). Theoretical underpinnings of geographical and environmental education research: hiding our light under various bushels. *International Research in Geographical and Environmental Education, 7*(2), 87–89.

Lidstone, J. (2005). The worst taught subject: International perspectives on Bill Marsden's guest editorial. *International Research in Geographical and Environmental Education., 14*(1), 61–62.

Lidstone, J., & Williams, M. (2006a). *Geographical education in a changing world*. Dordrecht: Springer.

Lidstone, J., & Williams, M. (2006b). Researching change and changing research in geographical education. In J. Lidstone & M. Williams (Eds.), *Geographical education in a changing world* (pp. 1–17). Dordrecht: Springer.

Marsden, W. (2005). Guest editorial reflections on geography: The worst taught subject? *International Research in Geographical and Environmental Education., 14*(1), 1–4.

Morgan, J. (2017). The making of geographical ignorance. *Geography, 102*(1), 18–25.

National Commission on Excellence in Education. (1983). *A nation at risk: The imperative for educational reform: A report to the Nation and the Secretary of Education, United States Department of Education*. New York: NCEE.

National Council for Geographic Education (NCGE) (2000). *A bibliography of geographical education*. Indiana: National Council for Geographic Education.

Natoli, S. (1986). The evolving nature of geography. In S. Wronski & D. Bragaw (Eds.), *Social studies and the social sciences: A fifty year perspective* (pp. 94–109). Washington DC: National Council for the Social Studies.

Rawling, E. (2004). Introduction: School geography around the world. In A. Kent, E. Rawling, & A. Robinson (Eds.), *Geographical education: Expanding horizons in a shrinking world*. SAGT with CGE: Glasgow.

Reinfried, S. (2001a). Curricular changes in the teaching of geography in Swiss upper secondary schools: An attempt to develop skills for lifelong learning. *Journal of Geography, 100*(6), 320–330.

Reinfried, S. (2001b). Ready for the twenty-first century? The impact of curriculum reform on geography education in upper secondary schools in Switzerland. *International Research in Geographical and Environmental Education, 10*(4), 421–428.

Reinfried, S. (2004). Do curriculum reforms affect classroom teaching in geography? The case study of Switzerland? *International Research in Geographical and Environmental Education, 13*(3), 239–250.

Rutherford, D. (2015). Reading the road map for 21st century geography education in the United States. *Geography, 100*(1), 28–35.

Sahlberg, P. (2012). *How GERM is affecting schools around the world*. Retrieved from https://pasisahlberg.com/text-test/.

Schrettenbrunner, H. (1990). Geography in general education in the Federal Republic of Germany. *GeoJournal, 20*(1), 33–36.

Slater, F. (2003). Exploring relationships between teaching and research in geography education. In R. Gerber (Ed.), *International handbook on geographical education* (pp. 285–300). Kluwer.

Solem, M., & Boehm, R. (2018a). Research in geography education: Moving from declarations and road maps to actions. *International Research in Geographical and Environmental Education, 27*(3), 191–198.

Solem, M., & Boehm, R. (2018b). Transformative research in geography education: The role of a research coordination network. *The Professional Geographer, 70*(3), 374–382.

Stannard, K. (2003). Earth to Academia: On the need to reconnect university and school geography. *Area, 35,* 316–332.

Stoltman, J. (1992). Geographic education in the United States: The renaissance of the 1980s. In M. Naish (Ed.), *Geography and education: National and international perspectives* (pp. 262–275). London: Institute of Education, University of London.

van der Schee, J., Schoenmaker, G., Trimp, H., & van Westrhenen, H. (Eds). (1996). *Innovation in geographical education*. Utrecht/Amsterdam: IGU and Centre for Geographical Education of the Free University of Amsterdam.

Wagner, P., Wittrock, B., & Whitley, R. (Eds.). (1993). Discourses on society: the shaping of the social science disciplines. Dordrecht: Kluwer.

Williams, M. (Ed.). (1996a). *Understanding geographical education: The role of research*. London: Cassell.

Williams, M. (Ed.). (1996b). Positivism and the quantitative tradition. In M. Williams (Ed.), *Understanding geographical education: The role of research* (pp. 6–11). London: Cassell.

Williams, M. (1998). A review of research in geographical education. In A. Kent (Ed.), *Issues for research in geographical education. Research forum* (Vol. 1, pp. 1–10). London: Institute of Education Press.

Williams, M. (2003). Research in geographical education: The search for impact. In R. Gerber (Ed.), *International handbook on geographical education* (pp. 259–272). Kluwer.

Wise, M. (1992). International geography: The IGU commission on education. In M. Naish (Ed.), *Geography and education: National and international perspectives* (pp. 233–246). London: Institute of Education, University of London.

Wrigley, T., & Kalambouka, A. (2012). Academies and achievement: Setting the record straight. Accessed March 21, 2013 from www.changingschools.org.uk.

Chapter 8
Research in Geography and Geography Education: The Roles of Theory and Thought

8.1 Context

This chapter discusses the connections between research in academic geography and in geography education, with particular reference to the roles of theory and thought. In this context it provides an overview of some of the key theories which underpin academic research in education and addresses the philosophical nature of theoretical constructions that inform research themes, methods and methodologies. As such, this chapter aims to help the reader understand how geographical theories and philosophical traditions may affect the nature of research in geography education. The importance of establishing clear 'philosophy-theory-methodology connections' (Couper 2015, Preface) when conducting research, such that abstract ideas or questions are successfully translated into appropriate research practices, has been largely established. However, the place, status and role of theory in research in education is still contested (see, for example, Carr 2006; Thomas 2007)—a situation that is perhaps well-known to those who conduct research in geography education.

Given the generally accepted requirement for researchers in geography education to base their work within existing philosophical, theoretical and disciplinary traditions—mindful of the knowledge sets within which they are working—this chapter helps to (re) establish the nature of the contribution of both theory and thought to geography education research. It exemplifies how research in geography education tends to operate across a range of theoretical settings, involving the application of ideas from a variety of disciplines and contexts (Lidstone and Gerber 1998). Particularly pertinent is the observation that the contextual origins of much published research in geography education (i) straddles both the discipline of geography and the field of education, and (ii) currently originates from within higher education institutions (HEIs). However, in typically direct fashion, Rawding (2013) asserts that there is little to be gained from researchers, and others, attempting to make the geography curriculum in schools simply ape developments at the cutting edge of university geography:

© Springer Nature Switzerland AG 2020
G. Butt, *Geography Education Research in the UK: Retrospect and Prospect*, International Perspectives on Geographical Education,
https://doi.org/10.1007/978-3-030-25954-9_8

the wholesale adoption of the latest research in university geography is neither as straight-forward as it might appear nor always particularly desirable. Recent developments in the Academy have seen an increasing fragmentation of the subject and a tendency towards spe-cialisation over synthesis. These trends within the subject have been exacerbated by the modularisation of many degree courses, which in the worst-case scenarios leave graduates with a perception of the subject that's little more than a list of topics which have been covered during their degree (pp. 287–288)

8.2   Introduction

As with all forms of research in education, theory plays a number of crucial roles in shaping the ways in which research questions are established and subsequent research activities are planned, enacted and understood in geography education. Theory pro-vides the frameworks and structures within which appropriate research questions can be posed, methods selected, methodologies organised, fieldwork undertaken and findings discussed. Theory enables ideas to be formed, tested and understood—indeed, it stands at the very foundation of the creation of new knowledge. In some research—arguably that which is most successful and has greatest impact—theory can help the researcher to not only avoid merely describing what has been found, but also enable him or her to undertake deeper analysis and fuller discussion. This process may possibly involve developing existing theories further, or even helping to develop new theories. As such, theory is important to all significant and rigorous research activity in geography education.

Morgan (2013), in attempting to unpick what it means to 'think geographically', returns to the basic question of 'what is geography'? This apparently simple enquiry has caused much confusion and debate, with leading academic geographers admit-ting that even they are unsure (see Bonnett 2008). Morgan asserts that in recent years this question has been reformulated in geography education into a consideration of what it means to think geographically—alongside considerations of the nature of the 'geographical imagination' (Gregory 1994). Giving such questions an educational 'frame of reference' begs us to consider what geography teaching and learning should look like in schools; here Morgan believes that most geography education implicitly assumes that a 'real world' exits, to be studied by geographers—with the fruits of their labours (that is, geographical knowledge) then being passed on to students. Given the work of sociologists in education from the 1970s onwards, not least Michael Young, we must question whether this assumption is correct, for the forms of geog-raphy taught in schools surely reflect the particular positioning, beliefs and values of the geographers and geography educators who convey them. This situation is, of course, true of all subject-based education—not just geography. Our subjective interests mean that all geographical knowledge is socially constructed (just as all historical knowledge, knowledge of art, or knowledge of economics (say) is simi-larly constructed). Such knowledge, and the disciplines from which it arises, have accumulated and tested ideas within a community of practitioners and experts over many years; however, this does not remove the contention that knowledge is essentially socially constructed and subjective. Young's work, with Johann Muller and others, has attracted considerable attention from geography education researchers in recent years. This has proved to be a fruitful association, challenging those involved in geography education to think more clearly about the key concepts and knowledge sets that underpin their field, and about what is important (indeed 'powerful') to pass on to learners (see Chap. 3). Based within these debates are considerations of the social constructedness of knowledge, alongside the more recent endorsement of a social realist approach—contentions that Young has considered over the past 50 years, in a personal intellectual journey that has taken him from one philosophical persuasion to another. This reminds us that the nature of knowledge shifts over time, along with its attendant structures and functions, according to societal norms, pressures and needs (Harvey 1984; cited in Morgan 2013).

Geography is still, arguably, bound by three core concepts—place, space and environment. The emphasis given by geographers and geography educationists to each of the elements of this trinity has shifted over time, as the discipline of geography has responded to changes in contemporary thought, theory, research findings and—more prosaically—the economy, politics and society. But the core concepts remain. Even though geography has passed through numerous developmental phases, changes of emphasis and, on occasions, rather more radical paradigm shifts, these core concepts endure. Essentially, the study of geography still exists at the centre of the overlap between these concepts—even if the discipline has passed through phases linked to exploration and discovery (which witnessed the founding of the modern discipline), to a regional focus, to human and physical perspectives (with systematic and integrative approaches), to more recent specialization, fragmentation and 'boundary crossing' (Matthews and Herbert 2008). Peter Jackson, in his oft quoted article 'Thinking Geographically' (Jackson 2006), is heavily influenced by the more recent connection between geography and the other social sciences. Here the nexus between morality and belonging help to illustrate the ways in which people and places are connected geographically. Morgan (2013) articulates the relationship between disciplinary thinking and pedagogy—he argues, in contrast to Young, that teachers should be given licence to start from 'students' experience of 'everyday life" (p. 280) as a way into deeper enquiry into substantive themes: 'analysis should proceed from empirical enquiry rather than relying on high theory from the start' (Morgan 2013, p. 280). This gives opportunities for many different 'readings' and meanings of geography to come to the fore, enabling the students themselves to explore their everyday experiences ('popular wisdom'), ideas and insights—downplaying the notion that academic geographers, and geography teachers, have a unique insight into what geography *is*.

8.3   Theoretical Perspective for Education Research[1]

All research in education should sit within a theoretical and methodological framework. As Greenway (2018) rehearses, qualitative and quantitative methodologies describe the two main—and, in their purest forms, generally considered to be opposing—ways of gathering research data in the social scientific research tradition (Creswell 2014; Parahoo 2006). Quantitative approaches emphasise the measurement, and possible analysis, of causal relationships between variables in terms of quantity, amount, intensity or frequency (Denzin and Lincoln 2013), which Robson (2011) contends can be 'hypothesis generating' or 'hypothesis testing'. But in most research in geography and geography education the generation of multiple representations of a concept preclude such empirical approaches. Geography education research only infrequently plans to *test* a hypothesis—it usually seeks to describe, understand and interpret experiences and phenomena (Parahoo 2006), or to contribute evidence and findings for the development of practice or policy (Green and Thorogood 2014). Such 'what', 'how', 'why' and 'how ought' information is not easily accessed via quantitative research methods, thus favouring the application of qualitative methodologies which scrutinise the socially constructed nature of reality, seeking to enquire into how social experience is created and given meaning. This approach often encourages the researcher to engage with human representations and perceptions as the primary data sources (Denzin and Lincoln 2013; Mason 2002; Greenway 2018).

According to Crotty (2003, p. 6), a theoretical perspective can be considered to be the 'philosophical stance that lies behind a methodology', with Creswell (2014) asserting that methodological choices are in turn based on the research aims and questions devised by the researcher. This is essentially an interpretivist[2] paradigm which seeks out 'culturally derived and historically situated interpretations of the social life-world' (Crotty 2003 p 67), while acknowledging that truth is socially constructed and mobile, as opposed to external, fixed and awaiting discovery (Butt 2015b). Adopting a particular theoretical position has implications for the subsequent interpretation of the data collected by the researcher; this may be limited by the chosen theoretical stance, be it sociological, psychological, philosophical or scientific (Hitchcock and Hughes 1995). There is therefore a generally accepted requirement for researchers to base their work within existing theoretical standpoints and disciplinary traditions—in turn mindful of the knowledge sets within which they are working. Creswell (2014) also suggests that researchers in education need to consider their ontological positions carefully, with ontology helping to outline the nature of reality based predominantly on two approaches: realist and relativist. As Greenway (2018) explains, the realist approach proposes that truth exists and can be quantifiably measured, whereas the relativist attests that truth cannot be measured, as reality cannot be judged (see also, Ross 2012). Epistemology refers to the

---

[1]This section draws significantly on work previously published in Butt (2015a, b).

[2]Crotty (2003) describes three strands of interpretivism: symbolic interactionalism, phenomenology and hermeneutics.

ways in which our beliefs can relate to theories of knowledge—which Butt (2015b) briefly describes as the process of discovering what knowledge 'is', and how it can be acquired and communicated. Wellington (1996) affirms that ontology and epistemology are closely linked—if ontology is the nature of reality, then epistemology is how we gain knowledge of that reality.

8.4 The Problem(s) with Education Theory

As we have seen, theory usually provides the foundation upon which research enquiry—and therefore ultimately new knowledge creation—is built. It helps us to highlight factors which require deeper investigation, to identify gaps in our knowledge, to aid our understanding and to predict future outcomes.

However, the respected education philosopher Wilf Carr (2006), cited in Thomas (2007), refers to education theory as 'the various futile attempts that we have made over the last hundred years to stand outside our educational practices in order to explain and justify them… we should now bring the whole educational theory enterprise to a dignified end' (p. 137). Similarly, Cohen et al. (2007)—no strangers to the application of theory in education—are also rather critical of educational theory (compared with, say, theories in the natural sciences), referring to them as 'only at the early stages of formulation and (are) thus characterized by great unevenness' (p. 13). Although these positions may be criticised for their characterisation of educational theory in direct contrast to practice—rather than the foundations from which we might understand and explain real life situations—they nonetheless reflect a concern about the ways in which educationists choose to apply theory. Thomas (2007) concurs, asserting that the use of theory in educational research is neither always justified, nor necessarily sensible, for achieving a better understanding of educational practice. He calls for the clearer conception of values, thoughts, reflections and ideas—rather than relying on theoretically rigid structures for enquiry—for in the hands of many education researchers he believes that "theory' may simply be a synonym for 'critical reflection" (p. 94). Critically reflecting is *not* theorising, just as all forms of intellectual work cannot be described as 'theory-making': the thinking that arises from everyday professional practice may therefore not have any explicit link to theory (Fish 1989). Thomas concludes by reflecting on the heady, exalted position that theory has gained within educational research, as follows:

> theory seemed to act as a kind of drug—a hybrid of intoxicant, hypnotic and hallucinogen—
> to otherwise sensible people in education, offering delusions about practical intervention
> (Thomas 2007, p. 7).

8.5 The Status and Place of Theory in Educational Research

Educational research has been under pressure in recent years concerning its usefulness and robustness—simply stated, it has been questioned with respect to its fitness for purpose (see Chap. 3). Thomas (2012) details how education researchers in many jurisdictions have been pressurised, both by their governments and research councils, to employ more scientific principles in their research designs—with a presumption that qualitative research neither offers robust enough evidence, nor generates sufficient reliable data, to answer educational questions. Shavelson and Towne (2002), cited in Thomas (2012), also refer negatively to the utility of educational theory, claiming that education research in the US is widely assumed to have 'failed', has an 'awful reputation' and 'does not generate knowledge that can inform education practice and policy' (Shavelson and Towne 2002, p. 28). We have seen that in the UK similar assertions are easily found, particularly within government circles and specifically with regard to research impact (Furlong and Oancea 2005; Tooley and Darby 1998; Hillage et al 1998). Not surprisingly, given that research in geography education is a subset of all education research, as a community geography educators are also not exempt from similar criticism (see Naish 1993; Butt 2002a, 2006; Firth and Morgan 2010). However, the assumption that the application of scientific theories and methods will ensure rigorous results—to be applied to improve teaching and learning in ways that are both repeatable and generalisable—reveals a fundamental misunderstanding of educational practice and process. Essentially, this raises our awareness that the knowledge accrued by the social scientist is different from that of the natural scientist.

So, what place *does* theory have in geography education research? Perhaps, unhelpfully, the most reasonable response is 'it depends'—essentially on what the researcher wants theory to do, and whether it is used appropriately. Some research in geography education is *a*theoretical, that is it does not require the application of any theory—only the correct use of research methods and methodologies (for example, in research designed to gauge the effectiveness of policies, strategies and projects the use of theory may be restricted, or even non-existent). Funded, evaluative work may have its research methods, methodologies and theoretical frame pre-determined for the researcher, particularly if it is government sponsored or directed by a funding body that favours the adoption of a specific theoretical or methodological stance. This raises questions about the openness of research—whether it is simply applied without a strong theoretical framework, say to test how effective a particular project has been, or whether it is driven by theory-building and a desire to explore new knowledge (Lidstone and Stoltman 2009). Such differences are important: the former type of research may not use theory, or may seek its application within a tightly constricted frame, while the latter is more emancipatory, encouraging knowledge discovery and creation.

8.6 Types of Theory

One might, as Thomas (2007) argues, look to simply define different types of theory—as well as the common uses to which they are put—along a 'theory-practice continuum'. Here Thomas, slightly tongue-in-cheek, offers the pragmatic view that 'theory is anything that isn't practice'; although this tends to ignore, or fail to recognise, the persistence of a 'theory-practice gap' in applied fields such as education and nursing (Greenway et al. 2019). The phrase is widely used, but often without definition or description of its underlying concept—meaning that it is inconsistently applied and its attributes, antecedents and consequences are unclear. As Gallagher (2004) asserts, the term 'theory practice gap' has become a 'useful and convenient shorthand' (p. 44) for a number of complex educational problems and situations. Most frequently the gap is referred to as being 'bridged', 'breached', 'avoided' or 'negotiated' without recourse to its characteristics, confusing our common understanding of the concept. Following a conceptual analysis of the term, using research literature, Greenway et al. (2019) offer this brief definition:

> The gap between the theoretical knowledge and the practical application of (x), most often expressed as a negative entity, with adverse consequences (p. 1)

They conclude that striving for a definition of the term is important, given that the theory-practice gap is not tangible—it represents a metaphorical void that is felt or experienced, but which is not easily measured or quantified. In attempting to provide a 'standardisaton' of the concept, to secure common meaning and relevance, the researchers applied Rodgers (2000) framework to help define its attributes. In addition, they describe situations in which practice fails to reflect theory; theory is perceived as irrelevant to practice; evidence-based practice is used; or ritualistic practices dominate.

Schon (1987) comments on the status of theory, noting that in education research (and elsewhere) there is a 'high ground of theory' which is usually contrasted with the 'swampy lowland of practice'. With others, he is also concerned about the uncritical, 'promiscuous' use of theory in education—asserting that to describe almost anything that is not 'practice' as 'theory' is unhelpful. Thomas (2007) ironically concurs, highlighting that theory can be applied well or badly to denote explanation, to gather personal reflections, to orientate principles, to aid epistemological presuppositions, and to develop arguments. With admirable conciseness, Robson (2002) defines theory as 'a general statement that summarises and organises knowledge by proposing a general relationship between events' (p. 18).

In attempting to provide an overview of different types of theory applicable to geography education research the important relationship between theory, epistemology and ontology needs to be explored. Epistemology refers to our theories of knowledge and how knowledge is constructed—which naturally impacts on the theoretical perspective adopted in the selection of research design and methodology (see Crotty 2003). As geographers and geography educationists, we understand that research in the social sciences look at 'reality' in different ways to scientific research. Our comprehension of the social world is, however, built on assumptions—which may be defined as ontological (concerned with the essence of the social phenomena being considered, whether these exist independent of the individual in the 'real world' or are a construction of one's mind), and epistemological (what knowledge is, how it is acquired, and how it can be communicated). Ontology is a branch of philosophy that deliberates on the nature of reality and the form that this takes, employing contrasting theories which are nonetheless closely aligned to associated epistemologies. According to the researcher's view of knowledge—whether it is considered to consist of objective facts, or is personal to the individual—the theoretical position and research approach is determined. In effect, knowledge-seeking disciplines and practices are shaped and influenced by the reality they seek to describe; this reality helps one to judge the validity of all research findings and the appropriateness of the methodology adopted to create knowledge. Here Crotty's (2003) definition of the theoretical perspective of research is helpful when he describes this as 'the philosophical stance lying behind a methodology' (p. 66). It is possible to research a particular question in geography education from a variety of theoretical perspectives and methodological approaches, but some will obviously be more suitable than others. In academic geography, until the 1970s, the preferred theoretical perspectives from the quantitative revolution were predominantly positivist and objective with concomitant effects on the research methodologies employed. This is still the case in much research in physical geography, although human and social geography tends to adopt a more interpretivist, subjective stance—however, the nature of the discipline of geography means that in much geographical study the positivist and interpretive approaches may operate together and cannot be easily separated, as both the physical and social worlds exist side by side (Crotty 2003; Taylor 2009). This may create tensions, by simultaneously pulling the researcher in different directions. Interpretivist research also covers a wide spectrum of epistemological positions—here researchers see truth as socially constructed and mobile, rather than external, fixed and awaiting discovery. Research into geography education therefore faces many of the same tensions as research into academic geography, although some researchers argue that no single epistemological position can underpin either field of research. In the geography education community research within the interpretivist paradigm has traditionally been strong, particularly in the UK, but debate about the contribution of theory to research in geography education has been rather limited (Butt 2002a; Gerber and Williams 2000; Williams 2003).

Within interpretivist research, theory tends to be generated as a consequence of enquiry, which Robson (2002) refers to as 'hypothesis generating' (as opposed to 'testing'). Here data collection and analysis may proceed 'hand in hand', where the process of analysis in turn suggests what new data needs to be gathered. Theory generation and refinement may be non-linear and uncertain, with ill-defined outcomes. This may not necessarily be as big a problem as it at first suggests: a mechanistic, methodologically 'tight', theory-driven piece of research may look good in terms of its design and expected results, but may not yield much valuable data. As Thomas (2012) asserts, 'we seek to capture regularity, and its encapsulation goes under the title of *theory*' (p. 31, emphasis in original)—but generalisation is not always possible, or indeed the main point of research, in the social sciences. The correct balance is difficult to achieve: the adoption of too many insecure assumptions within our evidence, methods and theory—often because of the number of variables in play and the changeable conditions in which they occur—mean that generalisations, theory-building and model construction can be dangerous projects in geography education research. Robson (2002) notes that adopting a purist approach to the research theories and methods we choose to adopt, from either one tradition or the other, may ultimately prove unrewarding. He suggests that a more sensible 'real world' approach is afforded by bringing the empirical and interpretivist traditions together in some way—as achieved through the use of a mixed methods research design (see Onwuegbuzie and Leech 2005; Weis et al. 2009).

In geography education, as indeed elsewhere, theories are only effective if they are applicable—within their given parameters, caveats and sets of conditions—to the majority of situations they are called upon to explain; indeed, to function well theories should be able to make strong claims to truth and generalizability within their chosen context. However, theories often include rather guarded statements about their expected degrees of certainty—a 'strong' theory may have good predictive powers ('in these circumstances, given these factors, this will result'), making explicit links between cause and effect. 'Weaker', or more tentative, theories may not be able to make such bold and defensible claims. Research in geography education should always aspire to adopting theoretical positions that are applicable to the circumstances they are used in and be strong enough to resist counter claims. Nonetheless, it may be the case that a theory's generative mechanism, rather than the theory itself, is robust and enduring (or 'strong') and therefore observable across different contexts. Theories can be classified as being *normative*—which suggests how things should be, and what we can expect given a particular set of circumstances or interventions; or *explanatory*—which suggest how things work. Many different types of theory obviously exist—here I choose to categorise the main theories used in geography education (as indeed elsewhere in subject-based research) as 'grand' theories, empirical theories, interpretivist theories, and critical theories. These tend to describe the collections of accepted theoretical positions that researchers in geography education favour. Other, more specific, theories (such as grounded theory, socio cultural activity theory, feminist theory, etc.), will fall within these parameters.

8.7 'Grand', Empirical, Interpretivist and Critical Theories

'Grand' theory attempts to construct a meta-narrative of related concepts, structures and empirical findings on a particular theme, which can be presented in a defined domain of study. If, as is often the case, empirical 'evidence' cannot be presented to support and illustrate all aspects of the theory this may involve some speculation or uncertainty. As such, grand theories are often ambitious and open to criticism, particularly as they aim to encompass everything in a particular field.

Empirical theories try to create deductions and generate laws (nomothetic), which can then be tested and hopefully verified. With repeated hypothesis testing (and rejection) confidence in these theories grows, following attempts to falsify or verify their propositions (Popper 1968). In this way each theory is amended, or replaced, by a better theory as knowledge grows—importantly, the types of evidence that can either confirm, or refute, aspects of the theory are made increasingly clear through this process. Exceptions to the original theory can be built into new theories which then ameliorate, or even replace, the old one. New theories should be broadly compatible with those that have gone before, being more empirically sound and moving in the direction of greater generalizability. There must be internal consistency in the theory, it should have explanatory and predictive power and be able to respond to anomalies (see Cohen et al. 2007). In essence, positivist and empirical theories differ from those with interpretivist roots because they look to understand phenomena through two different lenses:

> Positivism strives for objectivity, measurability, predictability, controllability, patterning, the construction of laws and rules of behaviour, and the ascription of causality; the interpretive paradigms strive to understand and interpret the world in terms of its actors. In the former observed phenomena are important; in the latter meanings and interpretations are paramount (p. 26)

Due to the inherent difficulties in meeting the conditions of positivism, the use of empirical theories in geography education research tends to be modest. However, this is not to deny that valuable research within this tradition is produced, nationally and internationally, every year. Interpretivist approaches allow for some non-conformity when compared with 'pure' theoretical perspectives, as they refute claims to universal truths that are 'out there' and simply waiting to be discovered. Interpretivism therefore adopts contrary positions to positivism, essentially by looking for 'culturally derived and historically situated interpretations of the social world' (Crotty 2003, p. 67). Because of this, interpretivist theories are open to a variety of readings—not least because of their tendency to adopt broad, context driven, inclusive perspectives—but are adept at exploring complicated meanings and useful in considering 'the whole' rather than 'the parts'. The epistemological underpinnings of interpretivism highlight the differences between people and objects, with researchers using various theories and methods to understand the complexity and richness of human action. So, positivism aligns with scientific enquiry, and empiricism, where only observable phenomena can be accepted as knowledge and where the gathering of 'facts' can lead to the creation of 'laws'. By contrast, interpretivism holds the position that the achievement of objective, unbiased research is unattainable in the social sciences (Bryman 2004). It is therefore unsurprising—given the nature of most research contexts in education, geography education and indeed the discipline of geography—that interpretivist enquiries are most strongly favoured.

Critical Theory, originating from the Frankfurt School, regards the positivist and interpretivist paradigms as incomplete, for they tend to ignore the ideological and political circumstances within which social behaviours occur. Critical theorists claim that many theoretical approaches accept, rather than question, the *status quo*; within the educational context this simply replicates the flawed systems of society and schooling, despite our research efforts. Theorists in this paradigm work towards the creation of a more egalitarian, democratic society within which individuals are free to act and express themselves. Essentially such research is transformative, rather than descriptive, striving to break the perpetuation and reinforcement of existing social status and process (Habermas 1988). The desire to make changes that will move us towards a more just society, rather than merely *describing* a situation that currently exists, marks out critical theory as a force for emancipatory action. Originally founded on Marxist ideas, researchers who adopt a critical position in their research cannot claim to be ideologically neutral—they question (rather than accept) given research agendas, and seek to interrogate and examine how social institutions work (Kincheloe and McLaren 2005). In the context of educational research this might include enquiry into the ways in which schools reproduce rather than mobilise social status, how they perpetuate inequalities and class divisions, and how curriculum decision-making occurs. Who gains and who loses from education—often with particular reference to their class, race and gender—is a common concern, as is the consideration of the place of knowledge in education. Here researchers question the very nature of knowledge and the taken-for-granted assumptions about ways of knowing (Firth and Morgan 2010). This has become a focus for much contemporary research in geography education—driven by questions such as: whose geographical knowledge is being taught? what counts as valuable geographical knowledge? and who advocates and supports such knowledge creation and transition? Young's (2008) conception of 'powerful knowledge', recognising as it does the social constructedness of knowledge and its cultural significance, has been a stimulus to recent research activity for many geography educators.

Research methodologies favoured by critical theorists, among others, include methods associated with action research, interpretivist approaches, and ideological critiques. Action research tends to be undertaken by practitioners researching their own practice within its particular context (its specific place, historical setting, social dimensions, power bases, etc.), and often claims to be emancipatory and empowering (Butt 2002b). However, change may *not* occur as a result of this research. This is the consequence of research being carried out by those who have limited power, or who are situated in very specific, local contexts—their study being restricted to individual classes, or a school, rather than impacting more widely across schools and society. The power of the individual teacher, even one who adopts action research methods, very rarely affects the real locus of power and the actions of decision makers. The extent to which engagement with critical theory research can change situations therefore requires verification. Despite its claims, has this theoretical position really led to much change? Has teacher and student agency increased and transformation occurred? The political drivers of this type of research are both a blessing and a curse—critical research seeks real change, but by taking an ideological position it is open to legitimate claims of partiality, bias and subjectivity (although researchers in this paradigm would assert that the very situatedness of other research means that it works towards maintaining the status quo and the reproduction of existing power relationships). Critical theory also implies the adoption of a critical pedagogy— where teachers refuse to teach a curriculum that simply replicates society, favouring one that is emancipatory, and which strives to destroy social inequality and liberate students.

8.8   Theory in Geography Education

Morgan and Firth (2010) remind us that the status of theory in geography education research is tenuous, currently experiencing a degree of 'anxiety' about its role. Providing an illustrative (rather than exhaustive) overview of the types of theories adopted in such research over the past 30 years, they concentrate on a number of texts that have been produced in geography education to support the application of theory. There is a keeness to point out that theories are human constructs and responses to events—that they are not easily dissociated from the people that produce them, nor the conditions within which they are produced. Particularly pertinent is the fact that the contextual origin of much research in geography education is from within higher education institutions (HEIs), and that the location for much of this work is the classroom—which may determine the theories that researchers choose to apply. The historical approach is a valuable one—Morgan and Firth (2010) describe the emerging field of research in geography education at the end of the 1980s as 'a gradual flowering of a variety of substantive themes underpinned by a number of theoretical perspectives' (p. 89). Two pathways for research became clear: one that focussed on curriculum development, and another concerned with teacher development and pedagogy (with a lens on 'what works').

Morgan and Firth (2010) note that 'since the early 1990s, geography education as a research field has been challenged by pressures that seek to downplay the importance of theory' (p. 89). Government conceptions of initial teacher education (ITE) have derided the use of theory as irrelevant for learning how to teach, with centralised notions of 'effective teaching' and practical 'tips for teachers' always trumping the pursuit of theoretical understandings. As we have seen, the 'what works' agenda always seems to supplant any recourse to theory to inform and understand process. This has led to scholarship in geography education being dominated by a focus on classroom-based problems, but with little recourse to theory (Morgan and Firth 2010). Firth and Morgan (2010) assert that as a consequence the geography education community 'requires a wider range of orientations to research, concerned as we are with … classroom practice, policy-making and future directions for geography education' (p. 109). In the light of expectations in the UK for research to have 'impact' and to be of high quality, they pose important questions about the value attributed to theoretically informed research and about which theoretical positions are considered worthwhile. They conclude that, 'without theory our data on the experience of schooling do not get very far and are unlikely to tell us much that is not already obvious' (Firth and Morgan 2010, p. 111).

Attempts to bring more theory into the geography classroom has had some limited success. The *Theory into Practice* series, published by the Geographical Association from 1999, was an explicit attempt to convey theory into the lives of practitioners, offering clear research and theory-driven perspectives on the issues geography teachers faced. This set of short research monographs sought to provide access for busy teachers to research findings that related to their own, classroom-based, contexts—although they were not all driven by a desire to explore 'what works' in the classroom (Butt 2002a).

8.9 Conclusions

Research in geography education, as in other subject-based contexts, must consider its theoretical underpinnings—without recourse to theory it is doubtful that research 'findings' can be validated as being either rigorous, or fit for purpose. It is possible that some enquiries and evaluations need not have a strong theoretical foundation—for example, literature reviews will inevitably cover thematic issues and questions, and the findings of others, and may not need to be *driven* by theory. But such reviews will engage with the theoretical frameworks that underpin previous studies—which may then influence the structure, organisation and analysis of future research. Any research findings or recommendations in geography education will not simply recount a set of 'facts' which have been discovered, but will comment on the theoretical framework within which these observations sit explaining their links and relationships with the data collected. In this way, theory not only helps us explain what has been discovered, but also indicates remaining knowledge gaps and questions that still have to be pursued. Theory should also enable the researcher to organise and interpret their data and indicate the imperative for new data gathering. It is certainly the case that research can be carried out without the consideration, application, or generation, of theory—which is undoubtedly quicker and easier than undertaking research that is theory-informed or –driven—but the outcomes may not constitute what is generally accepted as research enquiry. This may mean that the results are not considered to be significant, rigorous, robust and fit for purpose.

Reflecting on the development of research in geography education—as Kent did at the turn of the century (Kent 2000), and Morgan and Firth attempted after its first decade (Morgan and Firth 2010; Firth and Morgan 2010)—reveals that internationally research in the field has some limitations. Kent's (2000) analysis, which drew on previous work by Williams (1998), split research activity into three stages—incipient, intermediate and mature—positioning geography education research in the first stage. Although there are obvious examples of research in the field being 'mature', the majority of geography education research is still dominated by immediate practical concerns, rather than being directed by universal theoretical issues. Looking positively, one might conclude that geography education research is at an exciting point in its journey—youthful, challenging, with a range of possibilities and different avenues before it. However, there are dangers in occupying this position compared with more established fields of research—our engagement with theory still has some way to go. Research in geography education has not significantly advanced since Firth and Morgan (2010) characterised the research community as having 'overlooked theoretical traditions in ways of researching and analysing geography education' (p. 112). This charge might, of course, also be levelled at other education and subject-based research communities—but it still represents an uncomfortable position for geography education researchers to be in.

References

Bonnett, A. (2008). *What is geography?*. London: Sage.

Bryman, A. (2004). *Quantity and quality in social research*. London: Unwin Hyman.

Butt, G. (2002a). *Reflective teaching of geography 11–18: Meeting standards and applying research*. London: Continuum.

Butt, G. (2002b). Geography teachers as action researchers. In R. Gerber (Ed.), *International handbook of geographical education* (pp. 273–284). London: Kluwer Academic Publishers.

Butt, G. (2006). How should we determine research quality in geography education? In K. Purnell, J. Lidstone, S. Hodgson (Eds.), In *Changes in Geographical Education: Past, Present and Future. Proceedings of the IGU—Commission on Geographical Education Symposium* (pp. 91–95). Brisbane: IGU CGE.

Butt, G. (2015a). Introduction. In G. Butt (Ed.) *Masterclass in geography education* (pp. 3–14). London: Bloomsbury.

Butt, G. (2015b). What is the role of theory? In G. Butt (Ed.), *MasterClass in geography education* (pp. 81–93). London: Bloomsbury.

Carr, W. (2006). Education without theory. *British Journal of Educational Studies, 54*(2), 136–159.

Cohen, L., Manion, L., & Morrison, K. (2007). *Research methods in education* (6th ed.). Abingdon: Routledge.

Couper, P. (2015). *A student's introduction to geographical thought: Theories, philosophies, methodologies*. London: Sage.

Creswell, J. (2014). *Research design: Qualitative, quantitative and mixed methods approaches* (4th ed.). Thousand Oaks, CA: Sage.

Crotty, M. (2003). *The foundations of social science research: Meaning and perspective in the research process*. London: Sage.

Denzin, N., & Lincoln, Y. (2013). *Handbook of qualitative research*. London: SAGE.

Firth, R., & Morgan, J. (2010). What is the place of radical/critical research in geography education? *International Research in Geographical and Environmental Education, 19*(2), 109–113.

Fish, S. (1989). *Doing what comes naturally*. Oxford: Clarendon Press.

Furlong, J., & Oancea, A. (2005). *Assessing quality in applied and practice-based educational research. ESRC TLRP seminar series*. (December 1) Oxford University Department of Educational Studies.

Gallagher, P. (2004). How the metaphor of a gap between theory and practice has influenced nurse education. *Nurse Education Today, 24*(4), 263–268.

Gerber, R., & Williams, M. (2000). Overview and international perspectives. In A. Kent (Ed.), *Reflective practice in geography teaching*. London: Paul Chapman.

Greenway, K. (2018). Does a theory-practice gap exist in nurse education? Unpublished PhD. Oxford Brookes University.

Green, J., & Thorogood, N. (2014). *Qualitative methods for health research*. London: SAGE.

Greenway, K., Butt, G., & Walthall, H. (2019). What is a theory-practice gap? An exploration of the concept. *Nurse Education in Practice, 34*, 1–6.

Gregory, D. (1994). *Geographical imaginations*. Oxford: Blackwells.

Habermas, J. (1988). *On the logic of the social sciences*. Cambridge: Polity Press.

Harvey, D. (1984). On the history and present condition of geography: An historical materialist manifesto. *The Professional Geographer, 36*(1), 1–11.

Hillage, J., Pearson, R., Anderson, A., & Tamkin, P. (1998). *Excellence in research in schools.* London: TTA.

Hitchcock, G., & Hughes, D. (1995). *Research and the Teacher* (2nd ed.) London: Routledge.

Jackson, P. (2006). Thinking geographically. *Geography, 91*(3), 199–204.

Kent, A. (Ed.). (2000). *Reflective practice in geography teaching.* London: Paul Chapman Publishing.

Kincheloe, J., & McLaren, P. (2005). Rethinking critical theory and qualitative research. In N. Denzin & Y. Lincoln (Eds.), *The sage handbook of qualitative research* (pp. 303–342). London: Sage Publications.

Lidstone, J., & Gerber, R. (1998). Theoretical underpinnings of geographical and environmental education research: Hiding our light under various bushels. *International Research in Geographical and Environmental Education, 7*(2), 87–89.

Lidstone, J., & Stoltman, J. (2009). Applied research in geography and environmental education: Rethinking the 'applied'. *International Research in Geographical and Environmental Education, 18*(3), 153–155.

Mason, J. (2002). *Qualitative researching.* London: Sage.

Matthews, J., & Herbert, D. (2008). *Geography: A very short introduction.* Cambridge: Cambridge University Press.

Morgan, J. (2013). What do we mean by thinking geographically? In D. Lambert & M. Jones (Eds.), *Debates in geography education.* London: Routledge.

Morgan, J., & Firth, R. (2010). 'By our theories shall you know us': the role of theory in geographical education. *International Research in Geographical and Environmental Education, 19*(2), 87–90.

Naish, M. (1993). 'Never mind the quality—feel the width'—How shall we judge the quality of research in geographical and environmental education? *International Research in Geographical and Environmental Education, 2*(1), 64–65.

Onwuegbuzie, A., & Leech, N. (2005). On becoming a pragmatic researcher: The importance of combining quantitative and qualitative research methodologies. *International Journal of Social Research Methodology, 8*(5), 375–387.

Parahoo, K. (2006). *Nursing research: Principles, process and issues.* London: Palgrave Macmillan.

Popper, K. (1968). *Conjectures and refutations: The growth of scientific knowledge.* New York: Harper and Row.

Rawding, C. (2013). How does geography adapt to changing times? In D. Lambert & M. Jones (Eds.), *Debates in geography education* (2nd ed., pp. 282–290). London: Routledge.

Robson, C. (2002). *Real world research: A resource for social scientists and practitioner-researchers.* Oxford: Blackwell.

Robson, C. (2011). *Real world research: A resource for users of social research methods in applied settings.* London: Wiley.

Rodgers, B. (2000). Concept analysis: An evolutionary view. In B. Rodgers & K. Knafl (Eds.), *Developments in nursing—Foundations, techniques and applications* (2nd ed.). Philadelphia: W B Saunders.

Ross, A. (2012). The new pluralism—a paradigm of pluralisms. *European Journal of Psychotherapy & Counselling, 14*(1), 113–119.

Schon, D. (1987). *Education the reflective practitioner.* San Francisco: Jossey-Bass.

Shavelson, R., & Towne, L. (2002). *Features of education and education research: Scientific research in education.* Washington, DC: National Academy Press.

Taylor, L. (2009). *The negotiation of distant place: Learning about Japan at Key Stage 3.* Unpublished PhD thesis: University of Cambridge.

Thomas, G. (2007). *Education and theory: Strangers in paradigms.* Maidenhead: Open University Press.

Thomas, G. (2012). Changing our landscape of inquiry for a new science of education. *Harvard Educational Review, 82*(1), 26–51.

Tooley, J., & Darby, D. (1998). *Educational Research: A critique.* London: OfSTED.

Weis, L., Jenkins, H., & Stich, A. (2009). Diminishing the divisions among us: Reading and writing across difference in theory and method in the sociology of education. *Review of Educational Research, 79,* 912–945.

Wellington, J. (1996). *Methods and issues in educational research.* London: Continuum.

Williams, M. (1998). A review of research in geographical education. In A. Kent (Ed.), *Issues for research in geographical education. Research forum 1* (pp. 1–10). London: Institute of Education Press.

Williams, M. (2003). Research in geographical education: The search for impact. In R. Gerber (Ed.), *International handbook on geographical education* (pp. 259–272). XX Kluwer.

Young, M. (2008). Bringing knowledge back. In *From social constructionism to social realism in the sociology of education*, London: Routledge.

Chapter 9
The Prospects for Geography Education Research—What Are the Ways Forward?

9.1 Context

This chapter considers the prospects for the successful continuance of geography education research and for sustaining the aspirations, and careers, of its researchers. Building on the findings from the previous chapters it questions how, in their research community, geography educators might both strive to strengthen their field of research and make the most of the circumstances they now find themselves in. It concludes that in the context of numerous forces over which geography education researchers currently have little control—not least in terms of government, university and research council policy and, importantly, funding—the most pragmatic way forward is to pursue and promote what is truly distinctive about their research. There must be an honesty in recognising the increasingly insecure position of geography education research, as a subset of all education research, but also a clarity about what research in the field can offer to geographers, geography educators, other subject-based researchers and those beyond. This requires the geography education community, nationally and internationally, to communicate the unique contribution to knowledge that high-quality geography education research can make. However, it must also recognise that levels of confidence are currently low within the community—many might argue that self-assurance, or indeed complacency, is misplaced at this time given the set of circumstances faced.

But how should the conditions necessary to engender a genuine confidence about the future contribution of geography education research be achieved? The need for continuing a debate about the organizing concepts, themes and ideas that geography education research might engage with is recognized, for the 'future directions for geography education research and its impact on classroom practice and policy-making are not straightforward' (Firth and Morgan 2010). The agendas and themes for research in the field—initially outlined early in the 21st century by, among others, Gerber and Williams (2000), Roberts (2000) and Butt (2002)—have been considered in previous chapters. Here the task is to contemplate the practicalities of success-

© Springer Nature Switzerland AG 2020
G. Butt, *Geography Education Research in the UK: Retrospect and Prospect*, International Perspectives on Geographical Education,
https://doi.org/10.1007/978-3-030-25954-9_9

fully emerging from a set of circumstances that seem to hold geography education researchers, and those in many other subjects, in a condition of stasis or, more worryingly, act to reverse the advances they have made so far.

This chapter therefore 'reflects and projects' on the current research culture in education, the conditions for producing research, and the outputs of geography education researchers across nations and jurisdictions. A commentary is provided on what has been learnt from a generation of geography education research, and suggestions offered about the pathways this research might now take to strengthen its status, position and contribution. The conclusions may not be palatable to those who have worked 'traditionally' as geography education researchers in the academy, but if the goal is to strengthen research in the field then some radical solutions to current circumstances may need to be sought.

9.2   Introduction

In the concluding section of their edited work '*Geography into the Twenty-first Century*' Richard Daugherty and Eleanor Rawling find themselves reflecting on the 'similarities and common purposes' of geography education in the university and school sectors (Rawling and Daugherty 1996). They challenge themselves, and us, to consider the principle areas for research, exploration and further discussion—carefully pointing out that this is not just about teachers simply knowing more about the geographical research produced in universities, or vice versa, but about creating a shared understanding of ways forward. The six points they identify are repeated below:

1. Which areas from the research frontier are relevant and appropriate for school geography?
2. What distinctive knowledge, understanding and skills can geography contribute to vocational courses?
3. What are, or should be, the 'residuals' of a geographical education?
4. How can we better our students' ability to adopt appropriate modes of enquiry in their study of the subject?
5. How can we define and measure progression in geographical understanding?
6. How can we ensure that geography's contribution is recognised and understood by the general public and educational decision-makers?

Individuals may agree or disagree as to whether these points truly represent our recent, or current, research priorities; what is interesting is that most of these potential areas for research and development in geography education still remain essentially untouched, or unresolved, over twenty years later. This may be for four reasons: (i) they are not recognised by geographers and geography educationists as important areas for research and development, or (ii) they are recognised, but not agreed on, as requiring joint action, or (iii) these questions are not answerable in a definite and permanent fashion, the answers keep changing, or (iv) the capacity and capability to research into these areas has not previously existed, and still does not exist, meaning that they have not been substantially pursued. Elements of all four positions may be valid—but my impression is that the fourth reason for explaining a lack of resolution is the strongest.

Whether producing prioritised lists of research areas to tackle represents a sensible way forward for any research community is debateable. There is a sense that such lists are rather authoritarian, or indeed anti-democratic—especially for researchers who have traditionally valued, and defended, their autonomy and academic freedoms. Nonetheless, there is a sense that any research activity should be supported by structures, organisation and healthy debate about what directions our efforts should take, and how to help facilitate successful outcomes. The practical means by which we connect our research endeavours, whatever the circumstances faced, must be realised. This extends to opening up channels for communication and discussion about research—increasingly through the use of social media, the internet and blogs. In geography and geography education such networks already exist, but their content should always seek to include issues of geography education *research*—here is an expectation not only for the professional associations linked to geography education, but also for organisations dedicated to researching geography education [such as the Geography Education Research Collective (GEReCo)], and even for public examination and awarding bodies. One might hope that in different jurisdictions the government of the day, through its Department or Ministry for Education, would promote subject-based education research, not least through publications, websites and themed seminars and conferences. One of the links that needs strengthening across such networks, as we have seen, is the connection between schools and the academy. Making research findings available and accessible to others—from academic geography *and* geography education—requires the efforts of those who Healy and Roberts (1996) refer to as 'mediators' between the sectors. These individuals need to be recognised and supported; valued in both professional and career terms by their institutions and by the wider community of geography educators—a point also raised by Butt and Collins (2013, 2018).

9.3   Retrospect and Prospect

Let us first consider the 'backdrop' to the existing state of geography education research globally—for without a sound appreciation of the wider sphere of research in education, and of the challenges it faces, we cannot make sensible decisions about our future. Oancea and Mills (2015) helpfully offer a timely overview of the 'state of play' in education research in the UK—fortunately many of their observations also have some relevance to an international audience. The standout points, especially for smaller groups such as geography education researchers and many subject-based research communities, are salutary. They are listed below:

(i)   academic staffing in Education in HEIs employed on 'research only' contracts declined by one third since RAE 2008, resulting in less access to government and research council funding for research.

(ii) use of 'teaching only' contracts has increased, particularly among Russell Group universities.

(iii) in 2012–13 one third of academic staff in Education were aged 56 or over, an increase from one quarter in 2005.

(iv) higher degree research student numbers showed a slight increase from 2009 (to 4800), both on traditional PhD and EdD routes, with the number of part time students staying stable. Despite some HEIs having large numbers of doctoral students, over 40 only had 20 students each. Teachers and HEI employees make up the bulk of part time student numbers (two thirds). The majority of education doctoral students[1] are over 30 years of age (80%).

(v)   Russell Group and older established universities attract two thirds of all research funding. All institutions have suffered a reduction in research funding since 2009, a drop of almost one quarter. Sources of funding for educational research have also shifted dramatically, with government funding reducing by 42%.

Other contextual information is also valuable. Research funding is primarily locked into the 'pre-1950s' universities in the UK, but the distribution of this money—and therefore the concentrations of research staff that benefit from it—is predominantly within the London region[2] (with 32% of education researchers located in the South East of England). Those institutions that have previously performed well in research assessments have tended to become stronger, squeezing out any subsequent performance gains made by the 'post 1992' institutions. Additionally, the distribution of research funding in education—when compared to that in other disciplines—is considerably more 'lumpy' and unstable. As a consequence, higher education institutions in the UK are often no longer the primary location for conducting education research; Oancea and Mills (2015) describe the commissioning and production of education research as now occurring over:

> a diverse set of sites, collaborations and organisations (where) UK universities are only involved in a small subset of these policy networks. Third sector organisations, think-tanks and policy networks are increasingly important and influential funders, producers and consumers of educational research (p. 2)

The shifts in research funding for education have contributed to a decrease in the number of research staff employed in this field—particularly those on a part time, fixed term or contract basis—with these employees being unlikely to be replaced by universities once they leave their posts. This is of particular concern given the demographics of the population of active education researchers in the UK, one third of whom are currently aged 56, or above. With external research funding in flux—a

---

[1]One third of doctoral students receive some funding for their studies. The percentage of education students gaining Economic and Social Research Council (ESRC) studentships fell from 8% in 2009 to 4% in 2013, party as a result of open competitions held by Doctoral Training Centres.

[2]However, the resource to staff ratio is higher in the North East and Northern Ireland.

situation that does not seem likely to change any time soon in the UK—and with the strongest, mostly traditional, universities having the highest research capacity, staffing levels and dominance of the research market in education, this is a backdrop of impending cross sectoral research instability.

What does this mean for the prospects for geography education research in the UK, and beyond? The geography education research community, often represented by only one or two people in the higher education institutions that employ them, is positioned on the 'periphery of a periphery'—a genuine 'outlier' (Oancea 2017). Oancea's work in this area is significant and well-respected. She cannot with *certainty* describe the prospects facing geography education research and researchers (or indeed those of any other subject-based research community)—but her evaluation of the impacts of the Research Assessment Exercise (RAE) and Research Excellence Framework (REF), the institutional responses to research assessments, her interviews of REF panel members, her enquiry into the research prospects in other disciplines, and interviews with staff in over 30 HEI Education departments give us a strong steer about the overall 'direction of travel' for education research in the UK context. Despite some worrying messages, there are also many positives—for example, the highest-ranking publications in education are of an excellent standard, comparable with the highest quality research in other disciplines in the social sciences, nationally and internationally. However, the research outputs from subject-based researchers (including those of geography education researchers) have traditionally scored lower than other research areas in education, and may be aimed primarily at teachers rather than other researchers. This trend is not straightforward—subject-based research has customarily been considered to be within a subset of 'Curriculum' research in education; here Maths Education research has tended to score well, but this is not the case for outputs from researchers representing 'non-core' subjects. Much subject-based research is small scale and generally considered to be under theorised by reviewers, hence the subsequent award of lower research assessment scores to their publications. There are, of course, structural reasons that may help us to explain the generally lower quality of subject-based research in HEIs: practice-based research has tended to score lower than other education research, and many geography education research publications are written by ITE staff, who typically research into aspects of subject pedagogy.

9.4 International Perspectives

Despite its inherent difficulties, there is a sense that education research from the UK has successfully 'punched above its weight' on the international scene for many years (Oancea 2017).[3] The preferred medium of English language for academic publications obviously helps, while academics in other countries have traditionally

[3]See also Bagoly-Simó's (2014) longitudinal analysis of research publications in geography education between 1900 and 2014.

viewed education journals based in the UK and USA as having higher status, rank and impact—making them target publications for sharing their research findings.

In the US, and indeed beyond, the National Geographic's *Road Map for 21st Century Geography Education Project* (Edelson and Pitts 2013) attempted to set a course for the improvement of geography education and geography education research over the past decade. In total, three reports identified the requirement to encourage and support research that was both quantitative and qualitative in nature, but which would also be theory-driven and replicable elsewhere (Bednarz et al. 2013). Research was also encouraged to be large scale and assessment focused (Edleson et al. 2013), as well as working to support the professional development of teachers and the production of teaching materials (Schell et al. 2013). Lambert and Solem (2017) have considered, in the light of the influence of the Road Map Project in the US and beyond, how Research Coordination Networks (RCNs) have helped to build research capacity, support research management, and provide transformative research 'designed to support broad-scale advances in geography education theory, methods and practice' (p. 375). They highlight the often-stated issues regarding education research, particularly that which is generated from subject communities—including its low visibility of outputs, weak record of inter-disciplinary and international collaborations, poor rates of transfer and uptake of research findings into policy and practice, and the erosion of graduate programmes which prepare students to undertake education research in geography. This context is important for Lambert and Solem (2017), who recognise the significance of much geography education research in the States being 'positioned' in Departments and Schools of Geography (and other related disciplines), rather than in Schools, Departments or Institutes of Education. The reasons for this historic trend are linked to 'snobbishness' (Hill and LaPrarie 1989), a perceived lack of relevance of geography education research (Gregg and Leinhardt 1994; Brown 1999; Segall and Helfenbein 2008) and—interestingly—popular judgements made about the marginal worth that research in geography *education* has in gaining researchers academic tenure, or promotion. Contrasting this with a generally healthier, more productive, situation in research in mathematics and science education, the authors also note that there is little consensus on 'what works' research in geography education—not for any philosophical reasons, but because school-based research data in geography education simply does not exist in sufficient quantities, or quality, to support such decision making. A consequence is that major research sponsors, policy makers and education communities are generally disinterested in the products of geography education research in the United States.

The foundational concept which underpinned the Road Map reports was that future research in geography education should be *transformative*—that is, research which is high risk, high reward and which 'radically change(s) our understanding of an important existing … concept or educational practice or leads to the creation of a new paradigm' (NSB 2007). The Road Map project team therefore took very seriously the need for potential research in geography education to both renovate and change practice, as evidenced in their construction of two driving questions:

1. what areas of research will be most effective in improving geography education at a large scale?
2. what strategies and methodologies can relevant research communities develop and adopt to maximize the cumulative impact of education research in geography?

Importantly, they also considered what was needed in terms of institutional infrastructure to help answer these questions, concluding that a small community of researchers—as typified by those involved in geography education research—required support with coordination, collaboration, information sharing and funding:

> that breaks down research silos, which in many ways runs counter to long-held traditions of individualistic academic research culture Lambert and Solem (2017).

They recommended that a national institution be established in the US that could coordinate the implementation of geography education research, and its subsequent dissemination, to achieve greater knowledge transfer. This culminated in the Association American of Geographers (AAG) and Texas State University establishing the National Center for Research in Geography Education (NCRGE) in December 2013, from which a Research Coordination Network (RCN) for geography was launched.

9.5 Research Coordination Networks

RCNs are viewed by the National Science Foundation (NSF) as integral to establishing transformative research across Science, Technology, Engineering and Mathematics (STEM) disciplines, facilitating the ways in which researchers can work together and then disseminating their findings to interested parties.[4] The advantages of supporting research in this way, according to NSF, include:

- Fostering communications to solidify partnerships for interdisciplinary and international collaborations;
- Accumulating, sharing and curating a body of knowledge, including validated research instruments, databases, annotated bibliographies and other resources;
- Supporting data collection at multiple sites and across diverse educational environments;
- Building capacity for acquiring funding for larger, more complex studies;
- Connecting local researchers to a broader community of scholars and helping them to see connections between their work with that of others who share their research interests.

There are interesting comparisons to be made with the establishment and functioning of the Geography Education Research Collective (GEReCo) in England from 2007—GEReCo was originally launched in 2007 by a small group of research active

---

[4]See also, Gerber's (1998) account of the possible impact of Strategic Alliances in geography education research.

geography educationists working in higher education in England to support the production and dissemination of geography education research. In many ways the aims of GEReCo are congruent with those of NCRGE—indeed there was interest shown by American academics in the ways in which GEReCo functioned before their own institution was established. The main difference between the two organisations are the scale of their operations—although both purport to work at the national and, increasingly, international scales—and the fact that NCRGE is generously, centrally funded. In the States the creation of RCNs is predicated on the belief that such networks help to organise research nationally in ways which promote and support intellectual endeavour and aid the transfer of research findings to practitioners. The benefits of RCNs are not yet widely reported in geography education (but see Lambert and Solem 2017), however the claims about their potential achievements appear reasonable: they may help to foster communication and partnership both within and across disciplines, and at the international scale; aid data collection and management across various sites; set standards for data collection and quality; support capacity building; and assist the connection of researchers at a variety of scales. The creation of organisations and working structures in different countries that mirror those of RCNs might therefore be highly advantageous to geography education research communities.

From 2014, the NCRGE collected data on the extant research community in geography education in the US—by identifying university departments that offered geography education programmes, analysing relevant articles in reputable journals and dissertations/theses in geography education, and noting academics who had an interest in geography education—analysing their research and teaching in this area, and paying particular attention to the circumstances of early career academics in the field. Academics who had previously shown an involvement and interest in geography education were invited to join the RCN, resulting in the recruitment of 130 researchers across 52 institutions in the US, and the gathering of a handful of international collaborators in 12 countries by 2016 (see www.ncrge.org). Importantly the NCRGE charged itself with using their RCN to recognise the origins of potentially transformative geography education research in its earliest stages, rather than delaying for (possibly) years until the impact of such research became more obvious. Mindful of the persistent problems facing research in geography education, they also tried to distinguish and fund key thematic areas for research with paid research fellowships being targeted at themes identified by the Road Map Project. Money was also used to support face-to-face and virtual meetings (including funding workshops, travel grants and suggested collaborations) and to add research outcomes to social media, such that they were accessible to RCN members. To help ensure effective outcomes, the RCN's activities are aligned to the NSF publication 'Common Guidelines for Education Research and Development' (NSF 2013), components of which include: affiliation with key research questions; situating research in a problem context; focussing on core ideas, practices, knowledge and skills in geography; drawing from research in related themes/concepts in STEM; and developing sequential tasks, activities and experiments (measuring these against research objectives). This programme has a logical structure and framework, but achieving agreement on these themes was not straightforward. It was recognised that future research efforts had to be relatively 'fleet of foot', and that establishing set and enduring research themes was unhelpful in situations where issues and problems change rapidly.[5]

The RCN employs a number of Research Group Chairs, whose sub groups reflect the various research themes to pursue in geography education. These Chairs have a responsibility to ensure that their groups function correctly, that the principles of RCNs are upheld, and that the products of their labours are communicated with each Chair in the network. One of the contrasts with other organisations which support, or help to organise, geography education research in other countries relates to funding. The sponsoring of research groups is obviously highly beneficial to the facilitation of their work, supporting their projects and disseminating information about their outcomes—particularly where funding is assured for a five-year period, enabling more secure and detailed future planning to take place. The geography education RCN in the US has clear aims concerning the building of capacity to ensure sustainability of research activity, to ensure vitality, and to increase the visibility of research outcomes. Lambert and Solem (2017) identify the issues of doctoral uptake and completion in geography education compared with those in the discipline of geography in the States, reporting that from 2001 to 2012 only 32 theses were awarded in geography education,[6] compared with a little less than 2400 in geography. This situation is exacerbated by university geography departments not replacing those academic staff who either retire, or relocate to other universities, who have previously held responsibilities for geography education. There is a growing realisation that recruitment to the wider geography education community must also come from Colleges of Education in the US (which have traditionally been less prominent than in the UK) and that other locations of research activity need to be accessed.

In conclusion, the US Road Map Project has recorded significant achievements for geography education and geography education research in a relatively small period of time. The prospects for these to be mirrored globally are arguably hampered by a general dearth of funding elsewhere, but it may be possible for others to follow the principles, procedures and organisational structures adopted by the Americans in their own jurisdictions without the prerequisite of comparable funding levels. Importantly, American geography educators have realised a significant priority: to work *across* subject specialisms in creating research into learning pathways and progressions, understanding that this will inevitably be to the advantage of the geography education community.

---

[5]The RCN for geography education acknowledges this issue, which it seeks to counter through establishing good communication networks and using learning clusters. These clusters are monitored to determine levels of activity, interaction and communication online, along with monitoring ideas emerging, hopefully to sustain a dynamic working community.

[6]See also, Catling and Butt (2016) for a discussion of the wider issues of uptake and completion of doctorates in geography education.

9.6 Prospects for Geography Education Research in the UK

To compare and contrast historic developments in geography education research in the UK with recent developments in the US is, in many ways, cathartic. David Lambert, in contemplating the key issues and challenges faced by geography education research and researchers internationally over the past 50 years, concludes that the work of the geography education community has been characterised by 'long periods of contributing very little to what we know' (Lambert 2017). Indeed, in recent years in England this may be considered to reflect a degree of complacency—given that the pinnacle of research productivity in the field might be recognised as covering the period when three, large, funded curriculum development projects for geography education were sponsored by the Schools Council in the 1970s and 80s (16-19 Geography Project, Geography for the Young School Leaver (GYSL), and the Bristol Project). These projects saw substantial curriculum development, related assessment through public examinations, as well as associated research and publication activity—they also brought national and international recognition for English geography education research, sometimes beyond the immediate host academic community of geography educators. The late 1980s and early 1990s, when the attention of many geography education researchers was perhaps unhelpfully and unhealthily drawn to the development of the first geography national curriculum (GNC) in England and Wales, were arguably years of limited productivity for geography education research. For some, once the curriculum battle for geography to be represented as a school subject in the national curriculum was 'won'—famously Patrick Bailey referred to geography and geographers having now gained their 'Place in the Sun' (Bailey 1988)—it was as though there was 'tacit agreement that we finally knew what the geography curriculum is, and where it should be heading' (Lambert 2017).[7]

The period of development of the GNC in England was interesting in many ways for geography education researchers, not least because considerations of policy and practice were forced together (Butt 1997; Rawling 2001). The effects of the national curriculum on geography education are a matter of record, but an analysis of its effects on the community of researchers, and on their productivity, may still require further work. Lambert's (2017) contention is that the GNC took the focus away from curriculum development work in geography education for a generation—it was as though the GNC 'settled' the main curriculum issue, until it resurfaced again for geography education researchers over a decade after the start of the new millennium. The response to the development of the GNC, by a significant proportion of geography education researchers, appeared to involve 'sitting on the edge and critiquing it' (Lambert 2017), rather than becoming involved in debate, action and the creation of research-led, or research-informed, responses. Almost with a sense of claiming 'it's not fair'—rather than engaging with a clear rationale for curriculum development—the *field* of geography education research appeared to lack coherent

---

[7]Lambert does, however, note that some research in geography education at this time was significantly different—being innovative, and theory-driven. He highlights Frances Slater's work on Language and Geography Education (Slater 1989) as one example of such research.

leadership and a theoretical base. Lambert contends that much of the work undertaken by geography education researchers at this time consciously *disengaged* with theory—being essentially pragmatic, practical and inward looking—rather than being driven by more theoretical, or philosophical, questions. This contrasts with the work of researchers in history education; indeed, the national curriculum History Working Group had a genuine, theoretically and philosophically informed debate about the representation of their subject in the curriculum thirty years ago. This may be a consequence of better research leadership in history education, a stronger tradition of history teachers engaging with research, fuller experience of involvement with funded curriculum development and research work, and of the more democratic way in which the history working group was chaired and went about its work of curriculum construction.

Lambert's (2017) observations are *not* designed to casually criticise geography education researchers for a lack of productivity, or for the quality of their work— much of value has been produced and achieved in geography education research, which is recognised by our peers internationally. A more important consideration is the proportion of geography education research that is of high quality, in research assessment terms, and which has 'weight' within the wider sphere of education research. The danger is that beyond the narrow confines of geography education much of what we do seems to have little impact, or regard—a consequence of a limited engagement with theory, perhaps—such that geography educators and researchers are in danger of merely 'talking to themselves', without the benefits of others listening and responding. Indeed, geography education researchers might be accused of being rather inward looking, or self-congratulatory, in their efforts, having comparatively little engagement with their parent disciplines of geography or education. A test of the importance of geography education research is surely one that is *external* to its own research community—essentially this is a question of whether others in education research, subject-based or otherwise, regularly ask themselves: 'what are the geography educationists doing?' (Lambert 2017). Indeed, it is salutary to question whether anybody has *really* been interested in what specialists in geography education have to say about how their subject is taught, especially when there are so many other important educational issues to worry about.

9.7   The Importance of Structure and Organisation

There are therefore legitimate concerns when one reflects on the products of geography education research and researchers, nationally and internationally, over the last 30 years. The apparent lack of theoretical engagement, and production of new ideas, is worrying. Indeed, although the presence of a 'grand plan' for research may be damaging and unhelpful, the field of geography education research does reveal a lack of organisational coherence. This may be a consequence of its small scale, of it being too introspective, or of researchers simply not wishing to grapple with theory and ideas. The reasons behind (mostly) unexceptional performance are interesting and need to be confronted, understood and acted upon if geography education research is to move forward. It is possible that other fields of education research, and social theory, are considered not to be applicable to geography education research—a contention that is difficult to support—or simply that their ideas are too difficult, or tangential, to the endeavours of researchers. Maybe the lack of a core set of principles for geography education research has reduced the drive towards research productivity, or, more prosaically, structural changes in the academy have meant that prospective researchers have been hampered from pursuing their research ideas?

With respect to the importance of structure and organisation in the promotion of research in geography education, we have seen the importance of the Road Map Project in the States. In England, the development of a small, independent group of academic researchers in GEReCo has some significance. It is apparent that GEReCo now sees a more expansive future role for itself—not least as a consequence of its decision to merge with the UK Committee of the IGU CGE. With many of its members interested in powerful knowledge, inferentialism, critical realism, geocapabilities and subject didactics, the work the Collective has recently undertaken exhibits a recognition of the importance of theory in geography education and research. Rather than attempting to avoid theory, the imperative is to embrace what is intellectually difficult—to avoid succumbing to the temptation of merely simplifying complex ideas for the wider community. Geography education researchers must not avoid, or reject, new and challenging ideas to focus on easier, more pragmatic, content. This requires researchers to hold in balance their interests in both education *and* geography. Lambert (2017) notes how the University College London, Institute of Education (UCL-IoE) has been something of a pathfinder in this respect, with its Subject Specialisms Research Groups (SSRGs) helping subject-based researchers to combine in mutually beneficial ways. With a recent focus on subject didactics, and the increasing involvement of geography education researchers at the international NOFA[8] (Nordic Conference on Teaching and Learning in Curriculum Subjects) conferences, a renewed energy towards investigating the place of knowledge in subject-based education has widened the focus for geography education research.

One of the key issues for the achievement of strong geography education research appears to be how the field can embed significant, high quality, research outcomes into its wider communities—geography, geography education, education, and beyond. Firstly, geography education researchers need to offer a more valued research *product*. Secondly, appropriate structures and organisations through which their research

---

[8]The recent engagement of UK-based geography education academics, predominantly from UCL IoE, in the NOFA conferences is arguably a tangible recognition that subject specialist didactics are important. NOFA has offered a forum, for almost a quarter of a century, for education researchers, predominantly from Scandinavian countries, to explore issues of subject didactics. The rise in interest in general didactics (competences and skills), as opposed to subject didactics, in education departments has meant that initiatives such as GeoCapabilities has opened channels for researchers in a number of national contexts to develop their own projects based on similar ideas. For Lambert (2017), this signifies an imperative for geography education research in England to have an international focus. Significantly NOFA conferences are now held in English and attract a wide international audience—including academics from England, the Netherlands, Austria and Germany.

can be disseminated and discussed need to be established. Some of this infrastructure already exists—in the form of professional and academic journals; conferences, symposia and seminars; and research projects; Masters in Education modules, doctoral supervisions and scholarships—but more needs to be done. Importantly, as a third point, the geography education research community may need to achieve a greater awareness of the importance of sustaining what might be described as 'soft' organisational factors—enabling doctoral students to present to audiences of their peers, or subject experts; encouraging, or directly inviting, researchers to write for particular journals; realising that established researchers in geography education should publish not only in their usual, favoured, journals but also in disciplinary and generic education academic journals; and facilitating discussion and debate with experts from outside the field of geography education. This requires direction and leadership—and an expectation that as a research community geography educators will be able to stimulate a *reaction* from others, to help embed their ideas more extensively. Such leadership has previously been shown, most effectively, by (among others) the history educators—for example, Christine Counsell (see Sect. 9.8 below) has encouraged history teachers to publish theoretically informed material in journals such as *Teaching History*, which contrast with the articles typically offered in *Teaching Geography*. This point extends to the importance of practitioners being encouraged to read, and use, research evidence—from professional *and* academic journals. Indeed, with the future locus for initial teacher training being in schools, and the corollary that much education research will follow, there may be a need for collaborative research leadership: school-based research and researchers in concert with academics in HEIs. The issue is whether this will turn into a symbiotic, or parasitic, relationship—the only sensible resolution must be to strive for a symbiosis, based on a mutual respect and understanding of what each party brings to research in geography education.

9.8 Identifying Possible Ways Forward for Subject-Based Research

The question of whether subject-based research will be increasingly driven by education researchers employed by schools, rather than universities, is an interesting one. Christine Counsell recognises the issues facing subject-based research and concludes that this should logically become located in schools, mirroring developments in ITE: 'all the policy drivers from government at the moment are about putting schools in the driving seat; schools or communities of schools—trusts, federations, local authorities'[9] (Counsell 2017). With research into non-STEM subjects under pressure in higher education institutions, the prospects for history and geography education

---

[9]Christine Counsell—a renowned history educator whose career in the School of Education, University of Cambridge, spanned two decades—became the first Director of Education at the Inspiration Trust, in Norwich, in 2016. Her views, as a historian, have a particular resonance for geography educators, given the co-positioning of the subjects in much curriculum thinking (under a 'Humanities' banner), and the more widespread engagement with education theory among history educators. The

research being primarily based in universities appears weak. Counsell, like others, recognises that subject-based research was primarily undertaken by those involved in ITE, by researchers who in the near future may no longer be employed by universities. However, what might be referred to as the 'Cambridge model' for ITE—which insists that mentor teachers in partnership schools maintain a strong connection with subject related education theory and research findings—could provide a blueprint for future subject-based research endeavours. As Counsell explains: 'it is important to build a local hub for teachers to work together and to build subject teachers into a community that is both a producer and consumer of research' (Counsell 2017).

Basing her comments largely on the outcomes of recent developments in ITE, Counsell (2017) reminds us of the imperative that those involved in teacher training— be they mentors in schools, or academics in universities—are research literate, as well as agentive in both *producing* and consuming research. University-based initial teacher education, utilising its experience of research leadership, has served as a hub for the professional development of teachers in many local authorities. Here the successful preparation of trainee teachers crucially relied on the building of a cohort of vibrant *mentors* in schools: teachers who were knowledgeable about educational theory, who engaged with contemporary education research and developments in their discipline, and who were willing to become involved in research practice. Since the 1990s we have witnessed, in England, the uncomfortable fracturing of these mutually beneficial arrangements, as the two sectors have gradually moved apart; this has been paralleled by the unhelpful development of models of craft knowledge for teachers, with systematised subject and education knowledge being thought to reside only in universities. Both types of knowledge need to be shared and possessed by both sectors: for Counsell (2017), a 'difference of degree, but not of kind of knowledge'. As such, school-based mentors must read and understand the research material which trainee teachers engage with, or else these ideas are simply 'translating across two cultures' (Burn et al. 2018). The unfortunate consequence is that if trainees do not see their mentors valuing theoretical and research-based knowledge, they too will reject education theory—simply focusing on practical aspects of teaching in their preparation for the classroom. In a research-based model of ITE there is an expectation that experienced teachers have achieved an advanced appreciation of theoretical, practical and technical knowledge. A thorough knowledge of disciplinary, subject and educational research enables mentor teachers to offer informed opinions about the organisation of ITE courses and about the research that should underpin them.

The prospects for the relocation of subject-based research into schools is fascinating. But there are legitimate questions about the transferability of the model of ITE that underpins this relocation across the plethora of different schools involved in teacher training, and beyond—particularly concerning expectations for the consump-

---

significance of her move from academia, back to school-based work, was noted at the time of her appointment to the Trust. Counsell subsequently moved again, to serve on the Office for Standards in Education's (OfSTED's) curriculum advisory group, in 2018.

tion and production of research by teachers, in concert with HEIs. The Cambridge[10] model benefits from large numbers of its own Post Graduate Certificate of Education (PGCE) trainees finding jobs in local schools, participating in professional development and certificated education courses from their *alma mater*, and subsequently training to become subject mentors on the Cambridge programme. Arguably the academic standards achieved by these trainees, teachers and mentors is above the national norm; fostering individuals who recognise the importance of teaching being a research-engaged profession, and whose preparation for the classroom reflects this belief. The key question is whether this model of research-based teacher preparation could be repeated everywhere. Indeed, does it provide a future model for both subject-based education research and ITE (in school partnerships) in our current education system? Counsell (2017), as a previous Director of Education for the Inspiration Trust, presents a vision which replicates the benefits of the Cambridge model of ITE with respect to research engagement by teachers. Here vibrant hubs of research active and research-engaged professionals would 'lock into' a national profile of similarly minded teachers and researchers, countering the excesses of government education policy, and providing a bulwark against school senior management teams that constantly 'retreat to generic, policy driven areas of concern' (Counsell 2017). The importance of achieving a research-informed subject perspective on educational issues comes to the fore. Counsell (2017) observes that a 'huge vacuum in schools is the absence of senior curriculum leadership'—few deputy heads have responsibility for the school curriculum and, as a consequence, there is little recognition that some subjects (such as history and geography) need to maintain a close link with their cognate academic disciplines.

The question of what to research in education, and whether the academic freedoms enjoyed by university-based researchers mitigate against researching the major issues that *schools* face, has shifted somewhat given that research outputs should now have demonstrable 'impact'. But, for Counsell (2017), 'too few people in university-based research still think like teachers, like practitioners'—skewing the choice of research themes pursued. From the schools' perspective, senior managers need to understand the research evidence available to them and to be convinced that it has something important to contribute to problem solving in schools. There is an underlying issue that much subject-based research is either of low quality, is too esoteric, or avoids engaging with the questions that schools want answered. Like Stenhouse (1975), it is

---

[10]Counsell (2017) describes how, in spite of her involvement with the Department for Education (DfE) reshaping the future of ITE, there was no preferential treatment shown with respect to the numbers of PGCE Historians her course was allocated. Policy drivers hit history PGCE numbers hard in universities—reducing the Cambridge allocation from 20 to 11 students in just two years (and with the prospect of a complete removal of the history cohort in the following year). For Counsell, this revealed a lack of 'joined up' thinking in government—having assisted on the Carter review of ITE (Carter 2015), and with the Secretary of State for Education, Michael Gove, having praised the excellence of the History PGCE at Cambridge in parliament, Counsell faced the immediate prospect of her course being closed. The catastrophic effect on a community of mentors built up over the previous 20 years was particularly galling: 'they were hard won, they knew all the literature, and about half of them had published themselves—almost all of them had MEds and I was working towards 100%' (Counsell 2017).

possible to see classroom problems as *curriculum* problems, which can be theorised from a *subject* perspective. Indeed, Counsell believes that even some small-scale studies have outcomes that can achieve influence globally—research studies that have usually arisen from a practical, subject related, problem that has a curriculum focused solution.

9.9 Embedding Good Research Practice

For subject-based education research to matter, one issue to be resolved is how to *embed* its best work into school practice. This process arguably starts with the recognition that teachers are not predominantly researchers—that their *primary* role is to teach children, not to produce research. For Counsell (2017), this is an issue that the Inspiration Trust, and potentially all other large confederations of schools, addressed by appointing a central leadership team of senior subject specialists (n = 16) who were all exceptional practitioners, with strong backgrounds in research (enjoying publication credentials worthy of university academics), and who were capable of becoming respected senior leaders in schools. Additionally, these staff were high profile bloggers, effective public speakers and successful communicators across different media. Importantly, these appointees were, first and foremost, *subject* specialists.

It is worthwhile understanding the six 'workstreams' within the Trust to appreciate the central positioning of subject-based research within its work. All subject specialists are expected to deliver on these workstreams, which encapsulate:

(i) Teaching—all specialists must teach for at least one day per week in school, using their teaching strategically to undertake research, generate teaching materials, trial ideas and gather data. Teaching outcomes must benefit *all* the schools in the Trust, not just the school that is lucky enough to host that person, meaning that the effective communication of research findings is essential. Indeed, specialists move around schools, across both the primary and secondary school sectors, to embed change.

(ii) Continuing Professional Development (CPD)—CPD is free for Trust members, but a modest charge is levied elsewhere. The principle of serving a wider community of teachers and educators with the fruits of research outcomes is key.

(iii) Curriculum Development—although the nature of curriculum development varies between the primary and secondary sectors, History, Geography, Religion, Science and Art are central to forthcoming subject-based curriculum development (note, *not* Maths and English). Inter-disciplinary work is carried out, but not cross curricular topic work that has 'little frame or focus'. Principals in Trust schools have agreed a programme of curriculum development, as well as acknowledging the importance of promoting a knowledge-based curriculum that will move away from skills-based progression models. The curriculum is therefore based on 'substantive and disciplinary knowledge', with a balance between prescription and flexibility.

(iv) Research—the importance of teachers being both producers and consumers of education research is agreed. The Trust aims to employ research literate teachers who understand the place of research in their daily work. The question of what research knowledge is necessary for a teacher to become a successful mentor, head of department, curriculum developer, etc. is considered to be fundamental. The appreciation of research questions and methods—such as those involved in practitioner reflection and curriculum theorising—is seen as a function of being literate within one's professional community, with an awareness of recent and enduring debates in education essential. In research terms this calls on Trusts to curate a canon of literature linked to staff development. However, the *production* of research is considered equally important—such that bidding for funding, commissioning research, and approaching universities to be involved in research in the Trust, is encouraged. Research engagement is also considered to be a vehicle for training teachers to become 'research associates'; recent teaching appointees have had to present evidence of previous engagement with research. The employment of a full time Research Director is crucial to this workstream.

(v) Initial Teacher Education (ITE)—The decline in numbers of new teachers trained in university-led courses is acknowledged—and although Counsell recognises that some School Centred Initial Teacher Training courses (SCITTs) are 'fantastic' in terms of their teacher preparation, many have quality issues concerning subject input (this criticism could also be applied to some traditional, university-based, ITE courses). The vision is for the Trust to train their own mentors and trainees, as a SCITT, but with the active involvement of senior leaders in schools. This would establish that university-based teacher educators, and senior leaders in schools, were not 'two cultures'. Potentially this changes patterns of promotion, career development, mentoring and the recognition of the relevance of research. As Counsell states 'we need a gold standard of intellectual capacity—of knowledge base—for teachers and senior leaders, not a competence-based model'. A sustained link to university education departments becomes important to help define the knowledge base of teachers.

(vi) Inclusion—all staff receive training in special educational needs, which includes access to the latest research in this area.

The workstreams devised for the Trust—most of which place considerable emphasis of the contribution of subject-based research, and the active engagement of all teachers in research—are undoubtedly ambitious. If this model is successful, and taken up by other trusts and confederations of schools, what effect will it have on traditional university-based research in education? Do researchers in schools and departments of education simply become handmaidens to trusts? Indeed, will schools of education in universities exist in future, if much of their work is carried out elsewhere? Importantly, the success of this emerging model is not yet guaranteed—will all teachers 'buy into' this vision, and are all teachers capable of rising to the challenge of being genuine members of a research-engaged profession? Counsell (2017) argues that the organisation of research in schools:

> can't possibly replace Schools of Education; we can't possibly do everything – we are still fundamentally a *user* group, but are becoming better at setting up conversations with producers. This doesn't preclude university-based research, but so little of what goes on in research in Schools of Education is what we need!

In essence, this new model for a research-based profession needs to offer something different to what has (often unsuccessfully) gone before—not just with respect to research leadership, but also relating to whole school leadership. A sustained link to university education departments may remain important—to help define the appropriate knowledge base for teachers, based on subject research. A shift in the location and means of production of educational research will arguably prevent its ossification and the maintenance of shibboleths. Knowledge creation in this model is *not* fixed, either in terms of its location or its initiators; the prospective outcomes are empowering and emancipatory, creating a base from which the status of the profession can be raised. Not all teachers need to be original thinkers, but they must be able to understand, transact and mediate research knowledge—as Counsell concludes 'we have to capture, sustain and mobilise knowledge in our professionals' (Counsell 2017).

9.10   Conclusions

In the face of the apparently overwhelming obstacles that confront geography education research, nationally and internationally, it is important to remain positive—but realistic. Little is to be gained by simply identifying and celebrating the victories of the past, in terms of research productivity: although some might argue that we should be quicker to acknowledge that there *have* been victories! It is now important to try to establish how to move forward. The landscape within which all subject-based education researchers exists is rather hostile; one therefore need to be clear headed and pragmatic about what is needed—not only to survive, but to flourish.

It is vital for the geography education research community to see itself as others see it. To understand what is of value, both within and *beyond* the community of educationists and researchers—geography education researchers must recognise what, and how, to change to maximise their effective contribution. This implies being aware of how geography is understood (or indeed *mis*understood) by policy makers, government officials and education ministers, by professional associations, by the business community and employers, and by teachers, students and parents. Interestingly, all of these groups seem to agree that geography *should* occupy some part of the school curriculum. In a sense the continued existence of geography in the school curriculum is *not* the major issue. A key feature of all jurisdictions which score well on external measures of their students' educational attainment is that geography education is always present in the school curriculum. However, geography educators seem to believe that they are constantly being marginalised, with geography's existence as a school subject always questioned. From policy makers to employers, from parents to students, no one is placing geography under sustained attack, or pushing for its removal from state schools—perhaps geographers need to be a little more confident about who they are, and what they do?

There is a requirement to present an epistemological argument about what is meant by 'geography education'. Government ministers do not have the same understanding of the discipline and subject as educators and researchers do—this is the difference between holding an entirely functional and utilitarian view of geography as 'general knowledge', or seeing the subject's potential as 'powerful knowledge'. Lambert (2017) puts this succinctly: 'what it means to be educated in geography is the big question, rather than focussing on the justification for studying the subject'. Geography as a 'big idea' needs a much stronger theoretical base, higher level thinking and elevation beyond the pragmatic—all of which presents a case for future research.

The prospect for geography education research is therefore fascinating. It will, in all probability, involve fewer universities being engaged in subject-based research and, if we follow Counsell's (2017) vision, see more schools employing Directors of Educational Research to lead small teams of highly talented subject researchers. In this changed landscape education research is both produced and consumed by teachers—it is central to what they do—facilitated by subject leaders who have strong links to universities. This is an exciting model, where the teaching profession itself takes greater ownership of research and development. Here research leadership resides both within universities and schools, but must be symbiotic and sustainable. The prospect is that far fewer people will be dedicated to geography education research in universities, but far more in schools. It will be an evolving relationship—an exciting change. Ensuring a secure future for geography education research will require dedicated leadership: it will not 'just happen', particularly with respect to encouraging 'rank and file' teachers to understand the importance of engaging with subject-based research.

References

Bagoly-Simó, P. (2014, August 22). *Coordinates of research in geography education. A longitudinal analysis of research publications between 1900–2014 in international comparison*. Paper delivered at IGU Regional Conference, Krakow.

Bailey, P. (1988). A place in the sun: The role of the geographical association in establishing geography in the national curriculum of England and Wales, 1975–1989. *Journal of Geography in Higher Education, 13*(2), 144–157.

Bednarz, S., Heffron, S., & Huynh, N. (Eds.). (2013). *A roadmap for 21st century geography education research. A report from the Geography Education Research Committee*. Washington DC: AAG.

Brown, L. (1999). Towards a GeoEd research agenda: Observations of a concerned professional. *The Professional Geographer, 51*(4), 562–571.

Burn, K., Hagger, H., Mutton, T., & Thirlwall, K. (2018). *Teacher education partnerships: Policy and practice*. St Albans: Critical Publishing.

Butt, G. (1997). *An investigation into the dynamics of the National Curriculum Geography Working Group (1989–1990)*. Unpublished Ph.D., University of Birmingham, Birmingham.

Butt, G. (2002). *Reflective teaching of geography 11–18: Meeting standards and applying research*. London: Continuum.

Butt, G., & Collins, G. (2013). Can geography cross 'the divide'? In D. Lambert & M. Jones (Eds.), *Debates in geography education* (pp. 291–301). Abingdon: RoutledgeFalmer.

Butt, G., & Collins, G. (2018). Understanding the gap between schools and universities. In M. Jones & D. Lambert (Eds.), *Debates in geography education*(2nd ed., pp. 263–274). London: Routledge.

Carter, A. (2015). *The Carter review of Initial Teacher Training (ITT)*. London: HMSO.

Catling, S., & Butt, G. (2016). Innovation, originality and contribution to knowledge—Building a record of doctoral research in geography and environmental education. *International Research in Geographical and Environmental Education, 25*(4), 277–293.

Counsell, C. (2017, February 28). Interview with Christine Counsell at the Inspiration Trust, Norwich.

Edelson, D., & Pitts, V. (2013). *Charting the course: A road map for 21st century geography education*. Washington DC: National Geographic Society.

Edleson, D., Shavelson, R., & Wertheim, J. (Eds.). (2013). *A road map for 21st century geography education: Assessment*. Washington DC: National Geographic Society.

Firth, R., & Morgan, J. (2010). What is the place of radical/critical research in geography education? *International Research in Geographical and Environmental Education, 19*(2), 109–113.

Gerber, R. (1998). Strategic alliances: One approach to developing research in geographical and environmental education in the next millennium. *International Research in Geographical and Environmental Education, 7*(3), 179–180.

Gerber, R., & Williams, M. (2000). Overview and international perspectives. In A. Kent (Ed.), *Reflective practice in geography teaching*. London: Paul Chapman.

Gregg, M., & Leinhardt, G. (1994). Mapping out geography: An example of epistemology and education. *Review of Educational Research, 64*(2), 311–361.

Healey, M., & Roberts, M. (1996). Human and regional geography in schools and higher education. In E. Rawling & R. Daugherty (Eds.), *Geography into the twenty-first century* (pp. 289–306). London: Wiley.

Hill, D., & LaPrarie, L. (1989). Geography in American education. In G. Gaile & C. Willmott (Eds.), *Geography in America* (pp. 1–26). Columbus: Merrill.

Lambert, D. (2017, May 23). Interview with David Lambert at UCL Institute of Education, London.

Lambert, D., & Solem, M. (2017). Rediscovering the teaching of geography with the focus on quality. *Geographical Education* (vol. 30, pp. 8–15).

National Science Board (NSB). (2007). *Enhancing support of transformative research at the National Science Foundation*. Arlington, VA: National Science Foundation.

National Science Foundation (NSF) (2013). Common guidelines for education research and development. *National Science Foundation, U.S. Department of Education*.

Oancea, A. (2017, February 27). Interview with Alis Oancea at OUDES.

Oancea, A., & Mills, D. (2015). *The BERA observatory of educational research*. London: BERA.

Rawling, E. (2001). *Changing the Subject. The impact of national policy on school geography 1980–2000*. Sheffield: Geographical Association.

Rawling, E., & Daugherty, R. (Eds.). (1996). *Geography into the twenty-first century*. Chichester: Wiley.

Roberts, M. (2000). The role of research in supporting teaching and learning. In A. Kent (Ed.), *Reflective practice in geography teaching* (pp. 287–295). London: Paul Chapman.

Schell, E., Roth, K., & Mohan, A. (Eds.). (2013). *A road map for 21st century geography education: Instructional materials and professional development*. Washington DC: National Council for Geographic Education.

Segall, A., & Helfenbein, R. (2008). Research on K-12 geography education. In L. Levstik & C. Tyson (Eds.), *Handbook of research in social studies education* (pp. 259–283). New York: Routledge.

Slater, F. (Ed.). (1989). *Language and learning in the teaching of geography*. London: Routledge.

Stenhouse, L. (1975). *Introduction to curriculum research and development*. London: Heinemann.

Chapter 10
Conclusions

10.1 Context

The concluding chapter draws together a number of strands from previous sections of the book. By necessity, given the nature of the challenges currently faced by geography education research globally, it offers wide-ranging suggestions about the future prospects for the field. These acknowledge the relatively modest progress previously made in advancing geography education research in many countries, the likelihood of an insecure future for all education research, and consideration of whether the probable setting for subject-based research will still be within university schools and departments of Education (see Tierney 2001; Counsell 2017). By considering the lack of culmination, and indeed repetition, of certain questions in geography education research, this chapter makes frank recommendations for the role of subject and interdisciplinary research, with particular reference to the school curriculum; for school geography's links to the parent discipline in academia; and for the prospects facing the next generation of geography education researchers. In doing so, I consider where subject-based education researchers may be employed, the shape of their possible career progression, and the likely future profile of their work. This raises important questions: not only about what and how geography education researchers might research, but also about who such research should be *for*. Essentially, we must consider the nature of the questions that concern many geography educationists—regarding what to teach and how to teach their subject—and why these questions should matter when choosing how to educate young people for their future lives in our rapidly changing world. A key concern, previously raised by mathematics educators (Wiliam and Lester 2008, cited in Lambert 2015), will be whether geography education research will ultimately occupy a 'space of research for research's sake', or a 'space of ideas'. Lambert (2010) rightly contends that the use of public money to carry out geography education research must be accounted for and defended—the argument advanced here is that those intending to teach, and those subsequently requiring professional development, should still be guided by research funded by

© Springer Nature Switzerland AG 2020

G. Butt, *Geography Education Research in the UK: Retrospect and Prospect*, International Perspectives on Geographical Education,
https://doi.org/10.1007/978-3-030-25954-9_10

the state (Brooks 2010b). Education, as an enterprise, will not and must not stand still—it will always occupy a changing landscape, and require a particular focus on the continuing role of school subjects (Lambert 2009).

The dangers of obsessing about the details of geography education research, and about its particular themes, are all too apparent—especially if such considerations occur with little reference to other areas of education research. Geography educationists are perhaps guilty, as a research community, of having been too inward looking and of not fully considering the contexts within which education research occurs, both nationally and internationally. The influences of social, cultural and political factors on research are re-stated here, as these are powerful motivators for what is researched and how this research is carried out. This 'backdrop' is something that geography educationists and researchers need to be mindful of if they are to position their research more securely for the future. Geography education research—and similarly the representation of geography as a discrete subject in schools—is not necessarily a 'given entity'; it needs timely and strategic defence and support, but this should *not* become the sole, overriding, and defining feature of all it does. The wider educational focus is always important. What might in future appear to be relatively benign shifts of education policy and practice may have far reaching effects on the future of teaching, learning and research in geography education.

The need to find answers to these issues is reasonably urgent, for it is possible to conclude that the discipline of education is nearing crisis—and, as a consequence, so too is geography education research.

10.2 Introduction

In preparing to write this concluding chapter I reflected on one of the sub headings in the last chapter of my own doctoral thesis, written over twenty years ago (Butt 1997). This heading was in the form of a short, stark question: 'The future of geography – do we know what the discipline is anymore?' At a time when enquiries into the content of the school geography curriculum were prevalent—during the early years of the first national curriculum for geography in English state schools—this seemed like a valid research question, which reached back to the origins of the discipline and its expression as a school subject. The question was further refined into a series of investigations about the form and content of geography education in schools, all of which were driven by an 'insecurity about what geography actually *is* as a school subject' (Butt 1997, p. 302). Even today, the question about the relationship between the parent discipline and the school subject appears to be remarkably persistent. Perhaps it will always be so. Back in 1997, and through the application of a postmodern 'take' on academic disciplines, school subjects and their contribution to state education, there appeared to me to be a fundamental confusion about the status of subjects and about the underlying question of 'what to teach?' At the time I concluded that geography was—indeed I believe still *is*—composed of a constantly moving, 'restless' and changing set of ideas, making attempts at fixing it into a strictly defined and delimited subject order a risky business. Although it would be foolish to eschew any efforts to outline at least *some* core content for curricular subjects—we need to have a set of ideas about the guiding concepts and principles of subjects, and of their parent disciplines—the first iteration of the geography national curriculum immediately looked 'out of date' and laboured (see Graves et al. 1990; Lambert 2009). Indeed, official reports on the nature of geography as a school subject continued to mention it's 'tired and dated content' up to twenty years later (QCA 2005; OfSTED 2008, 2011). Faith in the old liberal-humanist certainties about 'passing on' the best of what we know to the upcoming generations was thrown into doubt; surely this would fix what is a dynamic, responsive and ever-changing subject in such a way that it would prove difficult to maintain its relevance and interest for young learners? It would also represent a fundamentally flawed, perhaps ultimately dishonest, interpretation of the discipline. Unfortunately, then as now, the research evidence concerning what geography in schools should look like—ideally evidence that would have previously been gathered by geography education researchers—was either unavailable, unconvincing or incomplete. We could neither persuade politicians of the need to change their thinking, nor offer enough data to effect a rapid change in the curricular form of geography. On reflection this situation represented an opportunity missed—there was clearly then, and still is today, a pressing need for research in geography education to help define the school subject in modern form, a task which must bring together the discipline of geography and the school subject through an act of curriculum making.

Geographers may at times wrongly assume that the immediate period of educational change is significantly different, complex and more challenging than any time that has gone before—Norman Graves, for example, reflecting on contemporary educational change in the mid-1980s, opened an article with the following statement:

> I wonder whether you feel like me and believe the present time to be one of the most confusing periods in the recent history of education? (Graves 1985, p. 15)

I would argue that, despite pressing concerns, the present is not so radically different to the past—it presents its own particular challenges and uncertainties, but so too has every other period of educational history.

However, we are definitely living in interesting times.

10.3 The Political and Disciplinary Imperatives

There have been frequent attempts by politicians and policy makers, nationally and internationally, to re cast education into forms which they believe best suit young people—as well as serving the interests of the state—in meeting the demands and challenges of the modern world. This process attempts to respond to the changing needs of society, against a backdrop of globalisation and increasing international competitiveness. Unfortunately, political responses to education matters are often more akin to a *reaction* than a more measured and reasonable consideration of evidence. I consider that most politicians believe that they act in good faith—they make genuine and largely honest attempts to do what they think is right for young people and for the state, albeit within the political parameters of their party ideals and diktats. But the nature of most political thinking is short term. Politicians are driven to affect change within the expected life of their government—and they often lose sight of, or do not fully consider, what the longer term aims of state education should be. Carelessly putting energy into pursuing the next politically-driven educational goals (and there have been many to choose from in England in recent years: individualised learning, enterprise initiatives, 'thinking skills', child centred education, life skills, vocational education, personalisation, choice, 'learning to learn', well-being, to name but a few) without the guidance of reasoned thought, principles, debate or research evidence will inevitably be costly (see also, Gerber 2007). Each of the exemplars of government intervention in education presented above has damaged the contribution of curricular subjects to children's learning and downplayed the significance of disciplinary knowledge. This has led teachers and learners on fruitless 'pedagogic adventures'—driving them away from the educational achievements promised for young people through studying the subject disciplines (Peters 1967; Young 2010). With regard to the creation of the national curriculum statutory orders there is much evidence—garnered from the reaction of teachers, professional associations and academics—that the Geography Working Group did not successfully achieve a (re) definition of the school subject in modern form, to the satisfaction of most stakeholders (including some of the members of the Group itself) (Butt 1997; Rawling 2001). The successful redefinition of geography as a school subject still eludes us. The flux occurring in academic geography in higher education at the time had some influence on the form and content of school geography, with concerns that any further severing of school and university geography would jeopardise the long-term future of the subject in both sectors. However, change is to be expected in any academic discipline, at almost any time: it is the nature of disciplinary enquiry, a feature of how disciplines advance their knowledge base, and a constant in the relationship between academic disciplines and school subjects.

It would be wise to acknowledge—but not to become fixated on—the fact that in both academia and in schools the presence of geography is not a 'given'. As Lambert (2009) reminds us, it is not just 'there': it has been, and continues to be, created by concerted human effort. To be able to draw from this resource of disciplinary knowledge, teachers need, in some meaningful way, 'to be engaged with it' (p. 4). The nature of this engagement is clearly important—teachers cannot be exposed to the whims and caprices of academic shifts, but must be able to keep pace with the main directions of change in the discipline. This is a responsibility for all those involved in geography education, not least those fortunate enough to be concerned with research, assessment and curriculum development. There is clearly a role for geography education research in helping to support the curriculum making process, although understanding the context in which such research will in future occur is complex. Indeed, it would be healthy and honest to recognise that for some considerable period of time geography in schools, and in further and higher education, has not achieved a 'consensus over philosophical, methodological and ideological issues' (Johnston 1985, p. 9). As the academic discipline continues to fragment and sub divide, school geography will inevitably struggle to find unity, or leadership, from this sector—indeed, Johnston (1986) has previously claimed that there is no necessity for a discipline of geography to exist at all, there being 'not one geography, but many geographies, created in response to circumstances specific to time and place' (Johnston 1986, p. 449). Achieving a conception of the whole discipline that is transferable into a manageable, workable curriculum for teaching the subject in schools is therefore highly problematic. This is particularly true when school and university geographies find themselves in a state of dislocation—a condition that some time ago was being referred to as a 'quiet divorce' of school geography departments from those in universities (Machon and Ranger 1996). If this coincides with a period when government policies have threatened the existence of certain school subjects, or when teachers have been focussed on re interpreting geography to fit a curriculum model of what politicians believe a school subject *should* be, then pressures are increased. We are, perhaps, fortunate in England that our geography national curriculum has now largely become a loose framework within which teachers can concentrate on 'big ideas' in geography, rather than an agenda for the delivery of lists of 'official' geographical content—much of which may be of dubious value, or utility. This creates the space for curriculum making by teachers—a considerable prize, but one which can only truly be appreciated if teacher preparation, professional development and research can combine to make this goal achievable (Brooks 2010c).

In conclusion, Furlong (2013) helps us to reflect on the notion that in the history of education there has been a constant, sometimes politically driven, struggle to come to terms with what knowledge is. This struggle is evidenced both in the field of education and within school subjects, influencing the ways in which knowledge is created, acquired, regarded and researched. Many involved in higher education have, not surprisingly, tended to favour the application of *academic* models, to describe the acquisition of subject and professional knowledge in education; governments, teachers and students have adopted more pragmatic, less research driven or theoretical, approaches. This entirely understandable division in the ways in which different stakeholders 'see' education, and the subjects taught in schools, must be at the forefront of our considerations of education matters—for it goes to the heart of why different stakeholders act in the ways they do. The lack of trust that governments place in university-based research in education reflects what is, for some, a broader and thinly veiled intention in government circles to drive universities away from all aspects of policy and practice in schools.

10.4   The 'Space of Ideas'

Lambert (2010) asserts that education researchers, in whatever fields they work, should occupy a 'space of ideas' within which they take intellectual risks. As such, research in geography education should not merely accumulate data (just because data collection is often easily done), but rather pursue more urgent, important, considered and 'wicked'—that is, not easily resolvable or possibly counter intuitive—

research questions. As a community of researchers, those engaged in research in geography education must take seriously the issue of where collective research efforts should in future be placed. This is a difficult matter to approach, as we have already rehearsed how simply listing a set of research priorities is both dangerous and arguably anti-democratic—the issues which require research resolution may also not simply affect geography education, but stem from wider afield. Essentially, the question of why geography education research is important, and for whom, must continue to be at the forefront of considerations. It is an expression of a desire to appreciate the advances made in the parent discipline of geography *alongside* research findings and thinking about how children learn (including their interests, motivations and issues of pedagogy).

Perhaps the best way forward for geography education research is to agree some key principles for action, rather than attempting to create definitive lists of topics, or questions to work on. Geography educationists will hopefully strive to remain mindful of recent developments in their parent discipline—here post modernity explains the recent fragmentation of the discipline into sub fields and the frequent crossing of intellectual borders, the re- formulation of boundaries, and the lure of interdisciplinarity. Where this leads to disintegration—or even what some may consider to be unhelpful, or potentially damaging, emphases on 'new' alliances and areas of study—the process should be countered. Getting the balance right is important—well reasoned criticism is always encouraged, but if criticism is stated too harshly, or is conducted with unhelpful references to (say) 'corruption' of the curriculum (see Standish 2007, 2009), the cause of geography education may be damaged. Nonetheless, we need to remain alert to the fact that in attempting to provide an 'education for all' we often end up achieving the reverse, forgetting—or down playing—the educational contributions and advances previously made by disciplinary subjects. This has led to some referring to geography in schools as being 'lost' and struggling to find, or rediscover, its identity (Lambert 2009). A key challenge for geography teachers—one that geography education researchers can help with—is to redefine the nature of their working relationship with the academic discipline. As Lambert and Morgan (2010) assert, teachers have to constantly re assess their connection with their subject *and* its parent discipline:

> teaching is an intensely practical activity, but is underpinned by intellectual effort. Without the latter it lacks efficacy (p. xi)

This is part of each teacher's role as a 'pedagogic boundary worker': linking the academic discipline and school subject in such a way that students understand it, as well as taking account of their learners' interests, social and community backgrounds, hang ups and personal geographies (Lambert 2011a; Lambert and Mitchell 2015). Care is needed. We can offer learners no immutable, future focused, 'grand narrative' of the world, and must be sceptical about looking back for any supposed 'golden age' when essential subject knowledge was taught.

10.5 The Role of the University

What do universities currently offer to support research, teaching and learning within the discipline of education? What makes their contribution distinctive and important? What will the future form of university-based research in education look like, and where will subject-based research 'sit' within this? These are difficult questions to answer given the complexity of modern universities, whose work is no longer based primarily on pursuing the single idea of the 'maximisation of reason' (Lay 2004). The function of universities is now predicated on numerous, sometimes competing, notions—we cannot claim that universities should simply pursue uncomplicated knowledge creation, because we understand that all knowledge is contested, problematic and partial (Barnett 2000, 2009). Knowledge is interpreted and used in different ways—these conditions of complexity and multiplicity of meaning are challenging, but must be embraced in preference to trying to identify one true way of interpreting the world.

Universities have traditionally provided the locations, contexts, organisational structures, funding and (importantly) staffing for the production of education research. This is not to say that other interested parties—businesses, industries, charities, governments, 'think tanks', schools—cannot, or should not, also undertake such research. Counsell (2017) provides us with a template for how educational research might, in future, be essentially school-based, but facilitated with the support of universities. Here the 'points of difference' concerning who conducts research, to what ends, and for whom, are important: universities currently hold a privileged position, and responsibility, to undertake research for society—even though their research is under pressure to compete alongside other increasingly well-placed research providers. These providers may have different primary aims and purposes; different to those of universities who are all about knowledge creation and transfer: to coin a phrase, 'it's what they do'. Research still defines the distinctive, core business of universities. Throughout this book is has been apparent that the pressures on higher education, and on the subsets of education and geography education research within it, have steadily increased—requiring fleet footed, sure-fired responses to sustain position and influence. In the university sector the casualisation of labour, the rise of new public management agendas, the increasing external controls on curriculum and research, the promotion of applied research, the fragmentation of disciplines into numerous sub specialisms, the growth of managerialism and neo liberal influence, the postmodern turn and its effects, have all impacted significantly on knowledge debates and research activity. This explains the current crisis of confidence about what researchers in geography education do—not helped by their reluctance to position themselves more securely, by providing clear statements of purpose and intent.

The nature of universities, and of the core work they do, is obviously changing. Across the world the effects of raised international competitiveness, globalisation and the upsurge in demand for the steady improvement of research performance and impact, have profoundly affected the ways in which the higher education sector functions. Marketisation of higher education has forced universities to become more competitive—encouraging them to adopt a wider, more business focussed, external gaze with respect to research and the attraction of larger numbers of students. Policy shifts have had profound impacts on the nature of academics' work—none more so than within education research and initial teacher education,[1] where state schools and departments of education now constantly juggle with forces that pull them in opposite directions. Worldwide, universities attempt to position themselves with reference not only to local, regional and national competitors, but also increasingly with an eye to international competition across the entire higher education sector. The massive growth experienced by universities over the last 50 years, not least in terms of the numbers of young people now expecting to be educated to degree level, is based partly on an economic assumption that everyone will benefit from more graduates entering the job market year-on-year. The question of whether this is socially and economically beneficial for the majority of young people, their country, and for the public at large is a moot one particularly following the global financial collapse of 2008—which led nations to question who should fund students' university fees and whether continued expansion of the sector was wise (Browne 2010). We must recognise that it is not just the field of geography education, nor education as a discipline, that may be under pressure within these university settings—the whole university project is under scrutiny:

> Education, as an applied discipline and as a discipline that was late arriving at the university high table, is perhaps no more than an extreme example of the difficulties that face universities as a whole (Furlong 2013, p. 161)

We have seen how a vision of the university—and of the schools, departments, faculties and institutes of education within them—can be formed around the idea of knowledge production; knowledge which is then disseminated through research, publication and teaching. Universities have been successful in this endeavour, producing new knowledge that can be tested according to the parameters of their cognate disciplines (even though postmodernity has questioned the production of objective knowledge and resolved that all knowledge production is situated). But alongside knowledge production has come specialisation—within any university education department academics may have only a shaky perception of what their colleagues do in the name of 'educational research', or indeed 'geography education research'. The lack of guiding, let alone universal, principles to shape our work leads to confusion and diversity. As Furlong (2013) asserts:

> If universities have no essential purpose, if there is no longer any means by which they can *insist* on their independence and if there is no longer any confidence in the knowledge that they provide, then why should governments take the university-based study of education seriously except where it has practical utility for their policies? .... But we should not fool ourselves; any valuing that there is, is fundamentally utilitarian (p. 166)

---

[1] As Furlong (2013) reminds us, 'Despite initiatives in England in recent years to move more and more initial teacher education into schools, initial teacher education is for the present at least, the core of the economy of most university departments and faculties of education. Without it, the system as a whole would be very much smaller' (p. 73).

Although the model of the University Training School (DfE 2010) in England—in many ways similar to Hargreaves' (1999) concept of the 'knowledge creating school'—has yet to progress beyond the initial establishment of two schools in Birmingham and Cambridge, these could provide a template for the future location of subject-based research. The wholesale shift of research production from universities to schools—given political perceptions of the former's inability to produce research that supports either teachers or policy makers—may in future be politically more acceptable than at present. However, currently there appears to be little government enthusiasm for funding more University Training Schools. Counsell's (2017) ideas—of creating consortia of research-engaged schools, rather than one off, isolated University Training Schools—are pertinent here. Collections of research-engaged schools would position teachers as the main knowledge creators in education—providing the necessary resources, time, opportunities for debate, reflection and critical engagement for research, alongside an expectation that teachers support informed experimentation and research in the classroom. Such an approach has support within the Department for Education—which has encouraged schools to drive their own research, rather than being wedded to the research conducted by universities. Arguably the best way forward is the creation of genuine research partnerships between schools and universities; much like the equal partnerships established in initial teacher education in the 1990s, when pre-service education courses were shared between higher education institutions and schools.

When considering the role of the university in shaping the future prospects for geography education research, it is valuable to reflect on the contribution of organisations dedicated to the promotion of geography education. These are often either based in universities, or have memberships composed of current or retired academics from higher education institutions. The restructuring of one such organisation in the UK, the Geography Education Research Collective (GEReCo), provides an interesting outlook on how geography education research might in future be supported. In 2018, the UK Committee of the IGU CGE agreed to amalgamate with GEReCo, to form 'UK GEReCo'—a body dedicated to the leadership and promotion of geography education research, at all levels and in all settings. The aims and purposes of this new collective—'to contribute to debates both nationally and internationally in order to strengthen geography education across the UK and beyond'—are undoubtedly ambitious:

- ensuring and promoting national and international links with cognate bodies;
- considering and identifying ways to promote and take action to support and contribute to the IGU Commission's programmes of work;
- considering and identifying ways to promote and disseminate research in support of the work of the COBRIG, the GA and the RGS-IBG;
- undertaking high quality research in geography education;
- developing original thinking, critical perspectives and new knowledge in geography education;
- pursuing research projects and funding from educational and/or governmental sources;

- organising and contributing to international conferences;
- holding a regular Research Forum;
- encouraging an international perspective in geographical education research;
- promoting publications and disseminating within and beyond the UK research in geographical education undertaken in the UK and in other nations.

These aims and purposes may offer a blueprint for the development of other, similarly minded, organisations in the future. The merger of the two parent organisations to form 'UK GEReCo' seems sensible, for although they had distinctive origins and purposes, they shared many of the same interests. There was also a tacit recognition among the members of both organizations that the research community of geography educators, nationally and internationally, is small—the proposed amalgamation of 'scarce resources (not least people-power)', would therefore serve to consolidate and exploit the complementarity of the two groups.

10.6 The Next Generation of Geography Education Researchers

The future for geography education research will be very different to the past. Its prospects are uncertain—inevitably tied to the fate of university departments of education, within which most of the high-quality geography education research, and researchers, currently resides. Without these departments flourishing, it is doubtful that the outlook for geography education research in universities will be secure. An added pressure is that most geography education researchers in the academy have traditionally been closely associated with initial teacher education in geography, particularly in the early years of their careers. With the physical shift of teacher preparation into schools, the employment prospects for teacher educators who are also geography education researchers appear unstable. Two further complications arise: firstly, even if universities continue to employ geography educationists, their job descriptions will change to reflect the emphasis on sustaining good performance in the Research Excellence Framework (REF) in England. As such, geography educationists in the academy—whose research is subject-based, and therefore often considered to be of lower status than that of other education researchers—have often been placed on 'teaching only' contracts, removing any prospect of sustained research activity in the future. This has happened particularly in Russell Group universities, even affecting senior members of academic staff who already have internationally respected profiles of research work and publications in their field. The impact on the next generation of university-based researchers in geography education could be brutal—reinforced by often insensitive, indifferent or short-sighted management systems in universities.

Secondly, for most geography education researchers in the academy—even those who retain a reasonable allocation of work time to pursue their research interests—there comes a realisation at some point in their careers that professional advancement will *not* be driven straightforwardly by excellence within their chosen field. So, the sensible response is to diversify: into areas of research and publication activity that are better regarded, better funded, more 'mainstream' and easily understood by other academics—both within and beyond the education department. Given the realisation that steady career progress also now requires the successful capture of research income—often from competitive tendering, as a Principal Investigator or Co Investigator—and that funding for research in geography education is extremely limited, many successful geography education researchers have been drawn into other areas of educational research. University-based researchers in geography education, as their academic careers have developed, have either tacitly or directly responded to these shifting and competing forces. This has resulted in geography education researchers living rather schizophrenic existences: juggling their desire to produce high quality research in geography education with the rather more prosaic demands of researching and publishing *outside* their chosen field. Anecdotal evidence, based on discussions with colleagues who, like myself, have researched and published in geography education for over 30 years, reveal that our careers have often advanced as a result of the impact of work published 'outside' geography education, or because of our contributions to management, administration and (in some cases) teaching in our institutions. It is of profound concern to me that the most highly cited piece of research that I have produced in my career is *not* what I would consider to be my best quality contribution to *geography* education—but a rather less significant 'think piece', co- written on the back of a large, well-funded and influential research project conducted in the early 2000s (see Butt and Lance 2005). The themes of this government funded research—concerning how teachers' job satisfaction and workload impact on recruitment and retention in the profession—are still important and topical, but significantly they lie *outside* what might be considered to be the narrower, more esoteric, focus of geography education research.

Powerful drivers are now forcing change. With initial teacher education becoming increasingly school-based, and schools showing a willingness to be more involved in educational research, the next generation of geography education researchers may well not be full-time employees of university education departments. The issue of whether future researchers will even identify themselves primarily as geography education researchers is a worry.

10.7 The Ideal of Geography Education Research

With regard to the future role and function of geography education research it is useful to have a vision of what an ideal might look like. However, this has to be tempered by a large dose of realism—it is very unlikely that subject-based research in education has any prospect of entering a period when it will be generously funded by government, or research councils, or charities, or professional associations any time soon. It will not attract the kind of attention among politicians that will displace interest in other areas of education research—which are often generic and respond to the wider needs of society, economy and education. Politicians will not prioritise subject-based research in terms of policy and funding. Schools tend to follow the political lead they are given—particularly where there is a possibility of gaining additional resource—with senior leaders in schools promoting activities that boost school and student performance, or help to deliver strategic plans. Universities will, in all probability, continue their rather grudging relationship with their own schools and departments of education, still wondering what they should do with these substantial parts of their own institutions—which often employ large numbers of academic and support staff, but which may perform in ways they consider to be far from exemplary.

Almost certainly, a more comfortable future for geography education research would involve researchers—be they university or school based—forging stronger alliances with academic geographers to advance better research and teaching in schools. Prioritising research into two great ideas—geography and education—would be productive. Ideally, funding would be attracted from various sources, and geography teachers would engage in active research and curriculum development. Teachers should be the key 'curriculum makers': informed by a profound understanding of the students they teach, their pedagogical knowledge and an appreciation of their parent discipline of geography. Each component requires an input from geography education research. Geography teachers, as well as teachers in all other curriculum subjects, would hopefully eschew the simple 'delivery' of chunks of geography content and be mindful that they are purveyors of a living subject, the content of which will shift and change as disciplinary research continues to advance.

One model for sustaining the future of geography education research might be school-based, in concert with some research support from universities. An ideal might be that teachers would be expected to have a much fuller—indeed a statutorily-driven—engagement with research in education as part of their role as modern professionals. Just as in Scandinavian countries, where teachers are expected to perform the role of public intellectuals, there might be an expectation that the school workforce is in future educated to Masters level, and beyond. Teachers would be rewarded for their intellectual engagement, and recognised as being part of a research-based profession. Here the role of educational research would be 'cradle to grave'—from the first steps the beginning teacher takes in their pre-service preparation, to the confident strides forward they might employ as senior educationists in schools. Recognition of the importance of subject-based research in helping to solve all manner of educational problems would be refreshing and enlightened.

But, one might argue, all I have presented here is a wish list. There is a need for a better overview (a map?) of the current prospects for research in geography education, nationally and internationally—which takes account of the progress made so far, the current challenges faced, and a realistic appraisal of what might lie ahead. The picture is complex—including consideration of policy, economics, ideology, assessment, culture and tradition. To create a credible vision for the future requires greater dialogue among stakeholders about possible 'ways forward': but all the talking in the world will not solve structural problems that essentially require greater political and financial support. It is always difficult to put aside day-to-day responsibilities and clear the space necessary for thought and action. Minds are inevitably focussed on the immediate and the urgent—students, assessments, research performance, management, leadership, or the latest political diktat on education—but more thought and action are definitely required. In essence, the current situation is just an extension of what has always been the case—a situation described by Daugherty and Rawling (1996) over twenty years ago:

> Research trends are not a pressing concern for geography teachers in schools; they are no more than a backcloth to the topics and themes listed in curriculum guidelines and syllabuses … the hard-pressed teacher is more likely to be concerned with managing statutory assessment requirements, contributing to the school's performance in league tables and meeting the demands of school inspections than with following up new ideas in geography (p. 360)

A position paper produced by GEReCo in 2017, *Curriculum Leadership and a Research Engaged Profession*, made some important points about the possible future for geography education research, and how its research base might inform teaching. This was an attempt by the Collective to create a wider research focus: one which has something to say about the relationships between education research and the teaching of specialist subjects in schools including, but not restricted to, geography education. The paper was underpinned by arguments advanced by Tony McAleavy's report for the Education Development Trust (EDT) in 2016, asserting the need for 'evidence-informed professionalism' to support the work of the 'research-engaged school' (as distinct from a narrower, evidence-based, or evidence-led, pedagogy). The paper recognised that teaching would never be *entirely* based on findings from academic research, arguing that such a situation would be undesirable. But it was unequivocal in stating that the products of education research are vital to the successful functioning of schools. As always, context is everything—the transferability of research findings to different educational settings is fraught with dangers and this highlights the limitations of relying on supposed 'what works' research evidence (see Biesta 2007). Indeed, the contention is that very little, if any, high quality educational research exists that can simply be transferred by teachers into classrooms in the form of blueprints to improve practice.

10.8   Re-Imagining 'Touchstones' for the Future

One way forward is to suggest points of principle and action for the future of geography education research. These 'touchstones', influenced by Daugherty and Rawling (1996), will all require the support of research evidence in geography education. Essentially, this approach enables us to consider the prospects for geography education research from a different angle—by seeking to identify what such research might be *for*, it helps us to recognise the themes and questions which need concerted effort to secure their resolution. These might include:

i.   bringing together the geographical content which forms the focus of geography education in universities and schools.

ii.   increasing levels of understanding between the two sectors—that is, expanding
      the knowledge and understanding of academics about what goes on in geogra-
      phy teaching in schools, and vice versa. Both sectors have a shared commitment
      to educating young people through geography and would clearly benefit from
      an improved comprehension of what the other does. Both have a distinctive
      educational mission, but also a commonality borne from their shared disci-
      plinary roots. There is more that pulls both sectors together, than draws them
      apart. A key responsibility is for academic geographers, geography educators
      and researchers to talk to each other—as Daugherty and Rawling (1996) stated,
      'How often and how effectively do geographers in university education depart-
      ments communicate with their counterparts in geography departments in the
      same institution?' (p. 370)

iii. striving for a greater 'internal coherence' for the subject of geography—such
      that geography is recognisable in terms of its core identity across all sectors.
      Essentially this entails striving for agreement, or at least better recognition,
      of the core elements of content, concepts and principles in geography—and
      an acknowledgement that even though research frontiers expand, and subject
      boundaries may be crossed, there is still such a thing as 'geography'. The
      search for greater disciplinary and subject coherence would also benefit the
      main stakeholders and 'consumers' of the subject—including students, parents,
      general public and employers.

iv.   clarifying the educational pathways for students who show an aptitude for, and
      desire to continue, the study of geography (as they progress through schools,
      further and higher education institutions). This clarification would also embrace
      the possible links between academic and vocational aspects of geography edu-
      cation. The process would recognise, but not be dominated by, the increasing
      demands by both government and society for more subject-based education—
      not only to deliver subject specific knowledge, but also *some* aspects of transfer-
      able skills suited to vocational needs, social and personal education, and well-
      being. Research into the most successful ways of solving this tricky conundrum,
      without having a damaging impact on the subject itself, would prove fruitful.

v.   sharing knowledge, understanding and skills relating to pedagogy—this is as
      important for higher education as it is for schools, given the changing cohort of
      students from different academic, social and economic backgrounds who now
      proceed to university education. As Roberts (2013) recognises, the challenge of
      teaching and learning geography in the twenty-first century requires increasing
      teacher expertise, to enable educators to interpret and (re) present their subject-
      rather than the adoption of a utilitarian view of teachers, merely as technical or
      managerial workers.

vi.   recognising that progression, as an aspect of teaching and learning geogra-
      phy, is under researched in universities and schools—here further research
      is required to help students transfer between sectors, with the intellectual
      gains from experiencing geography education at each level being identified.

vii. achieving greater agreement on what should be researched, both in academic geography and geography education.

These touchstones should not just be about the relationship between (research in) geography and geography education. As Slater (2003) has previously observed, there is still a dangerous 'one way' relationship between disciplinary research and the subject content taught in schools. Additionally, the contribution made by research in geography education rarely impacts upon its parent discipline, even in matters of pedagogy:

> Knowledge, concepts and ideas flow into research projects in geography education and also directly into geography teaching and its practices. I cannot however, think of examples of where ideas generated in the research and teaching of geography in education flow back the other way (Slater 2003, p. 295)

Elsewhere, Slater (1996) comments that most research in geography education is conducted by teachers, often in pursuit of further qualifications, and therefore tends to be unfunded, small scale and non-generalizable in nature. She makes an important point that even the best of this research usually goes no further than being offered for consideration for some form of certificated award (such as PGCE, PGDipEd, Masters in Education, EdD, or PhD):

> Too few completed research projects are published. There are not enough examples of promising research being followed up, either by taking the earlier research to provide a priori ideas for further research, or simply by replication in new places, at other times, by different researchers (p. 289)

Concluding this section on future 'touchstones', Chap. 5 revealed how a report by the Royal Society and British Academy in 2018—*Harnessing Educational Research*—suggested a range of priorities and strategies by which governments, the United Kingdom Research and Innovation (UKRI) organisation, higher education institutions and educational organisations could harness research evidence more effectively to improve the educational experience of learners. It carries an important message for our future research endeavours in education, particular those which concern research-informed teaching:

> There is increasing recognition in the UK of the need for teaching to be a research-literate profession. However, teachers repeatedly indicate that their working conditions do not enable them to spend time reading research to improve their understanding or to determine how to use it to adapt their practice. These activities must fit around the day-to-day practice of teaching, without taking teachers away from their principal role of nurturing their pupils' development. Factors such as repeated curriculum changes, demanding systems of accountability and shortages of experienced teachers, also limit the amount of time that teachers can spare for research related activities (Wilson 2018, p. 5)

10.9 Conclusions

The future for geography education research, both in the UK and further afield, is inextricably linked to the future of departments, school, faculties and institutes of education in universities. There is a distinct possibility that education research will become more focussed in, and on, schools—rather than maintaining its traditional 'home' within the academy. This potential transition will impact considerably on the fate of university-based research, for it is unlikely that universities will maintain their investment in education research merely for it to serve as a handmaiden to the changing requirements of schools. The discipline of education has regularly been referred to, in recent years, as being 'in crisis'—or, in need of 're-tooling' (Furlong 2013). This suggests a requirement for education researchers, and the institutions that employ them, to achieve greater clarity about their aims, objectives and *raison d'etre*: not only in terms of their philosophical and theoretical mission, but also with regard to the practicalities of future funding, staffing and income generation. This is, of course, part of a larger debate about the purpose of higher education—which for some has traditionally centred on the 'maximisation of reason', but which for others is now more clearly driven by neoliberal goals. I would argue that the fate of research endeavours in geography education will run parallel with those in university departments of education. Both are presently closely aligned, with futures that are inextricably linked. Nonetheless, it is certainly possible for schools and departments of education to exist *without* the presence of geography education research; it is less likely that the reverse is also the case.

A key issue is that the next generation of geography education researchers will not be inducted into their field of enquiry in quite the same way as the previous generation. The traditional route into becoming a researcher in geography education: by first teaching in schools, then becoming involved in the initial teacher education of geography teachers—either as a part-time or full-time employee of a university education department—and subsequently developing one's skills as a researcher, has largely gone. With initial teacher education, in England at least, now firmly located in schools the need for university-based geography educators has declined. As Lambert (2018) observes:

> the research-active infrastructure in many sub-fields such as geography is fragile in higher education, having been undermined by the deinstitutionalisation of teacher education in recent years (p. 368)

The logic is that without many people being directly employed in *geography* teacher education—traditionally academics who were also the source of much high-quality geography education research and publication—the future of geography education research may be in jeopardy. This is not to say that research will stop completely, but it will certainly change; it is most likely that research outputs in the field of geography education will diminish. The impact of the radical re-orientation of initial teacher education on subject-based education research is profound, particularly because universities—aware of the direction in which the political winds are blowing—have already begun to populate initial teacher education courses with ex teachers from schools, often on a part-time basis, whose interest in research cannot be assured:

> universities and colleges increasingly recruit staff directly from schools. Some of these new staff members were permanent, bringing with them their own strongly professional rather than academic forms of expertise; others were temporary and part time, opening the door to the future casualization of university staff (Furlong 2013, p. 34)

In some institutions, where the traditions of geography education research have remained strong and staffing levels have been sustained, the future may be vibrant. If these institutions can sustain senior academic positions in geography education—at Reader or Professorial levels—the likelihood is that research will continue, despite other shifts in teacher preparation and education. Research outputs must remain strong: retaining a foothold in research assessment exercises is essential, for this sends an important message, internally and externally, about the value of geography education research, and researchers. We are caught, nationally and internationally, in a numbers game. With the number of dedicated geography education researchers declining, and their research outputs in the field dissipated by the need to produce (more significant?) publications in other fields, the scale of high-quality publications in geography education is reduced. The number of programmes and courses in geography education—from initial teacher education to doctoral level—is also decreasing in many countries, with a concomitant impact on lectureships. Geography educators, and researchers, now engage in a highly specialised, but marginal, suite of activities. It is salutary to speculate on the number of genuine geography education researchers now working in the UK, and internationally—at best, perhaps a few hundred—often employed within small, specialised units that lack sufficient heft to sustain a community of researchers to engage in intellectual debate and scholarship in the field. Geography education researchers are often spread too thinly, or are too isolated, to have real impact. Despite the valiant efforts of notable individuals and groups we cannot engage in real intellectual work on recognised problems in geography education research—or indeed on more general issues of method, methodology, theory, research design and epistemology in the field—without greater numbers of researchers. There is therefore a pressing need for research in geography education to be recognised as important, both by politicians and schoolteachers. Unfortunately, we have not yet successfully made this case—a dangerous state of affairs at a time when education policy is often ideologically driven, and when most of the findings from educational research are deemed by stakeholders to be 'an irrelevance' (Furlong 2013, p. 40).

I conclude by looking to the well-intentioned *Declaration on Research in Geography Education* (IGU CGE 2015) for further support and guidance on the prospects for geography education research worldwide (see also, Chap. 7). This document describes the purposes of research in, and relevant to, geography education as follows:

- to provide and distribute evidence and/or conceptually robust arguments and practices that will improve the quality of geography education in national settings and internationally;

- to encourage a 'research orientation' among geography teachers and educators that enables reflective and critical engagement with habitual practices and a professional habit of mind that demands improvement in the quality of geography education; and
- to strengthen the scientific status of geography education and consolidate it as an area of knowledge by developing and reinforcing working networks among researchers and educators

These are reasonable, indeed laudable, statements. They are difficult to disagree with—although the pathways to their resolution may at present be rather badly lit. What appears clear to me now is that the shape of educational research in higher education institutions in many countries will look profoundly different in the next ten years. With change and uncertainty in the economy, in society, in the university sector and in our schools, the prospects for successfully sustaining research into geography education will almost certainly be radically different in future times.

References

Barnett, R. (2000). *Realizing the university in an age of super-complexity*. Buckingham: Open University Press.

Barnett, R. (2009). Knowing and becoming in the higher education curriculum. *Studies in Higher Education, 34*(4), 429–440.

Biesta, G. (2007). Why what works, won't work: Evidence-based practice and the democratic deficit in educational research. *Education Theory, 57*(1), 1–22.

Brooks, C. (Ed.). (2010a). *Studying PGCE geography at M Level*. London: Routledge.

Brooks, C. (2010b). Developing and reflecting on subject expertise. In C. Brooks (Ed.), *Studying PGCE Geography at M Level: Reflection, research and writing for professional development*. London: Routledge.

Browne Report. (2010). *Securing a sustainable future for higher education*. Independent Review of Higher Education. London: BIS.

Butt, G. (1997). *An investigation into the dynamics of the national curriculum geography working group (1989–1990)*. Unpublished Ph.D., University of Birmingham, Birmingham

Butt, G., & Lance, A. (2005). Secondary teacher workload and job satisfaction: Do successful strategies for change exist? *Educational Management Administration and Leadership, 33*(4), 401–422.

Counsell, C. (2017). Interview with Christine Counsell at the Inspiration Trust, Norwich. 28 February 2017.

Daugherty, R., & Rawling, E. (1996). New perspectives for geography: An agenda for action. In E. Rawling & R. Daugherty (Eds.), *Geography into the twenty-first century* (pp. 1–15). Chichester: Wiley.

Department for Education (DfE). (2010). *The importance of teaching: The Schools White Paper*. London: The Stationery Office.

Furlong, J. (2013). *Education: An anatomy of the discipline*. London: Routledge.

Gerber, R. (2007). An internationalised, globalised perspective on geographical education. *International Research in Geographical and Environmental Education, 16*(3), 200–215.

Graves, N. (1985). Geography in education: A review. In A. Kent (Ed.), *Perspectives on a changing geography* (pp. 15–22). Sheffield: GA.

Graves, N., Kent, A., Lambert, D., Naish, M., & Slater, F. (1990). Evaluating the final report. *Teaching Geography, 14*(4), 147–151.

Igu, CGE (International Geographical Union- Commission on Geographical Education). (2015). *International Declaration on Research in Geography Education*. Moscow: IGU CGE.

Johnston, R. (Ed.). (1985). *The future of geography*. London: Methuen.

Johnston, R. (1986). *On human geography*. Oxford: Blackwell.

Lambert, D. (2009). *Geography in education: lost in the post? Inaugural lecture*. Institute of Education, University of London.

Lambert, D. (2010). Geography education research and why it matters. *International Research in Geographical and Environmental Education, 19*(2), 83–86.

Lambert, D. (2011). Reviewing the case for geography, and the 'knowledge turn' in the English national curriculum. *The Curriculum Journal, 22*(2), 243–264.

Lambert, D. (2015). Research in geography education. In G. Butt (Ed.), *MasterClass in geography education* (pp. 15–30). London: Bloomsbury.

Lambert, D. (2018). *Editorial: Teaching as a research-engaged profession. Uncovering a blind spot and revealing new possibilities. London Review of Education, 16*(3), 357–370.

Lambert, D., & Morgan, J. (2010). *Teaching Geography 11-18: A conceptual approach*. Maidenhead: McGraw Hill/Open UP.

Lay, K. (2004). Pedagogy of reflective writing in professional education. *Journal of the Scholarship of Teaching and Learning, 9*(1), 93–107.

Machon, P., & Ranger, G. (1996). Change in school geography. In P. Bailey & P. Fox (Eds.), *Geography teacher's handbook*. Sheffield: Geographical Association.

Mitchell, D., & Lambert, D. (2015). Subject knowledge and teacher preparation in English secondary schools: The case of geography. *Teacher Development, 19*(3), 365–380.

OfSTED (Office for Standards in Education). (2008). *Geography in schools: Changing practice*. London: HMSO.

OfSTED (Office for Standards in Education). (2011). *Geography: Learning to make a World of Difference*. London: HMSO.

Peters, R. (1967). *The concept of education*. London: Routledge and Kegan Paul.

QCA (Qualifications and Curriculum Authority). (2005). *Geography Monitoring Report 2004–5*. London: QCA/DfEE.

Rawling, E. (2001). *Changing the Subject. The impact of national policy on school geography 1980–2000*. Sheffield: Geographical Association.

Roberts, M. (2013). *Geography through enquiry*. Sheffield: Geographical Association.

Slater, F. (1996). Illustrating research in geography education. In A. Kent, D. Lambert, M. Naish, & F. Slater (Eds.), *Geography in education: viewpoints on teaching and learning* (pp. 289–320). Cambridge: Cambridge University Press.

Slater, F. (2003). Exploring relationships between teaching and research in geography education. In R. Gerber (Ed.), *International handbook on geographical education* (pp. 285–300). Kluwer.

Standish, A. (2007). Geography used to be about maps. In R. Whelan (Ed.), *The corruption of the curriculum*. London: Civitas.

Standish, A. (2009). *Global perspectives in the geography curriculum: Reviewing the moral case for geography*. London: Routledge.

Tierney, W. (Ed.). (2001). *Faculty work in schools of education: Rethinking roles and rewards for the 21st century*. New York: State University of New York Press.

Wiliam, D., & Lester, F. (2008). On the purpose of mathematics education researcher: Making productive contributions to policy and practice. In L. English (Ed.), *Handbook of international research in mathematics education* (pp. 32–48). New York: Routledge.

Wilson, A. (2018). *Harnessing educational research*. London: Royal Society and British Academy.

Young, M. (2010). Alternative educational futures for a knowledge society. *European Educational Research Journal, 9*(1), 1–12.

Bibliography

Bampton, M., & French, R. (1995). Improving geographical education: A modest proposal. *Newsletter of Association of American Geographers (AAG)*, *30*(1), 5.

BERA/UCET (British Education Research Association/University Council for the Education of Teachers). (2012). *Prospects for education research in education departments in higher education institutions in the UK*. London: BERA/UCET.

Berger, R. (2013). Now I see it, now I don't: Researcher's position and reflexivity in qualitative research. *Qualitative Research*, *15*(2), 219–234.

Biddulph, M., & Lambert, D. (2016). England: making Progress in School Geography: Issues, Challenges and Enduring Questions. In O. Oslvado, M. Solem, & R. Boehm (Eds.), *Learning Progressions in geography education: International perspectives* (pp. 35–53). Switzerland: Springer.

Blades, M., & Spencer, C. (1986). Map use by young children. *Geography*, *71*(1), 47–52.

Boschetti, F., Prokopenko, M., Macreadie, I., & Grisogono, A.-M. (2005). Defining and detecting emergence in complex networks. In R. Khosla, B. Howlett, & L. Jain (Eds.), *Knowledge-based intelligent information and engineering systems, Lecture notes in artificial intelligence* (Vol. 3681, pp. 573–580).

Butt, G. (2008b). *Lesson planning* (3rd edn.). London: Continuum.

Castree, N. (2011). The future of geography in English universities. *The Geographical Journal*, *136*(4), 512–19.

Cresswell, T. (2013). *Geographical thought: A critical introduction*. Chichester: Blackwell-Wiley.

Csikszentmihalyi, M. (1991). *Flow: The psychology of optimal experience*. New York: Harper Perennial.

Davis, B., & Sumara, D. (2006). *Complexity and Education: Inquiries into learning, teaching and research*. Abingdon: Routledge.

DCMS/DfES (Department for Digital, Culture, Media and Sport/ Department for Education and Skills. (2006). *Government response to Paul Roberts' Report on Nurturing Creativity in Young People*. London: DCMS.

Department for Education (DfE). (1995). *Geography in the National Curriculum (England)*. London: HMSO.

Department for Education (DfE). (2012). *Attainment at key stage 4 by pupils in Academies 2011*. Research Report DfE-RR223.

Department for Education and Employment (DfEE/QCA). (1999). *The National Curriculum for England: Geography*. London: HMSO.

Department for Education and Skills (DfES). (2003b). *Raising standards and tackling workload: A national agreement*. London: DfES.

Eaude, T. (2012). *How do expert primary class teachers really work?*. Plymouth: Critical Publishing Limited.

Finn, M. (2017). *Transitions to University. From the 'new' geography A level to the 1st year of a geography degree*. Draft paper. Unpublished.

Fonseca, J. (2002). *Complexity and innovation in organizations*. London: Routledge.

Forsey, M., Davies, S., & Walford, G. (2008). *The globalisation of school choice?*. Didcot: Symposium Books.

Geography Earth and Environmental Sciences. (2006). *Special issue on threshold concepts and troublesome knowledge*. Planet 17 (available at www.heacademy.ac.uk).

Gerber, R. (Ed.). (2003b). *International handbook on geographical education*. Dordrecht: Kluwer.

Glaser, B., & Strauss, A. (1967). *The discovery of grounded theory*. Chicago, IL: Aldane.

Gorard, S., & Taylor, C. (2004). *Combining methods in educational and social research*. Maidenhead: Open University Press.

Graves, N. (1982). Geographical education. *Progress in Human Geography*, *6*(4), 563–575.

Guba, E. (1981). Criteria for assessing the trustworthiness of naturalistic enquiries. *Educational Communication and Technology Journal*, *29*, 75–91.

Hardman, M. (2010) *Is complexity theory useful in describing classroom learning?* Paper presented at The European Conference on Educational Research. University of Helsinki.

Hargreaves, D. (1997). In defence of research for evidence-based teaching: a rejoinder to Martyn Hammersley. *British Educational Research Journal, 23*(4), 405–419.

Hargreaves, D. (1999). The Knowledge-Creating School. *British Journal of Educational Studies, 47*(2), 122–144.

Haubrich, H. (1992). *International charter on geographical education.* Commission on Geographical Education of the International Geographical Union.

Hicks, D. (2006). *Lessons for the future: The missing dimension in education.* Victoria BC: Trafford Publishing.

Hicks, D. (2007). Lessons for the future: A geographical contribution. *Geography, 92*(3), 179–88.

Hicks, D. (2014). A geography of hope. *Geography, 99*(1), 5–12.

Hirst, P., & Peters, R. (1970). *The logic of education.* London: Routledge and Kegan Paul.

Huckle, J. (1985). Geography and schooling. In R. Johnston (Ed.), *The future of geography.* London: Methuen.

Jeffrey, C. (2003). Bridging the gulf between secondary schools and university-level geography teachers: Reflections on organising a UK teachers' conference. *Journal of Geography in Higher Education, 27,* 201–15.

Johnson, N. (2007). *Simply complexity: A clear guide to complexity theory.* Oxford: Oneworld publications.

Johnson, S. (2001). *Emergence.* London: Penguin Books.

Johnston, R. (1994). Department size, institutional culture and research grade. *Area, 26,* 343–350.

Jones, S., & Daugherty, R. (1999). Geography in the schools of Wales. *International Research in Geographical and Environmental Education, 8*(3), 279–282.

Lambert, D. (2013). Collecting out thoughts: School geography in retrospect and prospect. *Geography, 98*(1), 10–17.

Lidstone, J., & Williams, M. (Eds.). (2010). *Geographical education in a changing world: Past experience, current trend and future challenges.* Berlin: Springer.

Lincoln, Y., & Guba, E. (1985). *Naturalistic inquiry.* Beverley Hils, CA: Sage.

Long, M. (1964). The teaching of geography: A review of recent British research and investigations. *Geography, 49*(224), 192–205.

Long, M., & Roberson, B. (1966). *Teaching geography.* London: Heinemann.

McElroy, B. (1993). How can one be sure? Ground rules for judging research quality. *International Research in Geographical and Environmental Education, 2*(1), 66–69.

Marsden, W. (1989). 'All in a good cause': Geography, history and the politicization of the curriculum in nineteenth and twentieth century England. *Journal of Curriculum Studies, 21*(6), 509–526.

Mitchell, D. (2013). How to deal with controversial issues in a 'relevant' school geography. In D. Lambert & M. Jones (Eds.), *Debates in geography education.* London: Routledge.

Mitchell, M. (2009). *Complexity: A guided tour.* Oxford: Oxford University Press.

Morgan, J. (2009). What makes a "good" geography teacher? In C. Brooks (Ed.), *Studying PGCE geography at M Level.* London: Routledge.

Naish, M. (1972). Some aspects of the study and teaching of geography in Britain: A review of recent British research. *Teaching Geography.* 18 Sheffield: Geographical Association.

Newby, P. (2010). *Research methods for education.* Harlow: Pearson Education Limited.

Office for National Statistics (ONS). (2012). Ethnicity and National Identity in England and Wales: 2011. http://www.ons.gov.uk. Accessed March 18, 2019.

Osberg, D., & Biesta, G. (2008). The emergent curriculum: Navigating a complex course between unguided learning and planned enculturation. *Journal of Curriculum Studies, 40*(3), 313–328.

Osberg, D., & Biesta, G. (2010). The end/s of education: Complexity and the conundrum of the inclusive educational curriculum. *International Journal of Inclusive Education, 14*(6), 593–607.

P.R.C. (1968) Review of Chorley R. and Haggett P. (1967). Models in geography: The second Madingley lectures. *Geography, 53*(4), 423–424.

Pykett, J., & Smith, M. (2009). Rediscovering school geographies: Connecting the distant worlds of school and academic geography. *Teaching Geography, 34*(1), 35–8.

QCA (Qualifications and Curriculum Authority). (2007). *The national curriculum key stage 3: Geography.* Available from: www.curriculum.qca.org.uk. Accessed December 12, 2011.

Rempfler, A., & Uphues, R. (2012). System competence in geography education: Development of competence models, diagnosing pupil achievement. *European Journal of Geography, 3*(1), 6–22.

Rempfler, A., & Uphues, R. (2010). Sozialokologisches Systemverstandnis: Grundlage fur die Modellierung von geographischer Systemkompetenz. *Geographie und ihre Didaktik, 38*(4), 205–217.

Richardson, K., & Cilliers, P. (2001). What is complexity science A View from Different Directions? *Emergence, 3*(1), 5–22.

Roberts, M. (2010a). Where's the geography? Reflections on being an external examiner. *Teaching Geography, 35*(3), 112–13.

Roberts, M. (2010b). What is 'evidence-based practice' in geography education. *International Research in Geographical and Environmental Education, 19*(2), 91–95.

Roberts, M. (2012) *What makes a good geography lesson?* Available from www.geography.org.

Scarfe, N. (1949). The teaching of geography in schools: a review of British research. *Geography, 34*(164), 57–65.

Schleicher, H., & Schrettenbrunner, H. (2004). Schädigung von Bäumen auf dem Schulgelände: ein Unterrichtsbeispiel erstellt mit SchulGIS. *Geography, 34*(2), 21–23.

Shin, E.-K. (2006). Using Geographic Information System (GIS) to improve fourth graders' geographic content knowledge and map skills. *Journal of Geography, 105*(3), 109–120.

Smith, J. (2005). Flow theory and GIS-Is there a connection for learning? *International Research in Geographical & Environmental Education, 14*(3), 225–30.

Stacey, R. (2001). *Complex responsive processes in organizations: learning and knowledge creation.* London: Routledge.

Taylor, P. (1986). *Expertise and the primary school teacher.* Windsor: NFER-Nelson.

Thomas, G. (1997). What's the use of theory? *Harvard Educational Review, 67*(1), 75–105.

Thomas, G. (2009). *How to do your research project.* London: Sage.

Thomas, H., Butt, G., Fielding, A., Foster, J., Gunter, H., Lance, A., Lock, R., Pilkington, R., Potts, E., Powers, S., Rayner, S., Rutherford, D., Selwood, I., & Soares, A. (2004). *The evaluation of the transforming the school workforce pathfinder project. Research report 541.* London: DfES.

Tolley, H., Biddulph, M., & Fisher, T. (1996). *The first year of teaching: Workbook 5.* Cambridge: Chris Kington Publishing.

Totterdell, R. (2012). What makes a geography lesson "good"? *Teaching Geography, 37*(1), 35.

Unwin, T. (1992). *The place of geography.* Harlow: Longman.

Witt, S. (2013). Playful approaches to outdoor learning. In S. Scoffham (Ed.), *Teaching geography creatively* (pp. 47–58). London: Routledge.

Wooldridge, S., & East, W. (1951). *The spirit and purpose of geography.* London: Hutchinson.

Wright, D. (1992). Perceptions on research imperatives for geographical and environmental education to the year 2000. *International Research in Geographical and Environmental Education, 1*(1), 57–60.

Yarwood, R., & Davison, T. (2007). Bridges or fords? *Geographical Association branches and higher education Area, 39,* 544–550.

Printed by Printforce, the Netherlands